Beyond The Last Blue Mountain

VIKING
Penguin India

Books by the same author

- *The Creation of Wealth: A Tata Story (1981)*
- *Encounters With the Eminent (1981)*
- *The Heartbeat of a Trust (1984)*
- *In Search of Leadership (1986)*
- *Keynote (edited with S.A. Sabavala, 1986)*

Beyond The Last Blue Mountain

A Life Of J.R.D. Tata

R.M. Lala

VIKING

VIKING

Penguin Books India (P) Ltd., B4/246 Safdarjung Enclave, New Delhi - 110 029, India
Penguin Books Ltd., 27, Wrights Lane, London W8 5TZ UK
Penguin Books USA Inc., 375 Hudson Street, New York, N.Y. 10014, USA
Penguin Books Australia Ltd., Ringwood, Victoria, Australia
Penguin Books Canada Ltd., 10 Alcorn Avenue, Suite 300, Toronto, Ontario M4V 3B2, Canada
Penguin Books (NZ) Ltd., 182-190 Wairau Road, Auckland 10, New Zealand

First published in VIKING by Penguin Books India (P) Ltd. 1992

Typeset in Times Roman by IPP Catalog Publications Pvt. Ltd., New Delhi
Made in India by Ananda Offset Private Ltd., Calcutta

To

Four People Ever Precious

my father, my mother,
my mother-in-law and
my wife, Freny

Contents

═══

A Note on the Illustrations

The illustrations on the interleaving pages preceding Parts I and II are by the British artist Henry Morshead. Illustration 1 shows the twin heritage of J.R.D. Tata: The stairway at Persepolis representing his father's heritage from ancient Persia and the Eiffel Tower, the French heritage from his mother's side. Illustration 2 is a sketch of the Boeing 707 and the Puss Moth drawn to scale. Illustration 3 shows the Tata crest as designed by the founder, Jamsetji Tata. The words *Humata, Hukhta* and *Hvarshta* in the ancient Avesta language mean 'Good Thoughts', 'Good Words', and 'Good Deeds'. J.R.D. Tata later redesigned the wings. Illustration 4 is an artist's impression of the young J.R.D. Tata. The brief Tata family tree which forms part of the preliminary pages is taken from the Tata Electric Companies' brochure, 1989.

Acknowledgements

An author collects bouquets and brickbats on the publication of a book. If there are any bouquets I would like to share them with those who have played a notable part in its making.

In the five years that I have worked on it I feel indebted to many. My special thanks to two colleagues who have assisted me almost from the start—Arvind Mambro who helped sensitively with the research, and Mehroo Mistry whose enthusiasm for typing and retyping drafts encouraged me to chisel each chapter.

My senior friend Kenneth Belden, author and publisher from England who read the manuscript and gave me his valuable comments.

David Davidar of Viking/Penguin India for his perceptive points on the manuscript and Sudha Sadhanand for her meticulous editing.

Captain K. Visvanath for reading the section on aviation.

P.K. Sen for his early research on J.R.D.'s letters and later for verification of economic and statistical data.

Robert Robin for taking me round Paris to sites associated with the birth and life of J.R.D. Tata and Raymond D'Souza whose two decades of association with J.R.D. has proved a fruitful source of ready reference.

Shirin Mulla of Alliance Francaise who translated letters of J.R.D. Tata's mother of the early 1900s from French to English.

Khorshed Divecha and Freny Shroff for help in selection of photographs.

Villoo Karkaria and Sebastian Mathias for secretarial help in the final stages of the book.

What has touched me is not only the assistance of many, but their sense of participation in the making of this book. For all that, my grateful thanks.

Bombay
December, 1991

R.M.Lala

Preface

The year is 1930. A tiny Puss Moth with a single engine is droning over the arid expanse of Iraq at about ninety miles per hour. The pilot is a young man of twenty-six, who is flying without a radio, without landing aids, without instruments except an altimeter and a speedometer. In the distance rise the large mountains and the land beyond the Euphrates and the Tigris. He presses forward in that little plane to reach "beyond that last blue mountain".

The year is 1990. The same man at eighty-six is seated comfortably in his well-appointed office with a physical map of the world behind him, also portraying the elevation of the mountains of Iraq he once flew over. This time he is talking of the twenty-first century. 'You know I would like to live to see the 21st century because by that time you will be able to travel from London to New York within $1^1/2$ hours. Planes will climb into space, fly at three times the speed of sound without friction before re-entering the atmosphere and landing in New York.' Forever young, this man approaching ninety still seems eager to go beyond the distant blue mountain.

J.R.D. Tata was born when the twentieth century opened its eyes. The year was 1904, when the Russo-Japanese war broke out, when work on the Panama Canal began and when Sir Francis Younghusband led the expedition that opened up Tibet to the world. It was a time when Edward VII reigned as the King-Emperor of India, when Mahatma Gandhi was trying his early experiments with truth in South Africa and a rather well-featured young man called Jawaharlal Nehru was being groomed in Allahabad to enter Harrow.

The world was still young. There were many mountains to climb and rivers to cross. Amundsen had yet to reach the South Pole and the peak of Mount Everest had to wait half-a-century before its snow experienced the first footprint of man. The century to unfold was to result in amazing discoveries and inventions. The aeroplane, radio and television were to make the earth a much smaller planet to live in. It was a century that was to experience two world wars and see the rise and decline of Fascism and Communism. It was a century that would witness the old colonial order shatter into a hundred independent nations. Till the middle of the century, as late as 1950, there were only four independent countries in the whole of Africa: Ethiopia, Egypt, Libya and South Africa. By the late sixties there were forty-nine.

J.R.D. Tata, who has lived through the entire span of upheaval and scientific progress, says he has had no part in any of the events. All he can claim, he says, is some influence on the development of the House of Tatas—which broadly synchronizes with the growth of India as an industrial power—and, 'in a small way on the history of civil aviation in India'.

In 1932 he pioneered civil aviation in India. In 1948 he gave India her wings abroad. Thirty years later, when he ceased to be the Chairman of Air-India, Air Marshal Noor Khan of Pakistan hailed him as, "an epic figure". J.R.D. had given India an airline Indians could be proud of.

Had he not been involved in anything other than aviation, his claim to recognition would have been secure. But he achieved much more. For fifty years he headed the largest industrial group in India. When he took over as Chairman of Tata & Sons the group had fourteen companies. When he completed his half-a-century at the helm of Tatas on 26 July 1988 there were ninety-five enterprises which Tatas had either started or had a controlling interest in. Even so he bemoans lost opportunities, 'not for myself but for the country.'

For three decades—1955 onwards—he struggled under severe restraints. He has been a Prometheus bound by the red tape of bureaucracy, superimposed on the pink strands of a woolly socialism. He had great hopes when freedom came to India and Nehru spoke of "a tryst with destiny". The tryst somehow eluded him.

Speaking at the launching of his book *Keynote* in July 1986, J.R.D. said: 'My one sorrow and regret is that the Government had, from Jawaharlal Nehru onwards and at least up to a couple of years ago, not allowed many of us imbued with enthusiasm and hope to do enough. Today, things have changed and now the last sorrow of mine is that I have reached an age where I am not likely to be able to participate purposefully in the better things that are happening, the better opportunities and the quicker progress that I visualized.' He concluded: 'And I only wish that I'll be spared long enough to see that we are on the march.'

Behind this "tycoon", as *Life* called him, is a shy but nonetheless fascinating human being. It has been my privilege to have known him for the last ten years. Thanks to the nature of my profession—I am both a writer as well as a Director of a Tata Trust—our paths have often crossed. A petal opened at each meeting and I could view a little more of this intriguing man. Not many have had this privilege and I felt that those less fortunate than myself may wish to know him as I do.

The Greek dramatist Sophocles said: 'Call no man happy till he dies.' No man can be assessed until at least five to ten years have lapsed after he has gone. But while J.R.D. is with us and his memories are fresh, I have recorded them.

The portrait of the man that emerges from the book is not necessarily complete for, no matter how objective a biographer tries to be, he is only human and will project what strikes him sharpest. There is the story of the German philosopher Schopenhauer who was seated one day on a bench in a garden looking unshaven and unkempt. A policeman taking him to be a vagrant prodded him with a baton, and asked, 'Who are you?' The great philosopher, woken out of his thoughts, looked up and said, 'I wish I knew.' If a Schopenhauer didn't know who he was, is it not rather presumptuous for a biographer to presume that he can know fully the subject of his biography? He can only perceive glimpses and bring alive some facets of his life.

After giving me a couple of interviews for a biography in 1983, J.R.D. inexplicably played hard to get. I did not impose myself on him. Later I heard from friends that he was reluctant to have anything published in his lifetime. A couple of years later, thanks to the encouragement of his colleagues Jamshed Bhabha and Sharokh Sabavala, he agreed to resume the interviews. He protested more than once: 'I *don't* want my life to be written but if it is to be written, I'd rather have you do it than anyone else.' He agreed to cooperate with me on the understanding that nothing be published in his lifetime. He was liberal with his time and granted interviews whenever requested and made available two big cupboards full of his personal letters as he kept no diary.

Though he has cooperated handsomely with me, he has read only a few chapters and not the whole book. So I am alone responsible for any errors that may have crept in.

For five out of the six years I worked on the book, J.R.D. would occasionally say to me, 'Why are you wasting your time?' Once he said,

'You know it is not going to be published in my lifetime. And I'm not eager to oblige you by going earlier.'

We then arrived at a less drastic way to enable him to oblige. I suggested that I may be allowed to publish during his lifetime the first volume (from his birth till the death of Nehru), as most actors of the era were no more. He conceded. Only after he relinquished his last Chairmanship of a company—Tata Sons—in March 1991 did he agree to the publication of the full biography.

I have presented J.R.D. Tata in the perspective of the times he has lived in and in the setting of a group of companies he has headed for half-a-century.

The book is organized into four parts. Part I covers the period from his birth up to 1938, when he became Chairman of Tata & Sons. Though his forty-six years in aviation ran simultaneously with his most active years in industry, I have brought the entire aviation section together in Part II so that the reader can find in one place both his domestic and international contribution to aviation.

Part III covers his half-century as a captain of industry from 1938 to 1991. Till 1970, when the Managing Agency System was abolished, he was the all-powerful head of the whole of the Tata group. Later he became its constitutional head as the empire was converted into a commonwealth. Also, after 1970, he stepped out more as a statesman, giving his views on the national platform.

Part IV spans facets of his personality as a professional, a philanthropist, a citizen. He was the first Indian of note to propagate family planning and the Presidential System of government. His brushes with celebrities, a bitter-sweet friendship with Nehru, whom he knew for forty years, and vignettes from his life may enable the reader to know him better as a human being.

WE ARE THE PILGRIMS, MASTER, WE SHOULD GO
ALWAYS A LITTLE FURTHER: IT MAY BE
BEYOND THAT LAST BLUE MOUNTAIN BARRED WITH SNOW
ACROSS THAT ANGRY OR THAT GLIMMERING SEA....

WE TRAVEL NOT FOR TRAFFICKING ALONE,
BY HOTTER WINDS OUR FIERY HEARTS ARE FANNED:
FOR LUST OF KNOWING WHAT SHOULD NOT BE KNOWN,
WE TAKE THE GOLDEN ROAD TO SAMARKAND

-JAMES ELROY FLECKER

A SECTION OF THE TATA FAMILY TREE

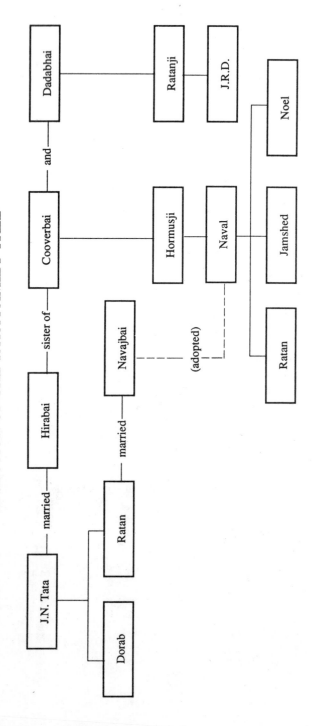

PART I

YEARS OF PREPARATION
1904—1938

CHAPTER I

The Heritage

A hot summer breeze blows gently over the plains of Persepolis. In the distance rise towering columns supported by a great terrace. The mountains provide a backdrop to the once magnificent palace started by Darius the Great and expanded by successive Persian kings.

As the taxi draws nearer, several fluted columns, each about sixty-five feet high, stand out like chimneys. Persepolis seems more bare than one expects but its grand central staircase is well preserved 2,500 years after it was built. On the staircase and the adjoining platform are carvings of rulers of adjoining kingdoms bearing gifts to the "King of Kings". The title was not inappropriate for the Persian Empire in its heyday extended from Egypt up to Sind in India, from the Mediterranean to the Arabian Sea.

Darius in an inscription near Persepolis declares: 'I am of such a sort that I am a friend to the Right. I am not a friend to the Wrong. It is not my desire that the weak man should have wrong done to him by the mighty, nor is it my desire that the mighty man should have wrong done to him by the weak. What is right, that is my desire.' On a carving at Persepolis is the image of the King slaying a monster symbolizing evil.

Zarathushtra, the prophet of ancient Iran, born 1,500 years before Christ, taught that life was a struggle between the forces of good and evil with good eventually vanquishing evil. He declared that man could be God's own ally in the struggle against evil and in doing so Zarathushtra bestowed upon the humblest peasant a sense of dignity that no Emperor could rob him of. It is, therefore, not surprising that the first Bill of Human Rights was discovered on a clay tablet of the reign of Cyrus the Great (it is now housed in the United Nations' building in New York.)

Today all that remains of Cyrus is a desecrated tomb at Pasargade upon which is inscribed: 'I am Cyrus, who founded the empire of the Persians. Grudge me not therefore this little earth that covers my body...'

3

After two centuries of rule, the Achaemenian Dynasty founded by Cyrus the Great was overthrown by King Alexander. Other dynasties followed. In the seventh century the Crescent and the Sword swept over the land of Cyrus and Darius the Great. The choice that confronted the Persians was to be converted to Islam or to be slaughtered. Some hardy souls escaped to the fastnesses of the mountains. A few adventurous ones set sail in little boats, treasuring their sacred fire, and headed for the west coast of India. They landed in Gujarat. More boats followed later.

When the first boat people came, the local Hindu raja sent a vessel full of milk to the refugees, indicating to them that the place was full to the brim and that there was no room for them. A Zoroastrian priest, who headed the group, poured sugar into the receptacle of milk which did not overflow. He requested the raja's representative to take it back to his master, to indicate that these refugees would mix with his people and enrich his state. The permission came to stay. For over a millennium thereafter the Zoroastrians (called "Parsees")—after the province of Pars in Persia—have flourished in India.

In Gujarat, some Parsees settled in a place called Nagmandal—land of snakes. The place reminded them of a town they had left behind in Persia called Sari. And so they called it Navsari (New Sari). Here, for 300 years, their sacred fire from Persia dwelled. Navsari became the centre of learning, tradition and the faith of the Parsee community. The tradition was kept alive by the priests. Even today there are records going back to the thirteenth century of all the young men of the town who qualified as priests, and among them is the name of Jamsetji Nusserwanjee Tata—the visionary and industrialist who ushered India into the age of the Industrial Revolution. It is he, who at the beginning of the century conceived for India the first steel plant, the first hydro-electric project and a University of Science, 'the like of which England did not have at that time.'[1] Jamsetji died in 1904 but he had laid the lines for development and by 1912 all the three projects were well advanced.

Just five years earlier in 1907, when Sir Frederick Upcott, Chief Commissioner for Railways in India, heard that Tatas were going to set up a steel plant, he replied:

> Do you mean to say that Tatas propose to make steel rails to British specifications? Why, I will undertake to eat every pound of steel rail they succeed in making.[2]

In the next decade Tatas supplied to the British Government for the First World War, 1,500 miles of steel rails which made possible the shifting of troops and war materials in Mesopotamia. Jamsetji's son, Sir Dorab Tata, dryly commented that had Sir Frederick Upcott kept his promise, 'he would have had some slight indigestion.'[3]

In recognition of the contribution of Tata Steel, Lord Chelmsford, the Viceroy, came in 1919 to Sakchi in eastern India where the steel plant was located. Standing on the steps of the Director's bungalow the representative of the British Crown said:

> I can hardly imagine what we should have done during these four years (of war) if the Tata company had not been able to give us steel rails which have been provided for us, not only for Mesopotamia but for Egypt, Palestine and East Africa, and I have come to express my thanks... It is hard to imagine that 10 years ago this place was scrub and jungle; and here, we have now, this place set up with all its foundries and its workshops and its population of 40,000 to 50,000 people. This great enterprise has been due to the prescience, imagination and genius of the late Mr. Jamsetji Tata...
>
> This place will see a change in its name and will no longer be known as Sakchi but be identified with the name of its Founder, bearing down through the ages the name of the late Mr. Jamsetji Tata. Hereafter, this place will be known by the name of JAMSHEDPUR.[4]

Today, Jamshedpur, the town Tatas created and still run, has 7,50,000, inhabitants.

In 1896 Jamsetji Tata offered half his fortune consisting of fourteen buildings and four landed properties worth three million rupees* to the British Government to establish a University of Science. For the last eight years of his life he tried to enthuse the British Government to make a contribution equal to his gift and pass an enactment to start a university. This university would offer post-graduate courses in electrical, mechanical and civil engineering, various aspects of the humanities, and would

---

* Around Rs 170 million in 1991.

enable students to do research into India's history, to study Indian archae-ology and a great deal more.

On 1 January 1899, a day after the arrival of the new Viceroy, a committee including Jamsetji Tata saw Lord Curzon. Lord Curzon inquired:

> Were they satisfied that the highly paid professors would attract a sufficient number of pupils? Were they satisfied that there were a sufficient number of posts for an accretion of highly trained scientists?[5]

A few months later Lord Curzon got the impression that perhaps Jamsetji Tata was angling for a baronetcy. Curzon delayed his approval of the scheme and two commissions were appointed to report on it. Whereas Jamsetji's benefaction fetched £8,000* a year the Imperial Government was not willing to go above £2,000 annually.

In 1903 the Secretary of State for India, Lord Hamilton, wrote to Lord Curzon that as the Imperial Government had not encouraged Jamsetji Tata's plan for the university, he would probably withdraw his benefaction and divert his resources to the iron and hydro-electric projects he had launched upon. Any other man would have done that but Jamsetji Tata did not. He had put his hand to the plough and he would not take it back. Unknown to the British rulers, in codicils to his will, he urged his sons not to touch his benefaction after his death and, if need be, put some more money from the other half of his wealth that he was leaving them, to make the university possible. He died in suspense not knowing if his grand scheme would ever see the light of day.

A year after his death, Lord Curzon agreed to the scheme for the University of Science: the share of the Imperial Government would be equal to the interest from the benefaction of Jamsetji. The university—to be called the Indian Institute of Science in Bangalore—became the fountainhead of scientific and technological manpower for the next half century and the mother of several of India's national laboratories that grew after independence.

When Jamsetji died in May 1904, the *Times of India* wrote in an obituary:

> His sturdy sense of character prevented him from fawning on

---

* Around Rs 125,000 million in 1991.

6

any man however great, for he himself was great in his own way, greater than most people realised. He sought no honour and he claimed no privilege; but the advancement of India and her myriad peoples, was with him an abiding passion.

The same paper spoke of Jamsetji's, 'quiet, strong, stern, unselfish determination to pursue his calling.' And in so doing, Jamsetji became the founder of modern industrial India.

Jawaharlal Nehru was to say of him:

When you have to give the lead in action, in ideas—a lead which does not fit in with the very climate of opinion, that is true courage, physical or mental or spiritual, call it what you like, and it is this type of courage and vision that Jamsetji Tata showed....[6]

Jamsetji created his wealth through textiles. In late nineteenth century when capitalism was at its roughest, he instituted a pension fund for his workers. In 1886 he instituted a pension fund for his mill workers and, in 1895, accident compensation. He was decades ahead of his time and miles ahead of his competitors.

Jamsetji established Tata & Sons in 1887 and took as his partners his eldest son, Dorab and a young cousin called Ratanji Dadabhai Tata (R.D.)—not to be confused with Jamsetji's younger son, Ratan, who became a partner a few years later. The father of R.D., Dadabhai Tata, had a sister Hirabai who was married to Jamsetji Tata. Five generations earlier the two Tata families had a common ancestor.

R.D. was born in Navsari in 1856 and his earliest recollection was of playing games barefoot with other children. He recalled his first attempt at experimental science as a child. At the age of nine, he and a young friend tied empty coconuts on a string, circled it like a lifebuoy and dropped it in a well. Then with a rope they lowered themselves into the water. What they thought would be a lifebuoy became a deathtrap. They cried out for help as they began to sink and elders came running to the rescue. R.D. was promptly hauled up before his father who gave him a sound thrashing.

In those days boys of Navsari played a game of cracking walnuts. The boy whose walnut cracked first had to pay a fee to the victor. R.D. was unbeatable. He had somehow procured an unbreakable walnut made of solid metal and raked in the money. One day, the father found out what the son was up to. A second flogging was duly administered.

R.D. came to Bombay to join Elphinstone College, which his cousin Jamsetji had attended earlier. After he graduated he took a course in agriculture in Madras. He was then whisked off to join his family's trade with the Far East. It is said that R.D. impressed his cousin Jamsetji—seventeen years older than him—with his handling of finance. Jamsetji who had started the Empress Mills in 1874 brought in twenty-eight-year-old R.D. as a partner in the textile enterprise. Jamsetji was struck by the way R.D. turned out and made him a partner in Tata & Sons when it was established thirteen years later. R.D. spent quite a bit of time in the Far East, mostly Hong Kong. He imported silks from China to India and exported rice from Burma to the Far East. Tatas were already established in Shanghai but R.D.'s enterprise opened branches in Kobe (Japan) and in New York.

R.D. was married at an early age to a Parsee girl from the Banaji family. She died childless not too long after the marriage. R.D. did not remarry till his mid-forties when the hand of fortune took him to Paris (the trip was probably also because he found it difficult to work with Jamsetji's eldest son, Dorab.)

In Paris R.D. hoped to trade in pearls (he had a fascination for gems) and silks. He wanted to learn the French language and Jamsetji recommended a teacher to him. At the teacher's home he met a very attractive young woman called Suzanne, just twenty years of age. She was tall and slim with blonde hair and blue eyes. He saw more of her. R.D. was by then forty-six. He confided to the patriarch of the family, Jamsetji, his affection for the young lady and his desire to marry her. In those days marriage outside the Parsee community was taboo and R.D. was prepared for an angry "No" from Jamsetji. Much to R.D.'s surprise and delight Jamsetji readily gave his consent. The patriarch attended the wedding in Paris in 1902 and even spoke at it. Soon after marriage, R.D. gave his French bride a Parsee name, Sooni (of gold), perhaps because of her golden hair.

After the wedding Jamsetji took R.D. and his bride to Britain and gave a lavish party at Kingston-on-Thames. It was, 'the largest gathering of Parsees which had hitherto been held West of the Suez Canal.' To make it a success Jamsetji Tata spared no expense. He wished his guests to see the countryside along the Thames and for this purpose he chartered a pleasure-steamer from Westminster to Kingston-on-Thames. An account of the occasion says:

> He played the host to perfection, though he deprecated, in courtly manner, the numerous expressions of thanks. His friends, Sir Jamsetji and Lady Jeejeebhoy, had cut short a tour of Scotland in order to be present; Sir Muncherji Bhownagree represented the House of Commons; Mr. Dadabhai Naoroji, the doyen of the Parsee residents in England, brought his family. Mr. R.D. Tata brought his French bride, who within a year, was invested in Bombay with the sacred thread and sudra of the Parsee religion. At Kingston, dinner was preceded by the prayers and blessings of the Avesta, the holy book of the Zoroastrians.[7]

On this occasion at Kingston-on-Thames, Sir Muncherjee Bhownagree, one of the three Indians elected to the British House of Commons at the time (the other two—Dadabhai Naoroji and Sir Shapurji Saklatvala, a nephew of Jamsetji—also hailed from the tiny Parsee community) in proposing the chief toast touched on R.D.'s marriage: 'I may recall as an example of enlightened sentiments of our host, that recently an event has happened in his family which, I am told, would have been impossible without his sanction and consent. I have the great good fortune to have on my right hand a lady of French nationality who is associated in life and fortune for the rest of her days with Mr. Tata. If I am rightly informed Mr. Ratanji Tata (R.D.) the lucky possessor of that bride had some misgivings as to how the projected union would be regarded by the head of the family. The fact that in spite of his many years of life and of what may be regarded as his orthodoxy, Mr. Jamsetji Tata gave his ready consent to the alliance is one more proof of his progressive tendencies and his interest in the social advancement of the community.'

This gathering was the last Jamsetji attended. When he returned to Europe in 1904, Jamsetji was looking forward to repeating the party in Kingston-on-Thames but it was not to be. He fell seriously ill in Bad Nauheim, in Germany. R.D. joined him while Jamsetji's son, Dorab and his daughter-in-law Meherbai, were still in Vienna. Jamsetji was very feeble by then but showed no signs of imminent collapse. From time to time he would rouse himself and talk to R.D. about his aims and what had

9

actuated his life and constantly expressed the wish that his sons and R.D. should together carry on his work. R.D. told him that to follow him was no small task and spoke of the lustre Jamsetji had brought to the name of Tata. Jamsetji replied:

> If you cannot make it greater, at least preserve it. Do not let things slide. Go on doing my work and increasing it, but if you cannot, do not lose what we have already done.[8]

On 17 May he suddenly took a turn for the worse. 'Where is Dorab?' he asked again and again,'Where is Dorab?' His son and daughter-in-law were hurrying from Vienna but hour by hour Jamsetji was sinking and was conscious only intermittently. During one of his lucid spells he handed over his Patek Philippe gold watch, which contained his mother's portrait and a lock of her hair, to his cousin R.D. It was found later in Jamsetji's will that he had bequeathed the precious watch to R.D.

On 18 May Dorab and his wife arrived. By that time the dying man was comatose, but he had made an effort to live until their arrival, and spoke to them a few affectionate words.'Where were you, where were you?' he said to his daughter-in-law, and he stroked his son's cheek. On the following morning he passed away in his sleep. It was 19 May 1904, the anniversary of the death of his political hero, William Gladstone.[9]

Two men were at his bedside when Jamsetji died—R.D. and Dorab. They did not always see eye to eye but once Jamsetji died they decided to work together and honour his wishes.

Jamsetji had dreamt and laid the lines along which India could erect a steel plant, a hydro-electric network and a University of Science. Neither the steel nor the hydro-electric company had been floated, nor had the University of Science been started. In the next eight years Dorab, R.D. and other loyal colleagues were to work as a team to make all the three dreams of Jamsetji come true.

Jamsetji was buried in Brookwood cemetery in England at the end of May 1904. The sadness of his going was assuaged when seventy days after Jamsetji's death, R.D. and Sooni's second child, a son, was born in Paris. They gave him the Persian name, Jehangir—"Conqueror of the World".

CHAPTER II

Childhood 1904–1915

The world that greeted Jehangir the little "Conqueror" in 1904 was one of hope and promise. Although earlier in 1871 Bismarck had defeated France and captured Emperor Napoleon III, by the turn of the century France had recovered from the shock of defeat. Between 1871 and 1904 France took long strides in Empire building. The whole of Indo–China had come under French rule, as had Morocco, Madagascar and Dahomey. France had the largest Empire in the world, next only to that of Britain. De Gaulle had proclaimed that France is not France unless it dreams of glory. When Jehangir was born, France was dreaming of glory again.

In Paris every eleven years a World Exhibition was held. That of 1889 coincided with the hundredth anniversary of the French Revolution. Alexander Gustave Eiffel, an engineer, lifted above Paris the Eiffel Tower, signalling to all the arrival of the age of steel. For forty years it was to remain the tallest structure in the world. From atop the Eiffel Tower the Seine appeared like a silver ribbon meandering its way through a gracious city, and beside the Seine, in the spacious boulevards at spring-time, chestnut trees blossomed, their white flowers looking like candles on a Christmas tree ready to be lit for a celebration. At the dawn of the twentieth century Paris was not an ordinary city. It was the cultural capital of the Western world.

A young British girl, Moma Clarke, who visited Paris about the time of Jehangir's birth describes the city she fell in love with. In her book, *Light and Shade in France* she says Paris is,

> surely the youngest city in the world, for all her age. She seems to me to be spring incarnate, a flaunting feminine creature swinging her skirts, scattering her laughter, endangering her virtue, so young, so careless, so ready to play... I

thrilled to the flying lights, the gay colours, that something in
the air which sets the pulse of youth racing and gives to every
ordinary happening of the day, romance. I longed to join in
the rhythm of the city's life, to be one with those self-assured,
busy Parisians who walked so well and had such bright,
challenging eyes.[1]

The Champs-Elysees was then,

the most perfect avenue in the world. Victorias bowled gaily
along, drawn by well-matched pairs of high-stepping horses.[2]

Moma Clarke revelled in the spectacle of the happy, luxuriant crowd
of people passing down the avenue with its magnificent Arc de Triomphe.
She found elegance in women with big feather-and-flower-trimmed hats
fastened to their cushions of hair by jewelled pins. As they walked, ladies
held their sunshades in tightly gloved little hands and men, as formally
turned out, escorted them. Bicycles, too, joined in the festive procession
of people but there were few motor cars, and no aeroplanes. Only captive
balloons and kites with papertails invaded the clear sky.

Paris had yet to have its underground Metro. Above the ground wheezy
steam-trams,

agonised down avenues and streets...Nothing moved swiftly
except the tongues of the Parisians.[3]

*

At the heart of the cultural life of Paris, then as now, stands L'Opera.
Today, on a sunny Saturday, scores of young people in their colourful
costumes sit upon its sweeping steps either licking ice-cream cones or
enjoying each others' company. To the left of the Opera is a spacious road,
Rue de Halevy.

The second house on Rue de Halevy is fairly modern and it must have
been almost new when R.D. Tata's family occupied it in 1904. It was there
that J.R.D. was born. It has a magnificent entrance, its curved doors almost
as high as Westminster Abbey's. To the right of the Opera stands the

Grand Hotel and on the ground floor is the Café de la Paix, spreading itself on the spacious pavement. There are three rows of small round tables, each with two or three chairs. At one table sits a couple in animated conversation; at another a lady in dark blue—face wrinkled, her eyes staring into the distance, talks to herself, oblivious of the cup of tepid coffee before her. Lonely she is, but at the Café she can at least see human beings, hear their voices, witness the flow of life and find some satisfaction in others living their lives.

At the Café de la Paix in the early 1900s R.D. used to sip coffee every day, watching the carriages roll by, while at a studio nearby, Auguste Rodin was delicately stroking his chisel on a stone. Four centuries earlier when Michelangelo was engaged in such an act he was asked what he was doing. He replied, 'I am releasing the angel imprisoned in the stone.' Rodin was about to release "The Thinker" in 1904 when Jehangir was born to R.D. and Sooni on 29 July.

In the two years between their marriage and the birth of Jehangir, Sooni had come from France to India twice. Within a few months of the lavish party of Jamsetji Tata at Kingston-on-Thames, R.D. and his bride had set sail for India by the *S. S. Imperatrix*. On board the British liner, the young lady, uprooted from her Parisian moorings, was homesick. She wrote to her mother on 12 December 1902: 'Every minute of the day I wish that you were here with me so that we could talk in French that I love so much and which I am having to abandon in order to speak English. I need sometimes to say silly things—foolish things but why say them because nobody will understand so I remain silent and sometimes sad. But this never lasts—I only have to look at Ratan (*mon petit*) and I am truly happy! My husband makes me feel safe, content, protected.' She said she felt so happy with him she felt she was dreaming.

On Christmas day, 1902, the *S.S. Imperatrix* steamed into Bombay harbour. To receive R.D. and his bride were Dorab and Meherbai Tata. On arrival at their home, says Sooni, 'before we crossed the threshold (of the house) a little ceremony of welcome (was performed). There are five servants in the house, one a Japanese called "boy".'

Sooni had come to a strange land with strange languages and stranger customs. Describing Bombay of the early twentieth century, Sooni speaks of 'Parsees with bizarre hats and dressed almost like Europeans; Hindus with turbans of every colour and form and they all have a red mark on their foreheads, red or white and different forms according to caste. They drape themselves in a sort of white cotton sari (*dhoti*) which is drawn up between their legs—a sort of bizarre trouser.... The women carry loads on their heads. There are also the Portuguese with their chic wives. There are Christians. I have met the wives of the Hindu bankers—they are always in very bright saris with so many bracelets on their arms and legs that they can hardly move. There are Jews and Arabs—an immense mixture of people.'

Within a week of her arrival, Bombay was readying itself for the coronation of King Edward VII, Emperor of India. On 2 January 1903, she wrote to her mother, 'Bombay is all lit up for the coronation. It is all like a fairy land. The Elphinstone College looked like a Gothic castle lit up with ancient oil lamps in all colours. It all sparkles like millions of precious stones and one imagines oneself back in the splendour of the Raj.'

There was no need for Sooni to look back. She was actually looking at the British Raj at the height of its glory. She comes back to her language problem in the letter: 'I hate not being able to express myself in the language of the country. I find myself gesticulating like a mad woman when I want a window to be opened or closed,' she tells her mother. 'I read, I write, I paint but what makes me really sad is that there is no one with whom I can talk French—I remain silent and often I feel like crying.'

Conscious of his wife's frustration, R.D. went to the French Consul in Bombay and requested him to find in Pondicherry an Indian woman who could speak French. Finally they found one who spoke pidgin French.

The warmth of India's people made up for the language barrier to some extent. Sooni found R.D.'s mother, 'a gentle and charming old woman.' She also looked forward to Sundays when she could be with her husband. For her it was, 'the best day in the week.' In early 1903 she wrote to her mother, 'Dear God, how I love him and how lonely I feel when he is not there.'

Sooni was expecting a child by March and both she and R.D. planned to sail for France. She wrote, 'We have lots of things to do for our future king; with muslin and lace we have to make him a lovely, lovely nest.'

Then she reveals to her mother the impending event that was to throw the orthodox Parsee community into turmoil. R.D. was a religious-minded Zoroastrian and wanted his bride to be inducted into the Zoroastrian faith. Conversion was not permitted under the rules and customs of the community. Sooni wrote to her mother: 'Dear Mama, I have very important news to tell you…only yesterday Ratan told me that it was impossible (to be converted) but that if at all it was possible, the decision might be made suddenly. I suddenly got word from Ratan's office that if I were to learn the necessary prayers in three or four days it could be all over. I was completely taken by surprise but I immediately started to learn the prayers and I recited all evening long the prayers in Pahlavi. Luckily, in Navsari, Ratan had already taught me two of the smaller ones. The official sanction had been given only this morning by the High Priest and Ratan wants the ceremony on Sunday…It will be attended by a whole lot of important Parsees and will take place at Mr. Sethna's.'

Excitedly she wrote that like the little girls in Navsari who have their thread ceremony performed, 'I will wear an *ijar* (an undergarment of silk worn to the knees) and will be wrapped in a white cashmere shawl. But what is most significant is that at the same time the priest will marry us and then nobody will give me another thought. I will be allowed to enter the temple (which I couldn't do before) or stay in a house where a Parsee lies dead. If a Parsee dies in a house, all non-Parsees, even if they are tenants in the same house, have to leave it until the body is carried away. I am a bit confused by all the stir that this is causing. Not in the last three or four hundred years has such a case as this been disputed. Luckily the press is for us. A *dastoor* (priest) is going to come and teach me the important prayers. I will recite them to you…Mama, and you will see how strange they sound.'

Another letter: 'Darling mcther, here I am, at last a Parsee. Everybody is happy for me and so am I. I spent five sleepless nights filling my head with prayers I had to learn—now I feel exhausted. Let me, however, try and recount the ceremony of my conversion and our marriage which took place at Mr. Sethna's big house.

'At 4.00 p.m. I was made to sit in a small room next to the huge salon in Mr. Sethna's house where the ceremony was going to be performed. A *dastoor* with the lower face hidden sat opposite me. I recited some prayers with him, ate a piece of pomegranate and then raised to my lips, in a gesture of sipping, a cup which contained the urine of the cow. It is supposed to purify but, of course, nobody really drinks it—not even touches it with their lips but it is a custom that has existed since the beginning. Ratan asked me not to tell you about this since he finds it distasteful—don't therefore talk of it. Then I had to go and bathe. Normally, a *dastoor* is present but this time he remained on the other side of the partition. The wife of the *dastoor* and the beautiful Meherbai Tata were with me.

'They dressed me in an *ijar* and confined my hair in a *mathu-banu* (head cover of muslin) and draped a white cashmere shawl around my shoulders. Then feeling very pale and nervous, and with my feet in *sapats* (slippers) I entered the drawing room where there were waiting at least 60 *dastoors* when only one was necessary. I was made to sit with my back to everyone facing the High Priest and I started to recite the prayers with him. After 15 minutes or so he placed my hands in the sleeve of the *sudra* (sacred white vest) and left—then all the Parsee ladies held up before me a white sheet to shield me from view. I put on the *sudra*, my blouse and a white sari with a silver border. When I was ready, the High Priest returned but this time we (both) stood—he standing just behind me. Then while I held on to his little finger, he tied the *kusti* (sacred thread) around me. Then seated again there were more prayers with the priest showering my head with pieces of pomegranate, coconut. Then it ended and I was led into the midst of all our friends who were waiting to congratulate me.'

But that was not the end. After most guests departed, close friends remained. The drawing room was rearranged for the wedding of the two according to Zoroastrian rites; it had to take place before sunset. Seated next to each other with two white-robed priests in front of them, the wedding ceremony began. 'I read out aloud the pledge of the Zoroastrian faith, in French, and then the ceremony began. Ratan and I were sitting side by side and the *dastoors* started to pray, showering us with rice. It took about 25 minutes. When everybody except the family and Mr. Kanga had left, we all drank champagne and then quietly we returned home. It all sounds so simple and everything went smoothly but until the very last moment we weren't sure that it would be allowed. There had been quite a few objections and some Parsees had not come since they had been afraid of trouble.'

The storm she feared did break out and was to result in a famous court case. But even as it did, Sooni was scheduled to sail for Marseilles. On arrival, Sooni left with her mother for Vichy to deliver her first child. "The king" she was preparing for, with "muslin and lace" for his nest, chose to delay his debut. Instead, a pretty daughter arrived, whom they called Sylla. Towards the end of the year, Sooni with her baby daughter returned to India. By March 1904 she was expecting a second child and sailed for France. This time R.D. stayed on in India to look after his business in Tatas.

Jehangir, later to be called Jeh, recalls his early years: 'My childhood and youth, so different from those of the average middle class Parsee, were mainly conditioned by the fact that my father had married a Frenchwoman, and we spent half of our early years in Paris and half in Bombay. My father loved life in France, French food and wines, and because my mother was at first not familiar with the English language, the language used by all of us was French. What I remember most vividly is that we always seemed to be on the move, and that my lovely and cultured mother had to uproot herself every two years or so to find a new home—alternately in France and in India. With servants and office help available in India, her task whenever we arrived in India or left was relatively simple. But in France, where our domestic help consisted of never more than a maid and a cook, the job of finding a new apartment, furnishing it while looking after her growing brood—there were five of us—represented a real chore which she accomplished with amazing efficiency and apparent ease but at the cost of much fatiguing work.'

One of the problems Jehangir faced early was of language, brought about by his mixed heritage: 'When I attended one of the government schools in Paris, the Janson Besailly, I was a much better student in French than I was in English at the Cathedral School in Bombay.'

The language barrier was considerable; in addition, Cathedral School in Bombay bored him. He saw no reason why he should learn British history. 'I used to ask "But what happened in India?" I have a good recollection of asking: "What about Aurangzeb?" and being smacked down. I don't know why I chose Aurangzeb!' Maths and Physics interested him.

'My first important memories from the point of view of a growing child, blessed with a fairly observant and inquisitive mind, were about cars and aeroplanes. My father decided that we needed a home of our own in which to spend our holidays, and he picked on a new and developing beach resort on the Channel coast of France, south of Boulogne, called Hardelot, where he not only bought a villa but later on built a number of villas and shops as a real estate developer. In fact, one of the two main streets of Hardelot was officially named Avenue des Indes.

'It happened that the legendary Louis Bleriot, who acquired world fame in 1909 by being the first to fly a plane across the Channel, also chose Hardelot for his family's summer resort. Bleriot built not only a fine villa close to ours but also a hangar near the beach. On the beach his small plane used to land much to the excitement of everyone there—

grown-ups and children, none more starry-eyed than myself. From then on I was hopelessly hooked on aeroplanes and made up my mind that, come what may, one day' I would be a pilot. I had to wait many years for that dream to come true.'

R.D. perhaps felt that his family members were more comfortable in France and kept all of them there. Sooni wrote almost daily to him and he—even with his busy schedule–dropped endearing notes and picture postcards to his wife and children. Sooni mailed her husband frequent postcards in French written in Gujarati script. No prying eye could understand it! After Sylla and Jehangir a daughter, Rodabeh, was born in 1909 and a son, Darab, in 1912. In 1916, Jimmy, her last child, was born in Bombay.

In his *A History of France*, Andre Maurois says that France before the 1914 war was as prosperous as any country of Europe, proud of her culture and literature. In every town and village there existed groups and societies avid for books and ideas, subscribing to the most modern journals published from Paris. An elite few journeyed every year to Paris from the countryside to see the season's new plays and to replenish their stock of ideas.[4]

Men of letters and scientists were more honoured in France than in any other country at the time. It was in this milieu that Jehangir grew up and he was to share its values. For all its plus points, on the eve of the First World War, France was a terribly divided country, politically.

When the guns of August 1914 boomed, R.D. was in Bombay, the children were on holiday in Switzerland with their Granny, while Sooni had remained in Paris.

Jehangir clearly remembers returning from their holiday with Granny. When their mother received them at the station, he was surprised to find her turn up in the white uniform of a nurse. Sooni had volunteered to work at the American Hospital in Paris.

Jehangir, by now ten, was an alert witness to the air raids on Paris. In those days there were no sirens. Instead the fire brigade charged all over the place clanging its bell and sounding the alert so that people could hide in their cellars. The first time the raids came, Jehangir was not in the cellar but on the terrace.

'We were in the shadow of the Eiffel Tower,' recalls Jehangir, 'and from the terrace I remember the stunning appearance of the first Zeppelin which tried to plant a few eggs (bombs) with little success on the cit There was the fully ineffective cannonade of the so-called anti-ai guns mounted on the Eiffel Tower, directed against the visiting acrid monster, lending to the excitement.'

J.R.D. adds:'I was not interested in or understood the grim and bloody struggle in the trenches which took the flower of France's and Britain's youth, but the great deeds of the knights of the air fired me with enthusiasm and I remember one day complaining to my mother that if she had the good sense to marry my father five or six years earlier I could have been a fighter pilot too. I fervently hoped at least that the war would last long enough for me to become one!'

He further adds: 'Oddly enough, some forty years later I was honoured, amongst other international airline bosses, with the French Legion of Honour. As our decorations were pinned on us by the then Minister for Aviation (who later became the Prime Minister of France for the usual short period) he addressed brief congratulatory remarks to each of us. When my turn came he said, "I shall not embarrass Mr. Tata by recalling his deeds of valour as a fighter pilot in the First World War." As I was only ten years old when the war started and fourteen when it ended, I was left with only two choices! Either to protest that I wasn't that hero—embarrassing the poor man in the process—or to be guilty of accepting an undeserved award. After about a second of mature consideration I accepted the decoration with as modest a mien as I could put on.'

As the war raged furiously in the trenches and the fields were soaked with blood, trains filled with casualties rolled into Paris. Sooni's work at the American Hospital was most demanding.

The strain of nursing and of looking after a family of four was too much for Sooni and she contracted tuberculosis in about a year's time. It was incurable then but a good climate and nourishing food helped the patient.

R.D. had to decide whether or not to take the risk and call the family to India. Submarines prowled the Mediterranean and the Atlantic waters and the risk was grave. R.D. decided to take it.

Winston Churchill was to say memorably in one of his wartime speeches:

> —ah yes! For there are times when all pray—for the safety
> of their loved ones.[5]

R.D., a religious man, must have prayed as his whole family sailed from wartime France to India. Fate was kind. They arrived safely.

CHAPTER III

Taj Mahal And Cherry Blossoms

In the past, whenever his wife and children came from Paris, R.D. had always rented a special house for them. On this occasion, in view of his wife's health, he did not want her to strain herself by running an establishment. He decided to instal the family and himself in a suite of rooms at the Taj Mahal Hotel.

Built by Jamsetji Tata in his own lifetime, the Taj overlooks Bombay harbour. On the waterfront, to the left of the Taj, stands the Gateway of India, erected to commemorate the landing of King George V, who in 1911 came to visit India—the first and last British monarch to do so. The harbour is dotted with launches and with ships waiting in line for a berth. In the distance, on the mainland in the shadow of the hills, burns the flame of an oil refinery.

When Jamsetji conceived of the hotel in 1900 neither the Gateway of India nor the refinery was dreamt of. Jamsetji had his reasons for erecting a grand hotel, the like of which did not exist between the Suez and Japan. As a young man he had read in Bombay's *Saturday Review*:

> The want of hotel accommodation was never so badly felt as
> at present. When will Bombay have a rest house worthy of
> the name?

The few tolerable hotels that existed catered exclusively to the "burra sahibs" and the "chhota sahibs". One day Jamsetji took a foreign friend of his to dinner. The doorman of the hotel told Jamsetji that his guest was welcome but, regrettably, he was not. The hotel was "For Europeans Only". That evening Jamsetji decided to create a hotel that would be the pride of India and would attract travellers from around the world to Bombay.

Normally Jamsetji planned the financial side of his operations carefully but in this case he decided that the sky was the limit. This hotel was his gift to a city he dearly loved. He planned every little detail of it. He purchased newly reclaimed land overlooking the Bay of Bombay. He toured Europe himself in 1902 and, in spite of a weak heart, personally walked the streets of Dusseldorf and other European countries and made purchases of soda and ice-making machinery, a laundry, lifts and an electric generator. When in Europe, he wrote to his son Dorab to have on the hotel premises a Turkish bath, a post office, a chemist's shop and a resident doctor (twenty-four hours a day) to attend to guests.

In 1903, facing Bombay harbour, the Taj Mahal Hotel rose in solitary grandeur. Ships from Europe were to witness the grand frontage of the Taj as a signal of their entry into Bombay harbour, somewhat like the much grander Statue of Liberty inviting to the New World the tired and the homeless from Europe. The day the Taj opened a crowd of a few hundred gathered at sunset to watch the spectacle of Bombay's first public building lit by electric lamps. Inside the grand hotel were only seventeen guests. Jamsetji, as usual, was ahead of his time.

Its impressive long facade, domed in the centre, had two wings running to the rear. The exceptionally wide corridors of marble were so designed to circulate the air and cool the guests in a humid climate. To negotiate one's way from the lifts to the corridor, one had to come down a few granite steps. Sooni, who was expecting another baby in 1916, tripped on these steps. She would have died of haemorrhage had not Jamsetji's foresight provided for medical attention night and day. Sooni was miraculously saved. The child, born prematurely in the seventh month, survived. He was named Jimmy, after Jamsetji. All her children were brown-eyed, but when Jimmy was born blue-eyed Sooni was excited. Jimmy was to grow up to be a well-built young man, but he did not oblige with blue eyes. His also turned brown.

Though Sooni was provided with better food in Bombay than she would have had in wartime France, the humid climate was unsuitable. T.B. sanatoria had just sprung up in hill stations like Panchgani, near Bombay, but as R.D. had good trading links with Japan and as its climate was temperate he preferred to send Sooni and the children there. The family stayed in Yokohama for all of 1917 and a good part of 1918. They hired a big house in Yokohama. Sooni liked Japan and the Japanese people. The climate suited her and they travelled extensively in Japan.

Rodabeh, then only ten years old, has clear memories of visiting the tea houses called "chaya" and the village inns of Japan, lying down on eiderdowns which she remembers as being pronounced "phtong". They went to Japanese plays they did not understand and their mother used to say, 'I want you to see those things and remember Japan.'

Sylla, eleven months senior to J.R.D., joined the Sacred Heart Convent school and J.R.D. went to a big Jesuit Boys' school. 'My main memories of Japan, apart from the beauty of the land and the strangeness of the Japanese people, their language, their clothes and their food, were the recurring earthquakes, to which we became quite inured. Fortunately, we left Japan long before the deadly 1923 earthquake which ravaged Tokyo and killed thousands. We did, however, experience the excitement of a typhoon in which our wooden house creaked and shook even more alarmingly than in an earthquake. We were perhaps lucky that it was a new and strong house, for others in our neighbourhood suffered badly. The one next to ours lost its whole roof which just took off to land not without damage in the middle of the adjacent race course, accompanied by a full-sized bathtub!'

The only English school in Yokohama, run by American Jesuits, 'left a deep impression on me,' J.R.D. recalls. He did not like the Jesuit teachers because they ill-treated a fat Jewish boy. 'They harassed the poor fellow. If he spoke out of turn they went out of their way to ridicule him. And I went out of my way to make friends with him because the other boys—mostly American—were anti-Semitic.'

He continues: 'There was a French family in Yokohama. The father had come out to Japan as a young man with his wife. He was a business-man and he gradually built up a family of five children and so the children and we were there together. The eldest son was about my size and age. We were very good friends but the friendship did not last beyond the time in Japan.' The daughter of the family was J.R.D.'s first girl friend. Over seventy years later he still remembers her phone number in Japanese. 'She was eighteen and I was fourteen. It broke my fourteen-year-old heart when she married an officer from the French Tank Corps. I didn't know they even had tanks in those days. A wonderful woman anyway.' J.R.D. did keep in touch with her family till as late as the 1980s and he knew her grandchildren.

The Tata family sailed for India in 1918 while the war was still on. Aboard a ship called *Hirano Maru*, fourteen-year-old J.R.D. spent his time learning typing on an old Remington machine in the lounge. The family disembarked at Colombo and the ship sailed on to Britain. They had made good friends aboard the ship, and were especially close to the captain and the purser. Some weeks after their return to India they got sad news of the *Hirano Maru*. It was torpedoed off the coast of England and sank within a few minutes, carrying with it the friends of the family.

Armistice was declared in November the same year. The following year Sooni and the children set sail from India for France.

CHAPTER IV

Sooni

The post-war France that Jehangir returned to was dominated by two great personalities, both saviours of their country—Prime Minister Georges Clemenceau and Marshal Foch.

Walrus-moustached Georges Clemenceau gave to France the indomitable will to fight Germany and win. Marshal Foch gave France the strategy to do so. Together, Winston Churchill said, they wrote a deathless page of history. The First World War took a larger toll of human lives than the Second World War—but once that devastating war was won, Clemenceau and Foch were set to tear each other apart. Winston Churchill, who was to save England in the Second World War, as Clemenceau saved France in the First, wrote about these two grand figures in his book, *Great Contemporaries*:

> As much as any single human being, miraculously magnified can ever be a nation, Clemenceau was France.[1]

"The Tiger", as he was called, waged inexorable war against those he saw as the enemies of the state, the monarchists, the swindlers and traitors within France and the Germans on French soil.

And there was the other France of Marshal Foch—ancient, aristocratic,

> whose grace and culture, whose etiquette and ceremonial has bestowed its gifts around the world. There was the France of chivalry, the France of Versailles, and above all, the France of Joan of Arc. In the combination of these two men during the last year of the War, the French people found in their service all the glories and the vital essence of Gaul. These two men embodied respectively their ancient and their modern history.[2]

In war, the two Frances stood together; came the Armistice, and Clemenceau and Foch began pulling in different directions. There were two Frances again.

Jehangir was too young to understand the struggle of the titans. In 1919, when the Tata family returned to France, the Versailles Peace Conference was being held. Clemenceau, with his fiery eyes and his quivering moustache, demanded that defeated Germany pay heavy reparations. Unknowingly thereby he, a victor of the First World War, sowed the seeds of the Second World War.

Jehangir was then only fifteen and, he says he had till then, 'done no real reading.' But the return of Alsace-Lorraine from Germany to France under the Peace Treaty, delighted him. (The area was captured by Germany from France in 1871 when Bismarck defeated Napoleon III.) Like every French schoolboy Jehangir had felt this area belonged to France and that it should come back to her.

Jehangir's next five years were to revolve round Janson Besailly which he says was, 'a very fine school, a great public school of Paris, but not in the British sense. It didn't have boarders but we were in the middle of the town, very near to where we lived, a hundred yards or so.'

He continues: 'In my class I had the reputation of being the fastest to run down the stairs and it was quite a long staircase we had in our school. I remember that in the French literature class I used to take pleasure in writing and perhaps I must have revealed some talent at it. I well remember our class in French literature, where the teacher used to call me for some strange reason, *L'Egyptien*. In front of me sat a big fat boy called Elkingon, a grandson of Napoleon's Marshal Ney. Our class had just written an imaginative piece on a subject given to us and I had written a very tearful story about how a woman's husband was killed in the war and how the poor lady was pursued by somebody. Announcing the results, the teacher walked down the centre aisle of the classroom. There were about 8 to 10 rows. The teacher said, "You will be surprised to know that the competition last week was won by..." and he paused. He was looking past Elkingon to me. Elkingon thought he was the winner. He rose with a broad smile and started bowing, when the teacher brusquely told him "No, not you. Sit down. *L'Egyptien*—behind you." Poor chap!'

J.R.D. was a naughty youngster: 'I was up to a fair amount of mischief and played practical jokes on people and I think I inherited my sense of humour from my French grandfather. In those days people used to wear hard straw hats in summer. And a favourite prank of mine used to be to stand on the balcony of my house (or my friend's) above the street and wait for a passerby wearing one of these hats. Then we would drop a pea or a pellet onto his hat... ping!... and the person would jump up with a start and look all around to see who had done it. But we would duck and be out of sight. On one occasion, however, we did get into trouble. The person on whom we dropped a pellet, happened to be a professor from our school, whom we had not recognised. He pretended as if nothing had happened and walked on a little further. He crossed the street and watched us from the opposite side of the street. Then he came upstairs to complain. Of course we didn't answer the door...we were hiding in the lavatory!

'In those days there used to be pennies with holes in them. So we would tie a long white thread through the hole and wait for someone to pass by; when he did, we'd quickly lower the penny onto the pavement... ping... and whisk it up just as quickly. And the person would think some coins had fallen out of his pocket, so he'd stop, shuffle around in his pockets and look for the coins...which he never did find!'

Almost seventy years after he had left school, the teenage editors of a school newspaper asked J.R.D. whether he was good at studies. 'I was just an ordinary boy,' he replied, with a shrug of his shoulders. 'I assume that I was reasonably intelligent....I was interested in politics and I was very much against foreign domination in India, but I didn't do anything about it. I was very interested in sports...hockey, football...every kind of sport. I wasn't interested in pretty girls then, I got interested in them much later. I found sports more interesting. But the real reason was that I was shy of girls. I had this impression that girls were silly, giggly persons. And girls used to make fun of boys. Girls also used to think that boys were stupid and this last bit, I believe is still correct!'[3]

Every summer in post-war France the Tata family went to their home on the beach resort of Hardelot. A golden sandy beach ran absolutely straight for a few miles south of Boulogne. Behind this little resort rose a forest rich in trees. Even in the early twentieth century Hardelot had the distinction of purifying its sewage before discharge into the sea. It was the site of the Field of the Cloth of Gold where one of the early summit talks were held between Henry VIII of England and Francis I of France.

Occasionally, as we've seen in an earlier chapter, a Bleriot aeroplane would land on the beach, to everyone's great excitement. 'It was flown by his chief pilot, Adolphe Pegoud, the first man to loop-the-loop.

Aeroplanes in those days did not have enough power or the driving speed to climb into a loop in what became the standard way later. So he did an outside loop! He just dived and kept pushing on the stick until he went all the way round, hanging on a belt! A brave but foolhardy thing to do. It was a miracle the wings didn't come off! As a result Pegoud became world famous in aviation circles. Once, when he landed on the Hardelot beach he hit a soft spot, the plane went up on its nose and broke its wooden propeller. The propeller was afterwards cut and he autographed pieces which were given away as presents. I treasured one for years.'

Professional pilots of the time made a living landing at various places and giving joy rides to people. One day Sooni decided to give her children an experience of flying. Jehangir was the first to be selected to fly, strapped into the extra seat behind the pilot in the small two-seaters.

*

Rodabeh was hardly eight years of age during this period. I asked her whether Jehangir was a shy child. She says: 'All of us are shy. Did Jehangir ever tell you he played the piano as a young man? Not brilliantly but with feeling and a beautiful touch when he was in his teens. I remember him playing the Etudes of Chopin. Jehangir became a businessman but he has always remained sensitive to poetry and music and he loved cars.'

When he was between fifteen and nineteen years of age, Jehangir had two good friends. One was Louis Bleriot Jr, and the other Zent D'Alnoys. Young Bleriot was his playmate on the beach, who joined in pushing the aeroplane into his father's hangar. He was to blossom into a fine aviator. In 1927, when Charles Lindbergh was making preparations for the first trans-Atlantic flight from New York to France, Bleriot Jr,. was planning to fly in the opposite direction, from France to America. Bleriot had built a bigger plane than Lindbergh. Just as he was ready to make his journey Bleriot was struck down with appendicitis; though operated upon, he died. Had he lived he may have been remembered for the first Atlantic crossing as his father is remembered for the Channel crossing.

Jehangir was even more friendly with Zent D'Alnoys but lost track of him for some time. Later, they began corresponding and even met once or twice. 'He got married and in the Second World War went underground. The Germans caught hold of him and sent him away to one of those camps. He never returned.' After a pause J.R.D. added: 'Those were the two close male friends. Now you'll ask me who my female friends were?'

'Why not?'

'Most of them have died. There were very few—to tell you the truth.'

*

Looking back, Jehangir feels that when he returned to Europe in 1919, his father should have sent him to England to brush up his English (so he could have entered a British university) instead of making him spend the immediate post-war years in France with his mother. But R.D. probably chose to send his eldest son to France to look after Sooni during her ill-health, as R.D. himself had to be in India most of the time.

The children adored their mother and as they grew up, they came to greatly admire her accomplishments. Her youngest daughter Rodabeh says, 'Mother was a person interested in everything very beautiful. She had golden hair and beautiful pale blue in her eyes, which I've been searching for in others ever since.'

Sooni was accomplished at tailoring and painting and could make ladies' hats. Her cooking, particularly her pastry, was delicious. A couple of British generals, the Head of the Medical Services, General Slogget, and a General Barter managed to get themselves invited to dinner at the Tata home in Bombay because they liked her French cooking so much. Indian kitchens were hot and stuffy, but in spite of her ill-health, she cooked the whole day for her guests. J.R.D. told me that at supper one evening, when one of the Generals, used to mess food, asked for some more salt, Sooni burst into tears. It was an insult to her French cooking. The other General reprimanded his colleague, 'You are not fit to eat in a pig sty.'

If the English Generals did not have an adequate appreciation of good food, Sooni's eldest son had a bit too much of it. J.R.D. was difficult with food. Recalls Rodabeh: 'He would find the meat tough. We would eat quickly and watch Jeh's reaction. One thing Jeh hated was celery. Mother used to say, you must eat everything. Jeh would resist, father would look angry, clear his throat and thunder at Jeh that he should finish his food. I remember one occasion when to teach Jeh a lesson father got up from the chair and Jeh got up and started running round the dining table saying, "No, Papa. No Papa."

'Mother was a sparkling personality, very witty and alert,' says Rodabeh. Jehangir admits, 'I was jealous of her. She was a strikingly beautiful woman.

Whenever we went into a restaurant, people would turn around to see this beautiful woman with golden hair and light blue eyes. I used to look daggers at men who used to stare at her and then turn to talk to each other. I felt they were insulting my mother.'

'Was your father jealous too?'

'No, on the contrary he used to be pleased at seeing her admired by other men, and she used to encourage his admiration for good looking women.'

In 1922, R.D. thought he should move his family from France to Bombay and send Jehangir to England to study. R.D. decided to give his French wife a beautiful home at Malabar Hill with a spectacular view of Bombay, a home in India that would remind her of France. Enthusiastically Sooni joined in the project. She selected the furniture in Paris, the door knobs and electrical fittings and the like and sent them on to India for their dream house.

Even as Sooni's spirit was buoyant and she looked forward to their future in Bombay, her health was flagging. In 1923 news reached R.D. first of her health deteriorating, and then of her critical condition. At the time R.D. was not only engaged in the struggle of Tata Steel (which was facing stiff competition owing to the dumping of foreign steel) but he was also fighting for the survival of the Tata Industrial Bank, the first of its kind in the country. Certain financiers were determined to crush the bank, and used rumours to start a run on the bank. 'Mother was dying and father knew that she was dying,' says Rodabeh. 'Every day, he said later, he was wondering whether he would arrive in time to see her.'

One can imagine the agonies R.D. went through, especially as there were no telephone lines those days and regular air travel to Europe had not yet started. Finally, the day came for him to climb aboard the liner at Bombay harbour. Just then a cable was delivered to him. Sooni had expired. She was only forty-three.

With that cable in his pocket he climbed up the gangway. His children needed him. According to Rodabeh, R.D. told the children in France, 'You know the day before the ship arrived at Marseilles your mother came to me (in a dream) and said, "I am so glad you are coming to look after the children."'

When that part of France which he was trying to recreate at Malabar Hill for Sooni was finally complete, he called the house in his wife's memory: "Sunita".

CHAPTER V

British School And The French Army

A flower shop called J. Poulain & Fils stands at the rear entrance to the cemetery of Pere Lachaise. As one enters its high walls one stops in amazement. The vista is that of a palatial garden. When the wind blows the leaves on the high trees seem to play with one another. In this vast expanse lies the "Who's Who" of France—La Fontaine, Moliere, Alfred de Musset... There lies Chopin, his fingers stilled, while his music still resounds round the world. In addition, there are distinguished foreigners. Around the corner from the Tata family grave lies Oscar Wilde.

Each grave has its own distinguished design. An occasional Grecian pillar, coloured marble and iron grills greet the visitor in that green garden of peace. People from France and from other countries visit the cemetery. They go around with a map, visiting the graves of their favourite author or composer or distant relative. Some arrange for flowers to be laid regularly at the tombs of those they cherish. Quite a few of those walking around the garden are young people who offer a pot of leaves or a bouquet of flowers. Chopin's grave seems generously endowed with different offerings.

In the very front row, as one enters from the flower shop end, lies the resting place of Sooni.

*

By the time R.D. landed in France, Sooni was already buried in the cemetery. All he could offer was flowers. R.D. spent a hectic seven days making arrangements for the family in France and boarded the next ship back to India to save the Tata Industrial Bank. Though a well-thought out venture, it had finally to merge with the Central Bank of India.

R.D. finalized preparations for all his children to return to India, all

30

except Jehangir whom he sent in October 1923 to a crammer in England. The purpose was to brush up Jehangir's English prior to his admission to Cambridge. Jehangir was keen to study Engineering at Cambridge and what could be better than Gonville and Caius College? His cousin Dorab had made a handsome donation of £25,000* to the institution some years ago.

The crammer was situated at Southwold, Suffolk, on the North Sea. It was one of the bleakest and coldest parts of eastern England. The school was to prepare him and other boys to take entrance exams to universities. Though most of the students were foreigners, there was a sprinkling of English boys who needed training for admission to British universities.

'It was a fairly useful time. I worked. I learnt. I boxed. And on one occasion I froze. It was a bright summer day so all the boys were marched off for a swim. With gay abandon I jumped in. But the North Sea even in summer is freezing cold. I thought I would die.' In spite of a year spent in England (1923–24) he could never quite understand the complexity of the English who venture out in the midday sun in the Orient but return home for Christmas to break the ice in the Serpentine for a heroic dip.

Called *L'Egyptien* by his teacher in Paris, Jehangir was called "Frenchy" in England for he spoke English with a marked accent. In the beginning he was made a little fun of for his pronunciation. Respect came to him from an unexpected quarter. At this school, his boxing instructor was an ex-army Sergeant. 'He found me a bit reticent to use my right hand so one day he stuck his chin out and said, "Come on! Hit me!" I clipped him with my right. His eyes rolled up and he staggered back, held himself and said "That's enough." I discovered that though I was thin as a reed, I had a very good punch.' (With young boys one punch speaks louder than a barrage of words.) Soon after, R.D. presented his son with the latest French racing bicycle, the like of which was hardly seen in England. Jehangir's circle widened overnight. Though he did not see much of the cycle, which was borrowed indiscriminately, he saw more of his friends.

Just when the crammer course was getting over and Jehangir was looking forward to entering Cambridge, a law was passed in France to draft into the army for two years, all French boys at the age of twenty. The

---

* Around £5,00,000 in 1991.

31

eldest of the family was to be given a concession and released, if he wished, in twelve months. As a citizen of France, Jehangir had to enlist in the French Army for at least a year and Cambridge had to be postponed for the following year, 1925. In between the crammer and the time in the army he had a brief spell at home in Bombay.

*

A few miles from India's peninsular coastline of the Arabian Sea, soar the Western Ghats. As one travels by train from Bombay to Pune, parallel to the railway line is a range of three magnificent mountains detached from the main range. One is a formidable rock called Parbal which means "strong" in Sanskrit. The second is Bawa Malang, named after a Muslim saint who lived and worshipped there. It is a place of pilgrimage and from a speeding train, pilgrims meandering up the hill path appear like a string of ants. After Bawa Malang comes a wide lush hill called Matheran, after the local word *Matha-ran* meaning the "forest-capped hill".

In 1851, a Collector of Thane called Hugh Poyntz Malet brought the 2,500-foot hill to the attention of the British rulers. Its salubrious climate later attracted Lord Elphinstone, Governor of Bombay. In the early days real estate in the area was parcelled out at a lease rent of five rupees per acre per year.

In the 1920s, R.D. and his family set out for Matheran. They first travelled on the Bombay–Pune line of the Great Indian Peninsular Railway (GIP) and changed at Neral. From Neral a privately owned match-box-sized train was hauled by a tiny engine. For two hours the engine puffed, spewed steam and blew countless whistles as it negotiated the sharp and frequent curves up the hill. When the train took the last curve through a thick forest, there was a perceptible drop in the temperature. As the passengers emerged from the station building they were greeted by two orderly rows of transport—one of horses and ponies and the other of colourfully painted, heavy wooden rickshaws, with three pullers for each rickshaw. No cars were allowed to operate on Matheran's deep red earth, so the family chose a humbler vehicle and made for their destination, a house on the crown of Matheran's steepest hill, owned by Sir Dinshaw Petit, a baronet.

Eva Wadia, a childhood friend of the Tatas, recalls the holiday in Matheran which she spent as a guest of the Petit and the Tata families.

The Petits, like Tatas, made their money in textiles but unlike Tatas, the Petit family kept expanding in the textile field. Tatas diversified. The families were close to each other. On this holiday trip were Fali Petit, the eldest son of the baronet and J.R.D.'s elder sister, Sylla, a beautiful girl. Jal Naoroji, the grandson of Dadabhai Naoroji, was coaching Sylla in tennis and she was to acquire the distinction of being one of the first Indian women to come up to the finals of the Western India Tennis Tournament. As Sooni was no more, R.D. had employed an Italian lady to be her companion. Eva recalls that one day the ladies heard footsteps on the roof and suspected someone was removing the tiles. Panic reigned as the women cried out *'Chor, Chor'* (Thief, Thief). It was not a thief but the future baronet—Fali and his brother. A few years later Fali Petit was to marry Sylla. Eva Wadia, at eighty-two, recalls that 'Jimmy (J.R.D.'s youngest brother) then only 7 years old would climb up to my lap (she was 19) and say: "I am going to marry you when I grow up." I had a snap of him sitting with me and others under a tree.'

Like all good things the holiday ended and soon J.R.D. had to catch a boat back to France to be conscripted. He recalls: 'If I had to join the army, I wanted to pick the cavalry. I had seen polo matches in Bombay and Poona. Polo seemed a very exciting game and I thought maybe I could play polo later but first I had to learn to ride properly. So when the time came to go into the army I pulled strings to join the cavalry, not too successfully as it turned out. My grandmother spoke to a General she knew. To my consternation, I was drafted in an Algerian regiment, an Arab regiment, where riding was very different from normal riding. You ride Arab horses—very exciting horses—on a different kind of saddle. This saddle had a pommel in front and if you jerked forward the pommel hit you in the plexus. If the horse galloped forward and you were not prepared for it, you lurched back where the rear saddle hit and bounced you forward, sometimes over the head of the horse.' So, to his dismay, far from learning to ride well enough to play polo he was constantly anxious of being pierced by the high pommel in front or being catapulted over the head of the horse. 'When riding you leaned forward in the saddle rubbing a part of the bottom different from the one on a European saddle. Within the first two days I had blisters. The Sergeant told me that when I got up in the morning I should go and sit in the horse trough. And for two days I did. It felt like fire but cured it.'

The regiment was called *Le Saphis* (The Sepoys) and was composed of Algerians, Tunisians, Moroccans and a few French; it was based in a dismal city called Vienne, south of Lyon. Brought up in luxury, Jehangir suddenly found himself with men from a totally different background. They were coarse in their manners and many of them were illiterate. Life in the barracks was pretty primitive. To his horror he discovered that most of his fellow soldiers never had a bath, save in summer. Jehangir was the odd man out because he went to a public bath once or twice a week. Many of the soldiers made fun of him for doing so.

Help came from an unexpected quarter. Jehangir's Captain discovered that an educated prodigy who even knew typing had joined the unit. 'The Captain, sporting the forbidding name of Massacrer, which in French literally means *massacrer*, promptly had me assigned as secretary in his office. One of the valued perquisites was that I did not have to get up at 5.30 in the morning to wash and groom my horse. The horse was ready for me. All I had to do was to climb the horse, sword in hand and exercise it. Another perk, so far as I was concerned, was that on my appointment to the good Captain's office, I was transferred from my highly redolent dormitory to a storeroom which I shared with an old Arab veteran called Guelool, who, unfortunately, smoked and coughed all night. I later discovered that the real reason for this transfer from the dormitory was to keep me out of sight for fear that I might be pinched by a higher regimental officer. It did not take long for the Colonel to find out that the First Squadron had an educated prodigy, who could not only read and write French and English but could even type. I was then transferred, for the second time, to the relatively luxurious office of the Colonel, and a bed was provided for me in a back room where I was gloriously alone.

'I became a favourite of officers applying for leave, whose applications I typed and who often graciously tipped me with a franc. During my year in the Saphis, I may have proved to be a good clerk, but I acquired little of military value for the possible defence of France. This was just 15 years before World War II broke out with its tanks and planes. We were given 6 rounds of ammunition, one live and five blanks. I fired a total of only 5 bullets. But I did learn to wield a sabre from horseback with reasonable proficiency. I remember feeling, and saying later, that if the Saphis were typical of France's Army, she would lose the next war. It nearly happened.'

Jehangir discovered that at the end of the twelve-month period of conscription, the educated soldier who volunteered for an extension of six months could be assigned to an Officers Training Course and after that, was eligible to attend one of the world's most famous riding schools at Saumur. 'I decided to apply for a year's extension of service. Fortunately, I had the good sense to consult my father,' he says.

R.D. angrily turned down the proposal. Though upset, Jehangir bowed to parental guidance, little knowing at the time that his father's decision would save his life. Shortly after he left for India, his squadron sailed to Morocco to fight the rebel chief, Abdel Karim. There it was ambushed and slaughtered to the last man, as the heroic garrison was in *Beau Geste*.

In the normal course, after his twelve months in the army, Jehangir would have proceeded to Cambridge for his engineering degree. 'But father decided that a college degree was not essential for a career in Tatas and summoned me to India. This decision is one I've regretted throughout my life and which caused me to have a long-lasting inferiority complex.'

On landing in India, Jehangir joined Tatas as an unpaid apprentice. It was December 1925.

CHAPTER VI

R.D.

When Jehangir returned to India after his time in the French Army, the house "Sunita" rang with laughter, the father setting the pace for the family's good humour. Perched at the top of Malabar Hill "Sunita" overlooked the sands of Chowpatty from where leaders like Mahatma Gandhi and Bal Gangadhar Tilak used to stir thousands with their call for the freedom of India. The bungalow, situated where an apartment building also called "Sunita" stands today, was furnished as Sooni would have liked it. Only occasionally some association of words or memory would bring about a tinge of sadness, as Sooni was remembered. It was silently borne and life moved on.

Bombay was a great and happy change for Jehangir. Almost overnight he had moved from the grimy barracks of Southern France to the shining taps and the gracious drawing-room of "Sunita". His yen for horses yielded to a desire for fast cars. Jehangir persuaded his father to invest in a blue Bugatti. When the car turned up R.D. discovered he had approved of a sports car without mudguards or a windscreen. Enthusiastic, Jehangir took his father for a ride in the Bugatti. A shaken R.D. showed no great anxiety to travel in it again.

Zal Taleyarkhan, one of his earliest companions in Bombay recalls spending a day with J.R.D. at an Altamount Road bungalow. J.R.D. had a rifle. Zal remembers going with J.R.D. to shoot crows and vultures. Huge vultures roosted on heavily wooded Altamount Road in close proximity to the Parsee Tower of Silence, attracted to the place by the dead bodies they scavenged on. Zal recalls J.R.D. aiming his rifle and firing and the loud report of a vulture falling in a compound. A Parsee from a neighbouring veranda yelled that dire consequences would ensue if the vultures were shot. Unconcerned, the boys proudly measured the size of the bird. It was seven feet long from wingtip to wingtip.

At "Sunita" every morning, at 8.30 sharp, R.D. would get into his office car. Rodabeh recalls Jehangir tearing away from the breakfast table and hurtling down the staircase, a piece of toast still in his hand, to join his father, for R.D. would not wait. Jehangir was soon to acquire his father's precision in matters of time. R.D. was the first to reach office and cleared his correspondence before visitors could call.

As Sooni was not there to receive the young children after school, R.D. made it a point to return home by 4.30 every evening. 'Father was very jovial,' says Rodabeh. 'He spoke French very well, but with an accent. He enjoyed life and had many friends. He was very helpful to people.' She pauses and adds: 'He loved people. We were blessed with two wonderful parents—human, intelligent, very much interested in all things; honourable to the nth degree; very lovable. Father was very quick tempered but he never slapped me. However, Jeh and Darab had developed a quickness of mind and agility to escape the ready slap directed at them.'

R.D. always spoke in French to the children and both Jeh and Rodabeh remember him asking them, 'Have you got your *sudra* and *kusti* on?'

'Father was religious, but in a big way. Granny Briere thought no end of him as an upright man,' says Rodabeh. Granny Briere was quite a character. After her daughter's death she felt all the more responsible for the children but stayed on in France where she looked after them on their occasional vacation to the continent. Rodabeh adored her. J.R.D. is less reverential: 'She was a very formidable lady. Her husband was a humorist and after some time with her ran away, as anyone would have done had he been married to my grandmother.'

'J.R.D. is unfair to her,' affirms Rodabeh. 'Granny was devoted to all of us.' J.R.D. admits she was, 'very, very fond of the children.' Granny lived on till the age of ninety-six, and died in 1946. Her hero was Marshal Petain, and even though Petain's reputation declined because of his collaboration with Germany in the First World War, to Granny, Petain ever remained the hero of Verdun. At that fateful battle in the First World War Petain told his soldiers, 'They (the Germans) shall not pass.' And they could not! Even when Petain surrendered in 1940, Mme Briere blamed the British, accusing them of wanting to swallow the French colonies. Admittedly, at ninety, she was getting senile. When they took over Paris the Germans would not allow people to leave Paris but Granny was permitted. 'I always say they allowed her to get out of Paris in self defence!' says J.R.D.

J.R.D. does not remember his maternal grandfather but says: 'In the 1940s, when the old man was about 80 or 90, I got a letter from him—I don't know how he knew me or of me—asking if I would be interested in operating a circus in Africa!' J.R.D. could not reply because the old man had forgotten to put an address on the envelope!

'Grandfather,' says J.R.D. 'was a joker, a humorist. He'd been in trouble all his young life. He'd been thrown out of almost every school he'd been in. He spent most of his military service in jail. I loved the stories about him. He was a wonderful fellow and he had a lot of fun.'

'Father,' says J.R.D. 'was a great tease, so we never knew what prank he would be up to. One day we were at the Ripon Club, where he had a lot of Parsee cronies. A friend was taking a drink at the dining table when R.D. smacked him from behind and the poor man spluttered on the table. There was almost a brawl. Then there was a dentist friend of father's who had palsy and his hands shook. Father went round loudly proclaiming he had pulled out the wrong tooth! A kind of childishness but he was good company.'

R.D. was about J.R.D.'s height and build; when he spoke, he sometimes stuttered. He had his flashes of temper—again a trait J.R.D. has. According to his son, R.D. was, 'very charming, a totally open, honest kind of man. He had no kind of pride in him. He was very unlike Sir Dorab who was pompous. R.D. spoke French, had lived in China, was cosmopolitan.' J.R.D. was fond of his father and admired him but, 'we never had time for a close and leisurely relationship. Father was not a great but a select eater. He loved French wine. Once he came home he threw away all care and over the weekends I don't remember him doing what I always do, having a box full of papers to clear.'

Some moments J.R.D. treasures, like the visit with his father to the steel city of Jamshedpur. He recalls staying at the Directors' bungalow and being included in all the discussions, though he was hardly twenty. He remembers meeting the fierce-looking, hard-swearing but kind-hearted General Manager of Tata Steel, T.W. Tutwiler (T standing for Temple instead of the "terror" he was.) Tutwiler knew how to produce steel but not how to handle British aristocracy. When the Viceroy, Lord Chelmsford, on a visit to Jamshedpur was being introduced, the Viceroy turned to him and said, 'And I presume you are Mr. Tutwiler?' The American barked back: 'You're goddamn right!' His Excellency took umbrage and Tutwiler was despatched by the Directors to New Delhi to apologize to the representative of the British Crown.

J.R.D. observed Tata Directors play poker with Tutwiler. Tutwiler could never understand how Indians could beat him at an American game but they invariably did.

Raw though he was at the time, J.R.D. was aware of the crisis that was brewing in Tata Steel. Flushed with the success of the war years, Tata Steel had set out to increase its production five-fold in order to meet the entire requirements of India. Just as they were pouring in funds for the expansion, steel was dumped into India from Belgium and England. The profits of the company fell from one year to the next, from ten million rupees* to 100,000** in the early 1920s.

J.R.D. can still picture the Director-in-charge of Tata Steel, John Peterson, clad only in pyjamas, sitting early one steamy morning on the veranda of the Directors' bungalow at Jamshedpur, typing a representation to the Imperial Government asking for protection to the steel industry.

On the recommendation of the Tariff Board, the Steel Industry (Protection) Bill was introduced in the Central Legislative Assembly in May 1924. It was a historic bill. For the very first time the British Government was introducing legislation to protect an Indian industry at the expense of a British industry. Would the Viceroy and his Councillors be able to withstand the pressures from home, especially given the fact that British trade was the prime purpose of the Raj in India?

Sir Charles Innes, the Commerce Member, moved the bill, 'to provide for the fostering and development of the steel industry in British India.'[1] Sir Innes said that one of the difficulties that confronted the government was, 'that the steel industry in India is represented by a single firm.' Silver-haired, a shawl thrown round his neck, looking somewhat like a Roman Senator of Caesar's times, Motilal Nehru rose to support the bill.

The debate was presided over by Sir Chimanlal Setalvad and a motion was moved that it be referred to a Select Committee. The names proposed for the Select Committee were those of Vithalbhai Patel (later Speaker); the eminent labour leader, N.M. Joshi, Chaman Lal, Pandit Madan Mohan Malaviya and M.A. Jinnah. As soon as Jinnah's name was proposed, a thin erect figure in a silk double-breasted suit rose to say that he had a small financial stake in the company but if it was all right with the House,

---

* Around Rs 240 million in 1991.
** Around two million rupees in 1991.

he would work on the Select Committee. Vithalbhai Patel interjected: 'Is it a big stake?' M.A. Jinnah, an affluent lawyer, turned to Vithalbhai Patel with a withering look and said, 'A big stake for Mr. Patel, but not for me.'

Pandit Motilal Nehru, whose name was also proposed for the Select Committee, rose to say: 'I have no interest in the Tata Company. The only interest I had at one time was that Mr. R.D.Tata had kindly put one of his motor cars at my disposal and I took good care of it...I had no other interest but do not despair of having some interest in future.' Looking intently from the gallery was a man of seventy with bright eyes and a twirled moustache. It was R.D. Tata.

As expected, the bill when debated was opposed by British trade interests in India, represented by the Associated Chambers of Commerce. It was also opposed by the Left, represented by those like Chaman Lal. Jinnah alleged that Chaman Lal espoused, 'the principles of socialism and Soviet doctrines.' Chaman Lal proposed nationalization of the steel company as did N.M. Joshi. Joshi insisted on a clause for the protection of the workers if protection was to be granted to the industry.

Jinnah fought for clarification: 'I want this House to carefully grasp the issue and not be led away by extraneous consideration.' The bill, he said, 'embodied a very important principle of state policy. For the first time protection was offered to an Indian industry. It is a mere accident that the Tata Iron and Steel Company is directly and very naturally affected. It was widely acknowledged that, during 1914–18, Tatas had supplied steel to the war effort at considerably less than the world price.'

After a heated debate the bill was passed by a voice vote in June 1924. A bounty was given on the output of railway lines and fishplates for three years and an import duty of fifteen per cent was slammed on imported steel. But for too long the company had struggled against great odds and it seemed that time was running out. As if other challenges were not enough, an earthquake struck Japan, the company's principal customer for pig iron.

The company was reeling under these blows when one day a telegram arrived from Jamshedpur that there was no money for wages. Sir Dorab and R.D. Tata went to the Imperial Bank and Sir Dorab pledged his entire personal fortune (which included his wife's jewellery) for ten million rupees* towards the loan of twenty million rupees** needed at that time.

---

* Around 250 million rupees in 1991.
** Around 500 million rupees in 1991.

He pledged his private fortune for a public company in which Tatas never had more than an eleven per cent interest. In addition to this burden and the Greater Extension Programme, the cost of which was rising three times above the estimate, there were debentures to be repaid. There was a period in 1924 when a good friend of R.D. Tata would call on him every day to ask when he was going to close down the works. Each day R.D. would reply:

> Ask me again tomorrow. We will be able to manage for today.[2]

At a grim meeting of the Directors someone suggested that the government be asked to take over Tata Steel. R.D. Tata sprang to his feet, and, pounding the table, declared that day would never come so long as he lived.

Fortunately, by the end of 1924, output had risen two-and-a-half times, thanks to the Extension Programme. The benefits of the Steel Protection Act were also coming in and Tatas were saved.

Throughout this baptism of fire, Tatas honoured their obligations. There was rationalization but no retrenchment. Every penny owed was paid on time but there were no dividends. At one of the meetings, where shareholders were anxious for dividends, R.D. pleaded: 'We are like men building a wall against the sea. It would be the height of folly on our part to give away any part of the cement that is required to make the wall secure for all time. That is why we and you have to use this money, to build up this great industry.'

The battle for the survival of Tata Steel and its wider impact on the nationalist movement, brought Motilal Nehru and R.D. Tata closer together and later Motilal stayed at "Sunita". At first R.D. had taken little interest in the freedom struggle but the battle in the Legislative Assembly brought him close to the nationalist cause.

Throughout the period of struggle to save Tata Steel, R.D. was also involved in saving the Tata Industrial Bank. 'Even as people were saying that Tatas were crashing he did not show the strain,' says Rodabeh. He went dancing in the evening and mixed with friends. But when the children had retired to sleep, in the still of the night, Rodabeh recalls her father pacing up and down on the veranda of "Sunita" overlooking Chowpatty. He was praying.

CHAPTER VII

Train To Jamshedpur

The Tata Empire of those days operated from Navsari Buildings. George Wittet, the man who designed the Gateway of India, was called upon to erect the new headquarters of Tatas off Flora Fountain in 1921. Called Bombay House, it was inaugurated in 1924.

In a room near R.D.'s, sat John Peterson. R.D. and Peterson came in close contact during the 1914–18 war when Peterson was Director of Munitions and R.D. had to ensure the delivery of artillery shells and steel rails on behalf of Tata Steel. A mutual respect developed. R.D. could not fail to notice that unlike others of his tribe in the Indian Civil Service (ICS), Peterson tended to be kind and more considerate to Indians than to his own countrymen. Given the problems confronting Tata Steel, R.D. felt a competent hand was needed to run the day-to-day affairs of the company. He asked Peterson whether he would consider joining Tata Steel. Peterson agreed, resigned from the ICS and was appointed the Director-in-charge of Tata Steel, stationed in Bombay.

Peterson's room, J.R.D. recalls, was very ordinary for someone of his rank. There was no padded furniture. Old-style wooden chairs and a solid wood table sufficed for Peterson. One day, soon after the arrival of J.R.D., in 1925, R.D. took him to Peterson's room. 'John,' he said, 'you know my son Jehangir. I would like you to look after my boy.'

That very day Peterson ordered a small desk to be installed in a corner of his room, and from that moment on, recalls J.R.D., 'Peterson never had a moment of privacy. Every single paper going to his desk was routed through me. I studied it before I sent it up. And I studied his comments before I sent them out. I must say that was a very formative and important time of my career, when I saw how a highly trained ICS administrator worked. I learnt a lot from that.' Except for one major break, this arrangement continued till 1931, when Peterson returned home.

After a few months with Peterson—and possibly in consultation with him—R.D. suggested that J.R.D. spend a year in Jamshedpur to understand the working of the steel company. In early 1926 J.R.D. went to Victoria Terminus to catch the Great Indian Peninsular Railway (GIP) train to Tatanagar, 1,000 miles east. As he journeyed across the Indian subcontinent for thirty-two hours, the romance of the steel city could not have missed him. Just a couple of years earlier, F.R. Harris had covered some of it in his book, *Jamsetji Nusserwanji Tata—A Chronicle of his Life*[1]. Through this book, and the first hand stories related to him by his father, the heritage of the House of Tatas came alive for J.R.D.

*

In the 1860s when Jamsetji Tata was about thirty, he went to a talk in Manchester by Thomas Carlyle. The steel age was just beginning. With vision Carlyle had declared: 'The nation that has the steel will have the gold.'

Ships were of timber at the time and so were bridges. The words of Carlyle hummed in Jamsetji's ears and some years later, in 1882, when Jamsetji was forty-three years old, he carried samples of coking coal and iron ore from India to Germany for testing. The ore was good but not the coal. The licensing laws of the British did not encourage private mining. Jamsetji postponed but did not abandon his plans. For the next seventeen years he kept a scrapbook of press clippings on minerals in India. In 1899 a report by Major (later General) Mahon was published. It said the time was ripe to establish an iron and steel industry in India. Mahon pointed to Eastern India where, at Ranigunj coalfields, coking coal had been discovered. Soon after, Lord Curzon wisely relaxed the rules for the licensing of mines. Jamsetji wasted no time and booked a passage to England. He arrived there in the twilight of the glorious reign of Queen Victoria. He saw Lord George Hamilton, Secretary of State for India, and told him of his desire to see the steel industry take root in India under Indian management.

Lord Hamilton, who held Jamsetji in high regard, encouraged him. Jamsetji told him that when he first thought of establishing a steel industry two decades earlier, he was young and ambitious. He would have undertaken the task for his own sake. Now he was sixty, and more than blessed with all he needed for himself. If, at this stage, he undertook a project as major as this, it would be for the sake of India. Hamilton was moved and said he would back him. Jamsetji replied that Viceroys come and go and a project as major as steel, with a long gestation period, needed a policy decision in writing. Hamilton promised to write to Curzon to support the Tata project. And he did.

Speedily Jamsetji cabled his office in Bombay to obtain prospecting licences for minerals and proceeded to the United States. He wanted the best technical advice. He studied coking processes at Birmingham, Alabama; visited the world's largest ore market at Cleveland; and in Pittsburgh met the foremost metallurgical consultant of the time, Julian Kennedy. Kennedy warned the enthusiastic, though ageing, Indian that even preliminary investigations would cost a fortune and there was no guarantee of returns. If, said Kennedy, a thorough scientific survey was made of raw materials and conditions, he would build the plant. He suggested the name of Charles Page Perin as the best man to undertake the survey. To Perin, Jamsetji went. Perin later described his encounter: 'I was poring over some accounts in the office when the door opened and a stranger in strange garb entered. He walked in, leaned over my desk and looked at me fully a minute in silence. Finally, he said in a deep voice, "Are you Charles Page Perin?" I said, "Yes." He stared at me again silently for a long time. Then slowly he said, "I believe I have found the man I have been looking for. Julian Kennedy has written to you that I am going to build a steel plant in India. I want you to come to India with me, to find suitable iron ore and coking coal and the necessary fluxes. I want you to take charge as my consulting engineer. Mr Kennedy will build the steel plant wherever you advise and I will foot the bill. Will you come to India with me?"

'I was dumbfounded, naturally. But you don't know what character and force radiated from Tata's face. And kindliness too. "Well," I said, "Yes, I'll go." And I did.'

Before Perin arrived he sent his partner, geologist C.M.Weld, to prospect for the raw materials.

Weld arrived in the summer of April 1903 and in the searing heat of Central India set out exploring with Dorab Tata and a cousin, Shapurji Saklatvala. Chanda district was one of the finest in the country for *shikar*. The party, however, was not hunting for tigers but for iron ore. They travelled by bullock cart over rough terrain. Clean water and food were difficult to obtain. They were often compelled to brew their tea with soda-water. As the days went by, the immensity of the task they had taken on began to dawn on the prospectors.

After many an adventure in prospecting a letter arrived from an Indian geologist , P.N. Bose, who had originally mapped the Durg area for ore. Now working for the Maharaja of Mayurbhanj, he had discovered rich iron ore in the state. It was within range of the Bengal coalfields and the ruler was keen to develop his state. In the wooden hills where elephants roamed and tribal Santhals eked out a precarious existence, the lofty Gorumahisani Hill rose to 3,000 feet. It was a superb storehouse of iron ore, later estimated at thirty-five million tons. The iron content was sixty per cent. Other neighbouring hills were also rich in ore. All the prospects were pleasing, but where was the water? A proposed reservoir had proved impracticable. Early one morning Weld and his Indian assistant, Srinivas Rao, plodded down a dry stream on their horses. It was heavy going through the sand. Said C.M. Weld:

> At length we came upon a sight which filled us with joy; a black trap-dyke, crossing the river diagonally, and making an almost perfect natural pick-up weir. It seemed too good to be true.[2]

Weld and Srinivas Rao clambered up the river bank shouting with excitement. They found themselves close by the village of Sakchi, near the meeting point of two rivers, Kharkai and Subarnarekha, meaning "gold-streaked", which between them, never run dry. A couple of miles away was the railway station of Kalimati (now renamed Tatanagar.) They had come to the end of their search in 1907. Jamsetji had passed away in 1904 but his vision was to outlast him.

A year after the site was found, a visitor to Sakchi village records:

> There were no roads from the station to the Sakchi camp. Job seekers had to find their way with much difficulty along the Susingeria jungle (where the Susingeria Gate now stands.) In the beginning, a few tents and thatched huts dotted amidst a jungle clearing, housed the small colony of people who were helping to lay the foundation of the Steel Works. The pioneers spent a hard and adventurous life, defying all dangers and

discomforts. No one could move out of his tent at night as wild animals from the neighbouring hills howled and prowled all around. The normal amenities of life were unknown. A cup of tea or milk was a luxury to us. Water was brought in wooden barrels from the Subarnarekha and distributed on a rationed basis. Sometimes, we had to go without water for hours, and there were occasions when we had to boil eggs and potatoes in aerated water![3]

Only three years later Lovat Fraser, an English journalist, noted:

I walked through street after street of commodious one-storey brick houses, all well ventilated, all supplied with running water and lit by electric light. Many of the houses had electric fans.[4]

It was still a hazardous undertaking to get to Sakchi and one could be waylaid by robbers even if one took one of the eight tongas which Pathans drove to the city. Then, as now, some trains arrived at odd hours and people preferred to sleep at Tatanagar station till dawn, before they resumed their journey. Each arrival was a great event. As many as ten men would accompany a friend safely to his train or collect him on arrival. Occasionally passengers were pressed into service at the steel plant. Recruitment officers, it is said, would crane their necks into crowded compartments and when they saw a well-built man they would invite him to stop off at Tatanagar, if he wanted a job as carpenter, welder or in a semi-skilled trade. 'You could not afford to be idle in those days,' recalls P.K. Chatterjee, who arrived as late as 1921. 'If a cluster of idle men were located, company officers who usually travelled on horseback would ask them to come along, give them a brass token and they started working that day or the next.'

Putting up the Works itself was a mammoth venture. The main machinery arrived by train from Bombay or Calcutta. An entire city was laid out, the first planned modern city of India (New Delhi and Chandigarh were to follow.)

In 1902 before the site of the steel plant was even located, Jamsetji when abroad, described his dream city of steel to his son Dorab in a letter: 'Be sure to lay wide streets planted with shady trees, every other of a quick-growing variety. Be sure that there is plenty of space for lawns and gardens. Reserve large areas for football, hockey and parks. Earmark areas for Hindu temples, Mohammedan mosques and Christian churches.'

Two decades after Jamsetji penned these lines, J.R.D. first visited Jamshedpur. The dream had come true. In the intervening years men of steel had raised a city out of a jungle.

In the years that intervened between 1908 and 1924 the steel plant won recognition as the mother of heavy industry in India. About a year before J.R.D.'s journey to Jamshedpur, Mahatma Gandhi had visited the place at the invitation of the Directors and the workers. The lady in charge of the new Directors' bungalow, where Gandhiji was to stay, was Lilian Luker Ashby, wife of the Deputy Superintendent of Police, Jamshedpur. There was much excitement at Gandhiji's arrival, with flowers and bunting in the streets and cheering crowds crying "Mahatma Gandhi *ki jai*". When Lilian Ashby showed him his room in the Directors' bungalow, Gandhiji was taken aback to see the British mattresses and pillows, drapes and carpets. Gandhiji requested her to remove them all. His wishes were obeyed but the room looked pretty barren. 'Will you be quite comfortable with so many things removed from the room, Mahatmaji?'[5] asked Lilian Ashby. 'I am quite comfortable with just a garment,' was the rejoinder.

There, at the heart of India's heavy industry, Gandhi installed his wooden *charkha*. After a short stay, Gandhi left one night to catch his train at Tatanagar station. The earlier floral decorations had withered, there were no cheering crowds, no flaming torches and only a few half-hearted cries of "Mahatma Gandhi *ki jai*" from the sparse crowd at the railway platform. But wrote Lilian Ashby, 'the Saint had passed that way.'

Not everyone who greeted J.R.D. at Jamshedpur was saintly. On arrival he shared a spacious bungalow at B. Road with J.K. Sondhi, the Town Administrator, Jehangir (called Jo) Ghandy, who was later to become the first Indian General Manager of the company and Jo's brother, Dinshaw (called Dinsi) Ghandy. Dinsi was the boss. 'Never before or since,' says J.R.D., 'have I heard anybody swear as he did.' Every Sunday after salary day he would line up the battery of servants as if they were guests at a Viceregal reception. Then he would go down the queue. As he ceremoniously handed an envelope to each, he selected the choicest Parsee–Gujarati swear words "*Sala mamai...*" "*Sala madar...*" and so down the line he would go. Admire as he did Dinsi's fluency, J.R.D. restrained from shaping himself after him. Instead he concentrated on absorbing the contents of a tome he had with him, *The Shaping and Making of Steel*.

'I had to go to different departments from time to time and I read in *The Shaping of Steel* the relevant chapter before I visited each unit of steel making.' He remembers showing Sir Purshotamdas Thakurdas, KBE, round the steel smelting shop No.2; just as Sir Purshotamdas neared it, one of the steel pitchers that tilts itself and blows hard, let fly a thick tongue of fire. Terrified, the knight of the British Empire, took to his heels and ran the whole length of the shop—his dhoti flying.

While J.R.D. was observing and learning at Jamshedpur, summer came along and R.D. left for France to spend some time there with his other children. Granny Briere was there to welcome them. The family went for a holiday to the Hardelot beach. It was a weekend. R.D. danced with Sylla on a Saturday night, came home and complained he was not feeling well. He went to the toilet and collapsed. It was a fatal heart attack. The news came to J.R.D. in Jamshedpur, contained in a telegram from the Bombay office.

'Though he was 70,' says J.R.D., 'to me he was a man who still had 20 years left in him. I had no inkling that I was not going to see him around but he seemed to have had a premonition about his death otherwise he would have allowed me to go to Cambridge and get a degree of some kind.'

Wistfully, J.R.D. adds: 'I remember him for his spendthriftness, no, it's a wrong word—generosity. He was always helping people, giving out money, spending money, building this white elephant, "Sunita", a beautiful house. So, although I am left with the impression of a very fine man, I would have liked to have known him more closely. We were separated so much of the time. It was always a joy to get together again but it didn't last very long.'

Regular flights by Imperial Airways from London to Karachi had not yet started, so J.R.D. could not get to France in time for the funeral. R.D. was buried next to his wife at Pere Lachaise in Paris. J.R.D. took the first train to Bombay. Happy memories and sad ones crowded his mind as the train sped westward—memories of father coming home after attending a board meeting and distributing the guineas he was paid in to his children; memories of Ripon Club; of Hardelot; "Sunita" and the happy times with father in Mahabaleshwar and Matheran, hill stations where R.D. travelled over weekends to be with his family. And when on one occasion when he saw his father arrive, J.R.D. did not stop play to run and greet him how upset father was! And how he sulked! It is said "Love knows not its depth—until the hour of separation." As the train neared Bombay, he realized the depth of his affection for his father. His responsibility as the head of the family began to dawn on him.

The stature of his father was evaluated by men such as Jamnalal Bajaj, the industrialist who was the treasurer of Gandhiji and one of the pioneers of the Swadeshi movement in India. He wrote in the *Nagpur Times*:

> If all businessmen in India would acquire half his love for things Indian, there is no reason why all our enterprises should not flourish.

The *Indian Mail* wrote in its obituary of R.D. Tata:

> His courage was such that those who worked with him were transported to the heights of the greatest altitude when they knew they had cause to be despondent.

CHAPTER VIII

Years Of Endurance

At twenty-two J.R.D. found himself the head of the family. Sylla was twenty-four and yet to be married. Rodabeh was nineteen, Darab fourteen, and Jimmy only ten. J.R.D. had still to find his feet in India. He had been to schools in France, India, Japan, England and had never had the anchor of residing in one place for long. Although he was no stranger to India, culturally he was more French than Indian. What is more, he had hardly a trusted relative or friend of R.D.'s he could turn to. Perhaps the struggle to save Tata Steel had taken a greater toll of both R.D. and Sir Dorab than either had realized. Sir Dorab, though living, was past his prime. He was afflicted with diabetes and had become a crotchety old man. The one person J.R.D. could turn to for advice was a solicitor called Dinshaw Daji of Crawford Bailey & Co. And given the state his father had left his financial affairs in, J.R.D. needed legal help.

R.D. was a prince of a man in his generosity. He also had the income of a prince. From Tata Sons he could draw upto Rs 200,000* per year, while Sir Dorab could draw Rs 300,000**. Taxation was low. Sir Dorab was careful with his money. R.D. was a spendthrift. R.D. had houses in Bombay, Pune and Hardelot. While he earned well from Tata Sons, he continued his private cotton business in the Far East and other businesses in London. In his last years R.D. was absorbed in the affairs of Tata Sons and his own private companies abroad were neglected. They ran up large losses. The London manager, an Englishman, used to speculate on the market with company money. J.R.D. says he passed on the losses to Tata Ltd. and pocketed the profits. Lion-hearted though he was, R.D. could also be gullible and was easily fooled. In the last year of this life R.D. and

---

* Around six million rupees in 1991.
** Around nine million rupees in 1991.

Sir Dorab had decided voluntarily to forego their dues from Tata Sons in view of the steel company's financial crunch.

R.D. had borrowed from Tata Sons and from Sir Dorab personally. On R.D.'s demise the Directors of Tatas—in view of his outstanding services—decided to write off R.D.'s debt to the firm. Sir Dorab, however, insisted that every penny of his personal loan be paid by R.D.'s family.

J.R.D.'s first priority was paying off his father's debt to Sir Dorab. "Sunita" was sold and J.R.D. moved into a suite at the Taj Mahal Hotel. The house at Ganeshkind was also sold, and later the property at Hardelot. All that remained intact were the shares of Tata Sons—a third of the total.

'It was a difficult period,' says J.R.D. 'because I did not have the background or experience. The Will said that I should get the first Rs 3,000 a month, Darab Rs 2,000 a month, Jimmy Rs 1,000 a month and then what was left should be divided among the five of us. In fact I would have got, I think, the first Rs 3,000 or Rs 2,500 and nobody would have got anything else at all. But we sorted it out easily. I remember consulting father's old friend Dinshaw Daji. He said, of course you could do what you like if you all (brothers and sisters) agree. So we ignored the Will. I decided the family income would be divided equally—1/5th each, including the shares. That was that.' The debts were soon paid off and R.D.'s offices in London and the Far East were wound up.

The one security J.R.D. had was that he inherited his father's position as a permanent Director of Tata Sons and the company sanctioned him a salary of Rs 750 a month to start with. The little desk adjoining John Peterson's gave him the base he could operate from.

*

For J.R.D. the friendship of Peterson and his guidance meant much. J.R.D. had earlier hoped to go back to Cambridge after a spell in India but now the die was cast. Like Winston Churchill who missed out on a university education too, J.R.D. determined to educate himself. Writing about Churchill's early years Ted Morgan says that he was a,

> self-created personality, who found ways to complete his missing parts. He lacked the well-rounded education of a University man. At Sandhurst the studies were martial. He

resolved to exercise his "empty, hungry mind" by reading works of history and philosophy.[1]

While in the army at Bangalore, in his early twenties, Churchill had his field exercises each morning. In the afternoons, when his fellow officers slept or played cards, he had a programme of five hours reading from Edward Gibbon's *Decline and Fall of the Roman Empire* and Thomas Macaulay's *History of England*—twenty-five pages of one and fifty of the other. Of this period of self-study Churchill said years later: 'First we shape our dwellings and then our dwellings shape us.'

Soon after his father's death J.R.D. had attacks of typhoid and paratyphoid. Rodabeh remembers seeing J.R.D. return from work, throw himself onto his bed at the Taj and pick up financial and business magazines. When she suggested he should rest he would refuse. 'I want to be worthy of Tatas.' So saying he would read on.

Over sixty years later J.R.D. still blames his father for not sending him to university and still refers to himself as "uneducated" or "semi-educated", an expression that upsets his close friends.

'What would you have taken up at Cambridge?'

'Mechanical Engineering.'

'You so wanted to go to Cambridge but it didn't work out. Could this factor account for your desire to educate yourself and be as good as anyone at your own work?'

He raised his eyebrows. I make the question more explicit: 'What has been the driving force in your life?'

J.R.D. replies, 'Honestly, if I had not been the son of R.D. Tata or the son of one of the main Tatas, I don't think I would have been so driven. I was driven by the fact that there was Jamsetji Tata in my life and so that is what urged me to do things to justify myself. I was very doubtful about my own capacity to follow these people. I had great admiration for my father, little for Dorabji Tata, enormous for Jamsetji Tata and what Tatas meant.'

The period immediately following the unexpected death of his father was the watershed of his life. The carefree young man was landed overnight with the responsibility of looking after his four brothers and sisters, with no close relatives interested in their welfare. He was also confronted with taking on the mantle of his father in business. The challenge, expressed in his own words, "I want to be worthy of Tatas" was to provide a motivating force in those formative years, which transformed him into a mature and responsible person at the age of twenty-two.

In those early days his main preoccupation was with the affairs of Tata Steel—an association that extended to sixty years. Between 1926 and 1931, the year when Peterson left India, the Scotsman was a major influence on J.R.D.'s life. Peterson kept J.R.D. out of trouble. One day J.R.D., then in his twenties, received a draft of a letter written by a Tata Sons Director, Burjorji Padshah. Padshah, a professor brilliant at Mathematics, had worked on establishing the Indian Institute of Science in Bangalore. He corresponded with Albert Einstein on abstruse subjects which the two of them seemed to understand but few others did. J.R.D.'s confidence in his newly acquired command of English was such that he slashed and corrected Padshah's draft. Peterson held back the enthusiastic youngster. 'Perhaps,' says J.R.D., thinking back, 'I was influenced by my father. He once told me that he had sent back a letter to Burjorji Padshah stating: "Please translate it into English!"'

'Peterson was one of the very rare people I respected both as a man and as one with an educated brain. There are very few men I know like that,' says J.R.D. 'I remember once driving with Sir Dorab somewhere in Vienna and he criticized my mentor, Peterson. I turned around and said to him, "Dorabji, it is not right that you criticise one of your colleagues and what is more, it is not true."'

The irreverent Tutwiler, the boss of Jamshedpur, at first refused to acknowledge Peterson as his Director-in-charge seated in Bombay. 'What does he know of steel?' Tutwiler would growl. 'A Government official! I will have nothing to do with him.' The first time Peterson was to arrive in Jamshedpur, Tutwiler sent Bombay a message to book his passage to America. Peterson's trip was postponed. The second time Tutwiler again tried to avoid him. The third time he agreed to meet Peterson as Tata Steel was struggling for survival and Tutwiler realized that Peterson was also needed if the plant was to survive. Consequently they became good friends. Speaking of Peterson J.R.D. says, 'He became a kind of father to me. It is from him I acquired many qualities.' J.R.D.'s ability to clear papers with speed and clarity came from Peterson, as also some of his passion for perfection. At eighty-seven he is as particular as ever. 'I know that aiming at perfection has its drawbacks. It makes you go into details you can avoid. It takes a lot of energy but that is the only way you can achieve excellence. So, in that sense, being finicky, is essential.'

French was his mother tongue and in those years he says he partly thought in French. As the decades rolled on, he no longer thought in French but he still counts numbers in French.

After his father's death J.R.D. realized that his place was in India. In the 1920s, he says, 'I had two personalities. I was an anti-British Indian and I was a little more of a Frenchman. Little more of a Frenchman than an Indian because French was my language.' But as the years unfolded he became more involved and integrated into the Indian scene and conditions. In 1929 he decided to surrender his French nationality. In those days you were allowed dual nationality and his father enjoyed both British-Indian and French nationalities. It was a wrench for J.R.D. to sever his French connection but after much thought he wrote to the Ministry of Justice, Paris, in 1929. He referred to how his father was made a Knight of the Legion of Honour by France at the end of the First World War, and how he happened to be born in Paris; he felt, however, that the time had come to renounce his French citizenship and his place in the French Army.

For J.R.D., switching from the French to the English medium of writing was necessary. When he commenced his career in Tata Steel he had a stenographer called Iyer, 'a very nice, very quiet old man. As I did very little dictating in those days, I used to write by hand. Then Peterson said, "Look, you must get into the habit of dictating otherwise one takes more time." Some people can't dictate at all and I certainly couldn't (in English.) I was terrified of dictating. So I'd send for Iyer to take a letter and read out from a letter I'd already written and concealed from view beneath the table, to give myself confidence.' One day a piece of paper J.R.D. was dictating from, flew off his desk (thanks to a ceiling fan) and Iyer picked it up. The game was over. Today he prefers to dictate than to write for in that way he can keep changing his draft and improving it, sorely testing his secretaries.

J.R.D. had a love of literature—and still does—but his earlier readings were mostly confined to French writing. Peterson not only instructed J.R.D. in his office work but instilled in him a love for the English language and poetry.

A beautiful red leather scrapbook with the name "J.R.D. Tata" inscribed in small gold type has been his abiding companion since the early 1920s. As he was far more conversant in French than English in the early days the first poems are mostly in French with Charles Baudelaire,

Victor Hugo, Alphonse de Lamartine and Alfred de Musset predominating. Some of the poems are illustrated rather attractively in ink. More artistic and by his own admission more difficult to draw than the Lindbergh plane (which also features in the scrapbook) are some of his reverse drawings. Occasionally, as in the case of three birds, the drawing has little connection with any poem and it seems the artist in him has urged him to attempt a drawing for his own pleasure.

André Maurois says of Benjamin Disraeli that in his early years, Disraeli was artistic but at the same time he was a man with a tremendous drive for action. In some ways J.R.D. has this combination of qualities. J.R.D. regrets that in his youth his travels from Bombay to Japan, Japan to Bombay and Bombay to Paris, made it impossible for him to have constant access to the piano and thereby to cultivate his taste for music but he did cultivate his love for poetry.

The greater part of the scrapbook was written in the 1920s. The most frequent themes of the excerpts quoted are death and love—in that order. 'I don't know why,' he said in 1987, 'although I am full of fun there is no doubt that in poetry, for instance, it is the more tragic things and death which appeal to me.' Having said that he went on to quote, from memory, a French poem by Baudelaire beginning *"O Morte"* ("O Death") and ending:

> *Let us go anywhere,*
> *I do not care where.*
> *So long as it will be*
> *different from this*
> *awful place we are in now.*

Just occasionally his voice cracked with emotion. Though such poems give an inkling of his mood and preferences at various times, he is careful not to give expression to his deeper thoughts, either in the scrapbook or elsewhere, save in some of his letters. He regrets he did not keep a diary that could have reflected his thoughts.

One poem which he likes to read and ponder is "I have a Rendezvous with Death...." by Alan Seeger:

> *I have a rendezvous with Death*
> *At some disputed barricade,*
> *When spring comes back with rustling shade*

And apple-blossoms fill the air
I have a rendezvous with Death
When spring brings back blue days and fair...

There is pathos in "Caprice" by Sarojini Naidu, "India's nightingale" whom he had met:

You held a wild flower in your finger tips
Idly you pressed it to indifferent lips
Idly you tore its crimson leaves apart...
Alas! it was my heart.

You held a wine cup in your finger tips
Lightly you raised it to indifferent lips
Lightly you drank and flung away the bowl...
Alas! it was my soul.

The poem on "Identity" amuses him today as much as it had sixty years ago, when he jotted it down. As he flips the pages of the scrapbook, he pauses and reads with a chuckle:

Somewhere—in desolate wind-swept space—
in twilight-land, in no-man's land
Two hurrying shapes met face to face
and bade each other stand.
"And who are you?" cried one shape
Shuddering in the gleaming light.
"I know not" said the second shape
"I only died last night."

Jotted at various times the inscriptions have no sequence. Sometimes there is a brief quote that once caught his fancy:

He took his misfortune like a man—blamed it on his wife.

Another:

I was sitting in jail with my back to the wall,

And a red-head woman was the cause of it all.

Shortly after this entry comes one from Tennyson:

But, O, for the touch of a vanished hand,
And the sound of a voice that is still.

Just occasionally a bit of advice creeps in:

To be kind to all, to like many and love a few, to be needed
and wanted by those we love, is the nearest we can come to
happiness.

Of the poems he refers to often in his eighties is one by Henry Shore
called "The Nomad":

When our ancestors found that wheat
Was a good bread to eat
They settled in Jericho.
All of us are settled now,
But in our souls there is great woe:
We don't know where to go.

I am settled in a fine place
I own a house, I live in grace,
I have a patio
But late at night when the wind laments
And the garden shivers—my soul is rent:
I don't know where to go.

One day when I say good-bye
To life and wife, and die and fly
Somewhere in a great flow
I shall be free to roam again
I'll try to find but try in vain
Where to go, where to go.

Our favourite poems are often a reflection of our own thinking at any given point of time. 'I was attracted to poetry because it sounds like music,' says J.R.D. He likes poets like Henry Shore and dislikes free verse.

Langston Hughes' poem, "Dreams" features in the scrapbook:

Hold fast to dreams
For if dreams die
Life is a broken-winged bird
That cannot fly.

Hold fast to dreams
For when dreams go
Life is a barren field
Frozen with snow.

Under the poem he has drawn a bird.

When I saw in his bookshelves a copy of the book, *The Miracle of the Bells* I recited a short poem—my favourite—at the start of the book. I picked up the book and showed him the poem. He read it aloud:

We're of one flame—all kin of stars and sun.
The brothel's beacon—altar's candle—one.
The sluggard's lamp—ambition's raging fire.
Saint—sinner—sage and fool—Life's deathless pyre.
The Christ who cried to One in agony—
The thief who cursed him from the neighbouring tree—
All God's—who out of Darkness ordered Light
And gave man's soul the miracle of sight![2]

'Leave the book here,' he said, 'I'll copy it down.' When I next saw the scrapbook, he had written down the lines.

CHAPTER IX

Meeting Thelly

In his father's lifetime, J.R.D. had asked for a motor bike but R.D. was dead against it. After R.D. died, J.R.D. bought a second-hand, two-cylinder motor bike for a hundred rupees. 'I forget what it was called except that it was British, and had a gearbox like a car.' After having run around on it for a while, J.R.D. decided to find out how it was built. 'I took it apart. When putting it together, I found it had a few surplus parts...' He was lucky to get the same hundred rupees he had originally paid for it.

In this twenties J.R.D. loved his Bugatti. It had no mudguards, no roof and he could cover Bombay to Pune (a distance of roughly 120 miles) in two-and-a-half hours which was probably a record at the time. J.R.D. admits that if he hadn't taken to flying, he would have loved to have taken up motor car racing. 'I like everything that is a little on the edge, on the verge of disaster—living dangerously....The car was the love of my life then.

'Unfortunately, the Bombay Police took a special interest in the car and I later learnt that they intended to nab me at all costs. They thought they had succeeded when, in 1929, they tried to involve me in an accident. After an evening at Juhu, Russa Mehta and I left in separate cars and decided that we would link up at Kemp's Corner. When I got there Mehta was not present, so I waited. After some time I got a message that there had been a major accident up Peddar Road.'

J.R.D. drove along Peddar Road slowly. In those days it was a beautiful road with one-storeyed bungalows, surrounded by compounds lush with trees. Bombay's traffic was slight and at sunset one could always find a man with a tall steel hook walking from lamp-post to lamp-post, lighting the gas lamps. Some lamp-posts were beautifully fluted and usually painted silver. 'As I drove along Peddar Road, I suddenly stopped to find that Russa's car had banged against a lamp-post. His companion's lifeless hand was hanging out. The companion was the brother of Kanga, a famous cricketer. The police thought this was too good a chance to miss and filed a charge against me, alleging that Russa Mehta and I were racing along Peddar Road.' J.R.D. needed a good lawyer and was advised to see Jack Vicaji, a top criminal lawyer. It must have been a Sunday, so he had to visit Vicaji's home behind the Taj. A second accident took place! J.R.D.'s eyes alighted on a niece of Vicaji. 'She was a very beautiful woman,' he says.

'At the court hearing, the police had trumped up witnesses who collapsed under the stiff cross examination of Jack Vicaji. For example, a policeman had claimed he could read the number plate of a car driving fast with its headlights on. A fine old magistrate called Jungalwalla dismissed the case and passed strictures against the police. The police were so annoyed, not only at their failure to nab me but more so at the strictures passed by the court, that they harassed me to a point where I decided to sell my beloved blue Bugatti.

'It was not till 30 years later, thanks to Tatas' association with Daimler-Benz, that I laid my hands on the wheel of a modern, high speed car, the Mercedes 300 SL. I drove it at accelerating speeds and without incident over the roads of Europe. The performance I remember most was between Stuttgart and Munich, when I averaged a 100 miles per hour, which means most of the way I was driving at 130 miles per hour, only having to slow down when passing other cars and trucks. A great regret in my life in the motor car field was that I was never able to drive, let alone own, a Hispano Suiza, the finest and fastest car in the world, produced just after World War I and the 12 cylinder Ferrari, which is till today, one of the most prized sports cars in the world.'

Even in his eighties J.R.D. regularly reads one magazine on motor car racing as he does one on aviation. And in the late 1930s he attended the La Mans car rallies whenever he could.

In the days after the court case J.R.D. who still lived at the Taj would, every now and again, glimpse the girl who had captivated him in Vicaji's flat which, fortunately for him, was directly behind his room. He was to learn later that Thelma (called Thelly) and her sister Kitty, whom Uncle Jack looked after, were daughters of Jack's brother and wife Muriel. Muriel was an intrepid English doctor who, in 1908, went across Russia in the Trans-Siberian Express. She married Thelly's father, Sohrab, the same year but the marriage broke up after some years. She made her own life in Italy, where she remarried. At this time Thelly's father was travelling around the world making a fortune and losing it. When he came to Bombay he stayed at the Taj, where he was called "Prince Vicaji" for his lavish taste. Prince or not, the Vicajis had princely connections. In the early nineteenth century, they were extremely wealthy.

In a book on the Parsees, Piloo Nanavutty writes:

> Parsis were (also) pioneers in the field of banking. The first Parsi bankers we hear of are the brothers, Vicaji (1781–1853) and Pestonjee Meherjee (1799–1854). They came from the port town of Tarapore in Thana district near Bombay. They were revenue contractors and bankers. Their success attracted the attention of the Nizam, Nasar-ud-Daulat of Hyderabad, who placed them in charge of his mint. In 1840, a silver coin was minted with the initials "P.V." and was called the "Peshtanshahi sikka". Two hundred and twenty thousand of these silver coins were minted and were in circulation. This is the only occasion when an Indian coin was named after a Parsi and a non-Prince. The brothers lived in grand style, and even advanced forty-one lakh rupees to the Nizam who refused to pay his debt. In 1851, the two brothers went to England and appealed to Parliament, but to no avail. They became insolvent and died within a year of each other.[1]

The ruling Nizam wanted to settle their loan in 1840 at five annas in a rupee, less than a third of the loan. The Vicajis had refused.

In 1880 Sir Salar Jung, the distinguished Prime Minister of the Nizam, realized that a great injustice had been done. As it was within his power, he allocated a pension to the Vicaji family, which continued for a few decades. But Thelly's father Sohrab and another brother of his, Rustom (not Jack), were headstrong characters, and they decided to take to a court of law, His Exalted Highness, the Nizam, known as the "faithful ally of the British Crown". Not only did the Vicajis lose the case, they lost the pension too!

Neither "Prince Vicaji" nor his estranged wife Muriel seemed to care much for their beautiful daughters and it was left to Uncle Jack to bring them up. He looked after them as if they were his own children and each year in the summer, when Bombay's courts closed for a long vacation—designed for British barristers and judges to sail home—Uncle Jack would go with his neices to Europe, mostly Italy.

Thelly was born in America and had some of her education in Italy. When Muriel left her husband the children were left with a governess in America and Kitty, elder only by a few years to Thelly, was to take the place of the mother in Thelly's life. All who knew Thelly in her twenties, say she was incredibly beautiful. She had studied at the J.J. School of Art and drew good sketches and painted portraits. She liked swimming and dancing. 'Always well dressed in skirt or sari, she never went out of her way to dress to kill,' says one of her friends.

The other Vicaji brother Rustom had married a Belgian lady and his marriage too had broken up. They had only one son, Bobby. Bobby's mother had left Rustom Vicaji and had settled in Belgium with her little son. Struck down with a fatal illness she realized she would soon die and, therefore, requested Kitty, then hardly twenty, to look after her son. Not long after that Mrs Vicaji died. Little Bobby was brought to Bombay and was entrusted to Kitty and Thelly in the home of their kind uncle, Jack Vicaji. When J.R.D. started courting Thelly, Bobby was a precocious youngster of twelve. Whenever J.R.D. arrived in the drawing-room of the Vicaji home, the others would make excuses and withdraw but Bobby would stay put. One day J.R.D. asked the little man politely, 'Have you nowhere to go?' 'No,' replied Bobby. 'Would you like to see a film?' asked J.R.D. 'Yes,' replied Bobby, 'but I have no money.' So J.R.D. gave him some. Bobby found his presence in the drawing-room a fruitful source of income. He began to look forward to J.R.D.'s arrival almost as much as Thelly did. At times when he was not in the drawing-room, Bobby would make his entry as soon as he had heard that J.R.D. had arrived. And J.R.D. would promptly dig into his pocket. A time came, Bobby says, when he would open the drawing-room door and J.R.D. would slip the money into his hand without a word.

On 15 December 1930, J.R.D. and Thelly were married. Muriel wrote to send her warm wishes to her daughter.

The young couple could have been better advised on their choice of place for their honeymoon. At the height of winter they went up to one of the highest hill stations in the Himalayas—Darjeeling.

Says J.R.D., 'The only important thing about our honeymoon was that it enabled us to see Kanchenjunga from close proximity. We went to Darjeeling in winter, December, not realising how cold it was. All tourists to Darjeeling go to Tiger Hill at 4 o'clock in the morning to see the sunrise on a little white cone at a distance, which we're told is Mount Everest. Instead of that, we went up on a two-day jaunt partly riding on horseback to Tongloo and Sandakphu. Sandakphu is 12,000 ft. up and there is nothing in front of you but a long valley and Kanchenjunga, there like a cathedral before you. In fact what was most extraordinary was that the whole of the valley, between Sandakphu where we were and the Kanchenjunga, was just 1,000 ft. below us. Right up to Kanchenjunga was an uninterrupted layer of cloud—so that Kanchenjunga was sticking up by itself and what looked like a few miles away was really about 40 miles away. It was so cold that we had to spend the night there with a wood burning stove. And we stuffed newspapers under our clothes to keep warm.'

Darjeeling was the summer capital of West Bengal. Government House stands towards the edge of the hill at Darjeeling's highest point where the cliff drops sharply to a plateau that runs flat for miles and miles. At the end of it one sees a magnificent unbroken range of white mountains, several of them 25,000 feet high and above. Almost in the centre, and presiding over the range, stands the most beautiful one—Kanchenjunga. Lord Hunt (leader of the first successful expedition to Everest) reckons that Kanchenjunga is more difficult to scale than Mount Everest.

In the late afternoons as the sun begins to set the entire range turns pink. As the sun lowers, a dark shadow creeps over the base of the mountains. The shadow slowly rises as the sun sinks in the west. The broad band of pink on the white snow recedes and a shadow overtakes mountain after mountain, peak after peak, till the twin cones of Kanchenjunga stand out, pink against the blue sky, before they too are overtaken and dusk descends on the majestic mountains.

A British Governor, with a fascination for freezing weather, turned up in Darjeeling that Christmas.

When returning from their honeymoon, after Christmas, J.R.D. and Thelly were driving down by car. His Excellency, the Governor, Sir Stanley Jackson, decided to come down the same morning. The Bengal Governor, next to the Viceroy, was the most heavily protected official of the Raj as it was fashionable for fiery Bengali rebels to take a pot shot at the Bengal Governor.

On their way home the police stopped the car of J.R.D. and Thelly. Even though it was a bitterly cold morning the police halted traffic for almost an hour. All this was to give the Governor's car precedence. J.R.D. and his wife decided to register their protest. They planned that when the Governor's car came, Thelly would step in front of it while J.R.D. would give his Excellency a piece of his mind. It worked. Thelly and J.R.D. boldly stepped in front of the Governor's car when it came up. When it stopped J.R.D. ran to the Governor's window and shouted: 'Who the hell do you think you are, keeping five hundred people, women and children, in the cold for an hour? You damn fool!' Unfortunately Thelly also wanted to give the Governor a piece of her mind, and rushed to the car window. In the process, the Governor's car shot off. Among those shivering in the cold was a British Anglican priest. He came up to J.R.D. and with all the dignity he could command, said, 'Sir, I do not approve of your language but I certainly approve of your sentiments.'

Recalling this episode, J.R.D. says, 'I was against British rule and to think that had I stayed on in the French Army, I would have been used to put down a colonial rebellion!' Though no instant action was taken against him, J.R.D. thinks that the British Intelligence Service must have recorded this incident as a black mark against him.

On returning to Bombay, J.R.D. and Thelly moved into a ground floor apartment at "Heliopolis", a gracious new building at Colaba. Bobby stayed with J.R.D. and Thelly, although he had also been invited to stay in Kitty's home. Looking back on his childhood days, Bobby appreciates J.R.D.'s generosity in having him stay with them, specially as Bobby did not make life any easier for J.R.D. At "Heliopolis", Bobby made another discovery.

When J.R.D. returned from office he wanted a bit of quiet each evening. Bobby remembers how 'In his very charming way, J.R.D. would ask me, "Would you not like to go out and play with the other children?" I would respond grumpily, "I can't play with them! They have all got cycles and I haven't!"' After a few days of this, Bobby says, 'We got down to negotiating. J.R.D. offered me an English bicycle but I insisted on an expensive French cycle! I won.'

In the months that followed marriage, Thelly started helping at the Swadeshi shop at Flora Fountain. But as time passed, J.R.D. became more and more the centre of her life. She did all she could to be a good wife, even pulling the veins out of mutton cutlets as J.R.D. liked them smooth.

'Was she fond of books?' I asked Rodabeh.

'No. She only thought of Jeh. "Where was Jeh? What was he doing? What was he saying?"'

Sir Dorab (left) and R.D. Tata with Sooni and
Lady Meherbai Tata (extreme right)

The house near L'Opera, Paris, where J.R.D. Tata was born

The beach at Hardelot where the Tatas had a home next to the pioneer aviator Bleriot

R.D. Tata (centre) with Sooni and Sylla (right) and
Jehangir and Madame Briere (left)

J.R.D. in a sailor suit

Sylla, Rodabeh and J.R.D.

J.R.D.'s favourite picture of himself
as a young boy

Sooni Tata with her children: (from left)
Rodabeh, J.R.D., Sooni, Jimmy, Sylla and
Darab

J.R.D. as a conscript in the French Army

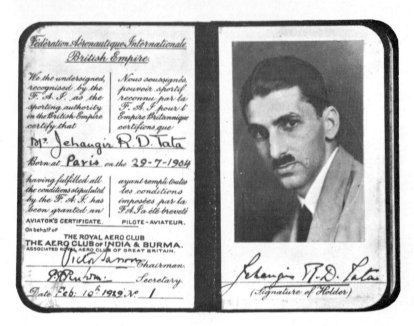

J.R.D.'s pilot's licence—Licence
No. 1

Arrival of the inaugural mail flight from
Karachi to Bombay: (from left)
Nusserwanji Guzder, Homi Bharucha,
J.R.D. Tata, Nevill Vintcent and two
officials of the postal service

The Leopard Moth in which J.R.D. made
a commemorative flight in 1962, on the
30th anniversary of the first mail flight
from Karachi to Bombay. Alongside is
an Air-India Boeing 707

J.R.D. with the Leopard Moth in 1982 on
the 50th anniversary of the inaugural mail
flight

First day cover released on the occasion of the
inaugural flight to London

J.R.D. at the controls of a Vampire jet

The De Havilland Leopard Moth DH-85 was a development of the Puss Moth and was used by Tata Air Lines between 1933 and 1934

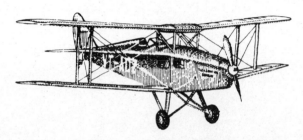

The De Havilland Fox Moth DH-83 was a four seater biplane which joined the fleet in early 1935

Tata Air Lines began a Bombay-Indore-Bhopal-Gwalior-Delhi service with the Waco YQC-6 aircraft in 1937

The Beechcraft Expeditor which joined Tata Air Lines
during the war

The DC-3 was the mainstay of Air-India's fleet in the
immediate post-war years. J.R.D has flown all the
planes on these pages

Petrol cans trundled up to the aircraft in bullock carts
in the early days of Indian aviation

Air Chief Marshal Arjan Singh with J.R.D. at the latter's investiture as Honorary Air Commodore

J.R.D. with Ratan Tata at the Boeing plant in Seattle

ARUL RAJ

An artist's impression of J.R.D.

A sketch of Thelly Tata by Homi Bhabha

La Traversée de l'Atlantique

"The Spirit of St. Louis"

Maintenant qu'il a fui, le merveilleux jeune être
Trop pur pour que personne ait pleuré devant lui,
Puisque nous sommes seuls dans un monde pâli,
Pleurons les Deux Amis qu'on a vu disparaître.

Sans ébauche, il n'est pas de chef-d'œuvre peut-être
Il faut au diamant quelqu'un qui le polit !
— Avant-coureurs d'un vol qui bientôt devait naître,
Pleurons sur Nungesser et pleurons sur Coli.

A distance plus tard, toute légende tremble.
Et quand on parlera, les évoquant ensemble.
Du vol que l'un protège et que l'autre promet,

La Gloire, confondant les noms dont on les nomme
Se souviendra qu'il fut dans une nuit de Mai,
Réussi par un ange et rêvé par deux hommes.

Maurice Rostand

(Juin 1927.)

A page from J.R.D's scrapbook. The drawing of 'The
Spirit of St. Louis' is by J.R.D.

An imaginative Air-India poster

A hoarding celebrating J.R.D.'s achievements in aviation

'She was Jeh-centred?'

'Yes.'

'That must have been a strain on Jeh,' I observed.

'A terrible strain,' said Rodabeh. 'Because Jeh is basically kind, he doesn't want to hurt people. But sometimes she was overdoing it—a kind of jealousy. I can't explain. Jeh is an independent character. He doesn't want to be...'

'Boxed in?'

'Yes.'

In the evening J.R.D. and Thelly would often go for a swim in the pool in the "Heliopolis" complex. J.R.D. was at that time wiry, fairly strong but very thin. He therefore decided to improve his physique and to this end stepped up his intake of milk and took to dumb-bells and weights for exercise. At the end of this regimen he had vastly developed his overall strength.

Every year or so J.R.D. and Thelly would visit Europe on a holiday and would be present at most sports events, especially tennis championships where the three musketeers from France—Réné Lácoste, Jean Borotra and Henri Cochet distinguished themselves. J.R.D. was particularly taken with Lácoste. Paris was then the venue for top tennis and J.R.D. did not worry about Wimbledon. Museums and art galleries did not attract him at that time as his taste for art was to develop much later.

After R.D.'s death, J.R.D. and Sylla, being the eldest, had to keep an eye on the younger brothers and sister. 'Darab was always a somewhat difficult child. Very obstinate,' J.R.D. says. Every year after the school holidays, he would solemnly and determinedly announce, 'I am not going back to school,' and every time Sylla and J.R.D., had to drag him to the railway station (he was at a boarding school.) 'Darab and I were fairly far apart in age. Oddly enough I got closer to my younger brother Jimmy.' Jimmy, too, was at a public school in the country . He was crazy about flying. By the time he was in his late teens, Jimmy was a big, strong man with blond hair. Looking at him no one could have imagined that he was prematurely born.

Bobby and Jimmy were about the same age and used to get into a lot of scrapes together. Bobby recalls an incident at Juhu. Jimmy was working in the Tata Hangar at Juhu airfield—then only an airstrip. Thelly, Jimmy, Bobby and J.R.D. were staying in a shack on the sea-front. One night, in high spirits, Jimmy and Bobby stole somebody's grand limousine. They

drove it round for a while and then, tiring of their adventure, parked it in
the same place they lifted it from. 'We felt conscience-bound,' says
Bobby, 'to report duly to our Lord and Master—Jeh. Jeh was furious at
the idea of a Tata hijacking a car! He went for us hell-for-leather. Jimmy
was big, J.R.D. was still skinny, and defiantly Jimmy lifted J.R.D. in his
arms and tried to rock him. With great agility J.R.D. slipped out and landed
such a clip on Jimmy's jaw that the tough guy swooned and fell flat,
immobile."O *mon petit frere*...O my little brother I have done this to
you..." said J.R.D., anguished.' When, after a little while, Jimmy opened
his eyes and recovered, J.R.D. resumed his tirade.

'Jimmy was an excellent flyer and navigator,' says Bobby. He was
training in a select Air Service Training School in England in 1936. Bobby
appreciated Jimmy's achievements more than J.R.D. did. If Jimmy didn't
do well in one subject, say Wireless Telegraphy, J.R.D. would question
and criticize him, rather than congratulate him for doing well in the other
subjects. A perfectionist himself, J.R.D. (according to Bobby) set his own
standards of excellence for his brothers and when they didn't or couldn't
live up to his expectations, he did not extend the understanding which they
needed. The same desire of J.R.D.'s that Darab should excel, affected
Darab for the worse, thinks Bobby. Jimmy was perfectly normal and
quietly took his hurt, he says, while Darab, slightly unstable to begin with,
became unsure of himself.

*

In the summer of 1936, the family spread out. Jimmy left to join his friend
Hans in London for a flight to Austria. J.R.D. was in Bombay, Sylla was
in the South of France setting up a new home for the Petit family, and Fali
Petit, with the children, was staying at a hotel in Paris with Rodabeh.

Fali Petit walked into the hotel one day and put a newspaper before
Rodabeh. A news item stated that an Indian called Jamshed Tata had
crashed in Austria. 'I could not believe it,' says Rodabeh. 'Jeh rang me
from India to ask "Are you going to Austria to fetch Jimmy?" I said,
"Yes," and the next morning I left on the train with the undertakers. I
wanted to bring back the coffin not in the train but all the way by car...
There was something very kind about Jimmy. He was in the flying school
in England and in the holidays he would come to Paris and stay with us.
His very good friend Hans from Austria wrote to say that he could not

find anybody to share the cost of the plane with him. "I have no friends who will come," wrote Hans. So Jimmy just told Fali that "I must go," and left.'

In Austria Rodabeh attended the funeral of Jimmy's friend Hans. She discovered that the two of them had been flying in a very small plane. Hans was a tall strapping young man and when he got into the plane in England, he told his friends, 'The plane is too small for my legs.' When Hans flew over his home in Austria and turned to wave goodbye to his family, the plane went into a spin and crashed into a garden wall. Jimmy was twenty-one when he died in May 1936.

When the Flying School got the news of Jimmy's death all his class-mates were on holiday but every one of them came from the UK to Paris for the funeral. It was a moving gesture and an indication of the esteem and affection in which he was held. 'The way they talked about Jimmy,' says Rodabeh, 'he was the best pupil.'

Jimmy's going was a terrible blow to J.R.D. They had argued as brothers, teased each other, but J.R.D. was perhaps fondest of Jimmy among all his brothers and sisters. This was certainly due to the fact that he was an excellent flyer but was also because he was a warm-hearted personality who could win people over easily. 'When he walked into a room,' says Rodabeh, 'everybody had to take notice of him. He radiated warmth wherever he went.' J.R.D. was planning that in the years to come Jimmy would work alongside with him and look after the airline. But fate had willed otherwise. 'The death of someone close to you is always traumatic,' he says philosophically, 'but you survive the shocks.'

The three Tatas—Sooni, R.D. and Jimmy lie beside each other at Pere Lachaise. The beautiful yet simple granite mausoleum is meticulously kept, its motif of torches signifying the Zoroastrian's respect for fire. At the same time the torch was the symbol of the French Revolution, of the time when the Bastille was stormed. The torch can also be interpreted as a symbol of freedom.

CHAPTER X

Sir Dorab And Sir Nowroji

In the early twentieth century, India was predominantly an agricultural nation. There were only four industries of note. The textile industry in Bombay was started by the Parsees and Gujaratis. The jute industry in Bengal was dominated by the British. The British were involved with coal mining, too, with which some Bengali families like the Tagores were associated. The tea plantations of the east and the south were the virtual monopoly of the British. It was in this setting that the successors of Jamsetji Tata set about the task of establishing the steel industry and a major hydro-electric project from their base in Bombay.

When Jamsetji died in 1904, Tata enterprises comprised three textile mills and the Taj Mahal Hotel. Under the stewardship of his son, Dorab, Tatas added an integrated steel plant, three hydro-electric companies, a large edible oil and soap company, two cement companies, a leading insurance company, an aviation unit and some other industries.

In 1907 Tatas were advised that the kind of capital needed for their projected steel plant was not available in India and that they would do better to tap the London market. However, London was going through a bad patch at the time and the British indicated their desire to exert control if they invested their capital. Tatas decided to try their luck in India. To Tatas' amazement, investors besieged their Bombay office at Navsari Building, from morning till late at night. In three weeks the entire capital was subscribed. A jubilant Dorab Tata wrote:

> For the first time in India's financial history, I have succeeded in raising for industrial purposes, such a vast sum from the hidden wealth of India for the development of our mineral resources. It was the first time that the raw materials of India did not go out and return as finished articles to be sold in the

country. Above all, it was a swadeshi enterprise, financed by swadeshi money, managed by swadeshi brains.[1]

The attainment was significant because it marked the end of the feeling of helplessness that had pervaded India since the Revolt of 1857. The fact that Tatas could raise the funds was a mark of the confidence returning to a subject land. It is interesting to note that Jawaharlal Nehru picked the year that Tatas raised these funds as the turning point in the country's struggle for independence.

As the first steel chimneys were rising in Jamshedpur, on the western side of India, giant pipelines were being riveted down the steep incline of the Western Ghats to generate hydro-electric energy. Up to that time (1910) hydro-electric power was harnessed only through generators placed under natural waterfalls. The scheme visualized by Jamsetji would house artificial reservoirs on the brink of the Western Ghats that would speed the flow of water down pipes to the power houses at the foothills of the Ghats.

The idea was one thing; executing it was another. Each pipe had to be perfectly laid on saddles hewn out of uneven and craggy rocks or earth. Even the artificial dam envisaged at Walwhan, off Lonavla, was a feat at that time. The dam was only a little smaller than the great dam constructed at Aswan by Sir William Willcock.

The object of the scheme was to supply cheap and clean electric power for the growing needs of Bombay. Tatas were convinced of the utility of hydro-electric power but many textile mill owners were wedded to working with coal and there was the risk that the capacity generated by Tata Hydro would not be lifted by Bombay's industry. Electricity had yet to be tried in Bombay. At first, only two textile mill owners guaranteed to lift the electric power. In faith, the hydro-electric company went ahead with its plans. The Governor of Bombay, Lord Sydenham, aware of this when laying the foundation stone of Walwhan Dam, said:

It symbolises the confidence of Indians in themselves.[2]

On the same occasion Dorab Tata recalled:

The anxious and depressing days through which we have passed.[3]

Sir Dorab added that to his late father, who proposed the hydro-electric scheme:

> The acquisition of wealth was only a secondary object in life;
> it was always subordinate to the constant desire in his heart
> to improve the industrial and intellectual condition of the
> people of this country.[4]

Having produced electric power it was a job to sell it. The biggest potential customers were, as we've seen, the textile mills of Bombay, each of which had huge steam engines which ran on coal. So for the next decade and more, Tatas had to persuade the mill owners to convert from steam to electric power. J.R.D. recalls a massive steam engine of Bombay Dyeing Mills which occupied an entire floor of the plant. He attended a farewell ceremony for the engine that had served the company so well for so long. It was a nostalgic occasion as if a popular staff member was retiring. Tatas often had to purchase a number of such engines from the mills so that they would agree to accept their power. They were paying the price of being pioneers.

Earlier on, in 1905, Lord Curzon had agreed to the Institute of Science in Bangalore. Preparations proceeded apace. The Vesting Order of the University came in May 1909 and spoke generously of Jamsetji, the "promoter and donor". The Indian Institute of Science opened in 1911 and its first Director was Professor Travers. To the credit of the British Government and its successor government of independent India, the autonomy of the institute and its academic freedom was maintained. From its inception the institute has been a tripartite venture run by the Government of India, Tatas and the Government of Mysore (now Karnataka.)

In 1912 the first steel ingot rolled off the lines of the Jamshedpur plant. Two years previously Dorab Tata had been knighted and his cup of joy was brimming over. He observed:

> Kind fate, has... permitted me to help in bringing to comple-
> tion his (Jamsetji's) inestimable legacy of service to the
> country.[5]

Apart from R.D., the men who had stood by Sir Dorab and helped him to fulfil his father's legacy were Sir Bezonji Mehta of Nagpur,

Burjorji Padshah and A.B.Billimoria.

The success of Tata Steel in the First World War period induced them to venture into sugar, electro-chemicals, construction and a lot more. Young J.R.D., drafting a speech for his Chairman, Sir Nowroji, in 1933, was to write of the 1920s, 'Those were truly mad days and perhaps the maddest feature was the supreme confidence of the public in Tatas and incidently, the confidence of Tatas in themselves.'

Tatas attempted to introduce technology and encouraged farmers to undertake sugar-cane growing as early as 1920. Their construction company broke the monopoly of British companies in undertaking large projects. The company and its subsidiaries executed such outstanding construction works as the Vaitarna Dam and several long railway bridges across the Indus, the Godavari, the Krishna and the Narmada rivers. The construction interests were sold in 1935 to Walchand Hirachand. The sugar corporation had to bow out in 1922. The New India Assurance Company started in 1919 was a great success. J.R.D. was a witness and a participant in some of these events.

Jamsetji's second son, Ratan, knighted in 1916, had joined his father, Dorab, and R.D. as a partner of Tata & Sons in 1896. A sensitive spirit, he had little interest in business and his main contribution was outside of Tatas. In early 1902 he supported a little known lawyer in South Africa called M.K. Gandhi, with the then substantial donation of Rs 125,000*. For a number of years Sir Ratan supported Gopal Krishna Gokhale as well with generous contributions. Able men had renounced their ambitions to serve in Gokhale's Servants of India Society. On 19 October 1909, referring to Sir Ratan's help to the Society, Gokhale wrote of, 'the deep gratitude I feel for your overwhelming generosity. There is no parallel to it anywhere in the country. I can only say that members of the Society will ever cherish your name as that of their greatest benefactor.' In 1912 Ratan Tata's annual donations to the London School of Economics (LSE) helped it start its Department of Social Sciences.

Sir Ratan lived well. In England he had bought the property of the Duke of Orleans at Twickenham and his receptions were written about in the London *Times*. When he died in 1918, Sir Ratan left a sizeable part of his fortune to a multi-purpose Trust named after him. His superb art collection went to the newly started Prince of Wales Museum in Bombay. His

---

* Around seven-and-a-half million rupees in 1991.

widow, Lady Navajbai Tata, took his place as a permanent Director of Tata Sons which she held till her death in 1962. As they had no children she adopted a grand-nephew of Jamsetji's wife called Naval. Naval's maternal grandmother Cooverbai and Jamsetji Tata's wife Hirabai were sisters. Naval was to become a Director of Tata Sons in 1941.

It was Sir Dorab's drive and R.D.'s that made Tatas a national enterprise. Sir Dorab's finest hour was in 1924, when he pledged his private fortune including his wife's jewellery for a loan to save a public limited company—Tata Steel. In the jewellery that he pledged to the Imperial Bank, was the celebrated Jubilee Diamond (245.35 carats) more than twice the size of the Kohinoor. Dinsi Gazdar, the well-known jeweller at the Taj Mahal Hotel, remembered Sir Dorab stating that every time his wife took the diamond out of their safe deposit vault in London, he was "fined" £200 by the insurance company.

Unfortunately, by the time J.R.D. joined Tatas, in the mid-1920s, Sir Dorab was already on the decline. He was once a strong and healthy sportsman, who excelled in feats of horse riding. He rode from Bombay to Kirkee (a distance of 110 miles) in nine-and-a-half hours—just double the time the train took. However, when J.R.D. began his association with Tatas, diabetes had taken hold of Sir Dorab. Its treatment was not so advanced then and consequently Sir Dorab was reduced to eating a tiny morsel of food when all the others at his table—and at his expense—helped themselves to tasty meals. J.R.D. recalls having lunch with Sir Dorab in Vienna at the latter's usual hotel-cum-sanatorium. Sir Dorab was in a very bad mood as he was not allowed to eat much. He had with him his wife, his secretary and his doctor. The food was tempting but all he got as an entreé was a tiny round piece of toast with an olive on the top. He gulped that down. When the waiter courteously asked Sir Dorab: 'Have you finished, sir?' Sir Dorab glared at him. He had eaten only that little piece of toast and he was furious. Even though there was nothing left on the plate Sir Dorab said in reply to the waiter's question that he had not finished. The waiter stepped away and came back after a time and repeated: 'Have you finished, sir?' Again Sir Dorab said 'No.' The third time the trembling waiter came and enquired: 'Have you finished, sir?' Sir Dorab got excited and shouted 'No! Can't you see I've still got to eat the plate?'

Apart from the tremendous industrial activity of his earlier days, the sportsman in Sir Dorab made him bring India into the Olympic movement and from his own purse he financed the Indian athletes who participated in the 1920 and 1924 Olympics. But in his last years illness had sapped his energy and interest. His ill-health had made him quite cynical, even with his co-Directors, and one day when a senior Tata Steel Director returned to his post after a spell in the Viceroy's Executive Council in New Delhi, Sir Dorab loudly noted to the discomfiture of all: 'The bad penny has returned.'

Dorab's wife Meherbai was an outstanding lady. Sir Stanley Reed, Editor of the *Times of India*, wrote that Sir Dorab's translation of Jamsetji's dreams into reality,

> would never have been accomplished if Sir Dorab had not seen his purpose with open eyes, if he had not always at his side a wife who was as staunch in the pursuit of these filial and patriotic duties as himself.

Sir Stanley also noted:

> Lady Tata was one of the clearest brains it has ever been my lot to find in a woman. No one could put her own case with more convincing logic.

Lady Meherbai Tata was one of the founders of the Bombay Presidency Women's Council. She brought India into the International Council of Women, as Sir Dorab had earlier brought India into the International Olympic Association.

In June 1931, Lady Meherbai died of leukaemia in North Wales. The following year her husband left for Europe hoping to study the possibility of starting a radium institute in India, and to visit his wife's grave. It was while in Germany that he passed away at Bad Kissingen. A few days later, almost on the anniversary of his wife's death, his ashes were laid to rest beside her at the Brookwood cemetery. The mausoleum in which they lie is modelled on that of Cyrus the Great of Persia. Inscribed are the words from the Zoroastrian faith:

Humata	*Hukhta*	*Hvarashta*
Good	Good	Good
Thoughts	Words	Deeds

The words are flanked on either side by Persian rosettes. Also drawn from an ancient Persian model, the single sarcophagus is of green bronze and stands on a marble plinth. It bears on its two sides the names: DORAB-MEHRI.

Three months prior to his death, Sir Dorab had formed a Trust to which he left his entire belongings, down to his pearl studded tie-pin. In place of a radium wing for a hospital he was planning, his Trust was to establish India's first cancer hospital, the Tata Memorial. His Trust also founded the Tata Institute of Social Sciences (TISS), the Tata Institute of Fundamental Research (TIFR) and the National Centre for the Performing Arts. J.R.D. was a trustee of the Sir Dorabji Tata Trust from its inception in 1932 and was to give direction to it for over half a century.

On the death of Sir Dorab, Nowroji Saklatvala, a son of Jamsetji's sister, Virbaiji, was appointed Chairman of Tatas. He was the seniormost Director of the firm and was highly respected. In his mid-sixties, when he became Chairman, Nowroji was, in J.R.D.'s words, 'a fine outstanding person, but not one with great intellectual capacity. He infused confidence, he took advice, he was a man of substance but he was not a real leader or a creator.' Fond of sport he was seen at all the major cricket and tennis engagements in Bombay with his tin of State Express 555 in hand. He smoked incessantly and worked hard.

Sir Nowroji trusted J.R.D. and J.R.D. respected Nowroji who was a true gentleman. Everyday at twelve noon, F.E. Dinshaw, the legal and financial wizard, would call to discuss problems with Sir Nowroji. 'F.E.' as he was called, was very shrewd and gave liberally of his acumen to Tatas. He was adviser to the Maharaja of Gwalior and was instrumental in bringing the Maharaja's substantial resources to the rescue of Tatas at a difficult time. Where he brought in finance, he did not hesitate to extract his price from Tatas, but his advice was sound and he had Tatas' interest at heart. J.R.D. was much impressed by him and calls him, 'one of the most distinguished intellects I've ever come across.'

Company shareholders' meetings in those days could be quite uproarious. There were shareholders who had made a profession of studying balance sheets and tackling the Directors. A shareholder called Shamdasani was brilliant at tripping up the Directors by picking holes in the balance sheet. There were not so many shareholders then and most companies held their meetings in their own offices after clearing the furniture. At a meeting of the Tata Oil Mills, the Directors were seated with a table in front of them and office cupboards behind. J.R.D recalls Shamdasani in his excitement pounding the table, then pushing it back in stages with the Directors retreating with their chairs, with as much dignity as they could muster. The climax came when Shamdasani finally jumped on the table, converted it into a platform, and harangued the Directors.

Shareholders' meetings in the 1920s and 1930s were colourful occasions. A contemporary recalls the variety of headgear in which Directors turned up; red *pugrees* from Kutch with pinnacles in their centre, wheel-like Bhownagarees, golden Khoja rings, black shining Parsee tophats. When Sir Shapurji Bharucha, a kindly old man, arrived at a Tata Board Meeting at Navsari Building he always created a commotion. Even before he collected his dividend he would fling a fistful of small coins resulting in beggars piling on top of one another to collect their booty.

'But all this has changed,' wrote the contemporary in 1946. 'Now there is regimentation in dress.... All (Tata Directors) are clean shaven, bare headed and in English dress, which again is mostly of the same cut, texture and colour. But there is hope. A day will come when lady Directors will grace the Board table, and then there will again be a riot of colour, and this time of the peacock variety.'[6]

*

After John Peterson left in 1931, J.R.D. attached himself to Nowroji. 'Feeling Nowroji needed help, I got a room next to his. He must have realised I was helpful. Soon the middle wall between the two rooms was broken and a glass panel was installed. He began to consult me more and more and relied on me.'

'Were you his executive assistant?'

'No, we never formalised the relationship. But I used to draft things for him or he showed me drafts of letters he intended to write.' J.R.D. continued, 'He was a fine fellow who had no time to think, moving as he did from one Board meeting to another. As he smoked too much, his hand was always shaking. He was an honourable man who inspired great confidence that Tatas were totally honest people.'

During Sir Nowroji's term as Chairman (1932–38) India was emerging from the Depression. Under his Chairmanship nothing significant was started except the Investment Corporation of India for the promotion of companies. It was to acquire holdings in major outside companies like Forbes, Forbes and Campbell and to underwrite capital issues like Henley Cables, Associated Bearings and others. Sir Nowroji's job was to hold on to what Tatas had. This he did well.

Sir Nowroji was married but the couple had no children. Sir Nowroji's only brother Kaikobad who could have been quite senior in Tatas, had a somewhat difficult temperament and had left the firm with ambitions to be a writer. Kaikobad's son Minoo* recalls how Sir Nowroji used to see them off at the railway station in Bombay when they left for their home at Kanpur. Even as a child Minoo was touched by the care of his uncle, who headed a large industrial empire but would take time off to bid farewell to his brother's family.

In 1938, while in Europe, Sir Nowroji died of a heart attack at Aix-les-Baines in France—the third consecutive Chairman of Tatas (Sir Ratan Tata and R.D. were the others) to die in Europe.

A meeting of the Board of Directors of Tata Sons Limited was held on Tuesday, 26 July 1938, at 11.30 a.m. at Bombay House, Fort. The minutes read:

> Those present were:
> J.R.D. Tata Esqr.
> S.D. Saklatvala Esqr.**
> A.R. Dalal Esqr.
> Sir H.P. Mody, K.B.E.
> Mr. S.D. Saklatvala was voted to the Chair.
>
> Mr. S.D. Saklatvala referred in feeling terms to the sad and untimely death of the Chairman, Sir Nowroji Saklatvala, K.B.E., C.I.E. and proposed the following resolution which was carried unanimously:

---

* Sir Nowroji was the son of Jamsetji Tata's third sister Virbaiji. Virbaiji's grandson Minoo Saklatvala is the nearest male descendant from Jamsetji's father. He inherited two settlements made by Jamsetji. Under the terms of the settlements, he changed his name to Minoo Tata.

** Sorabji D. Saklatvala was the son of Jamsetji Tata's fourth sister Jerbai.

The Directors, Tata Sons Limited, have received with profound sorrow the news of the sudden death of their Chairman, Sir Nowroji Saklatvala. The Directors feel that by his outstanding qualities and unique personality, he not only commanded their unstinted respect but evoked their warm affection. The confidence he inspired amongst his colleagues and staff and amongst all those connected with the Tata enterprises in various capacities as well as amongst the public, was of tremendous value in the maintenance of the traditions, and influence of the House of Tatas.

Pursuant to Article 118 of the Constitution of the Company, Mr. Sohrab Saklatvala proposed and Mr. Ardeshir Dalal seconded the proposal that Mr. J.R.D. Tata be appointed the Chairman of the Board of Directors of Tata Sons Ltd.

J.R.D was thirty-four.

PART II

EYES ON THE STARS

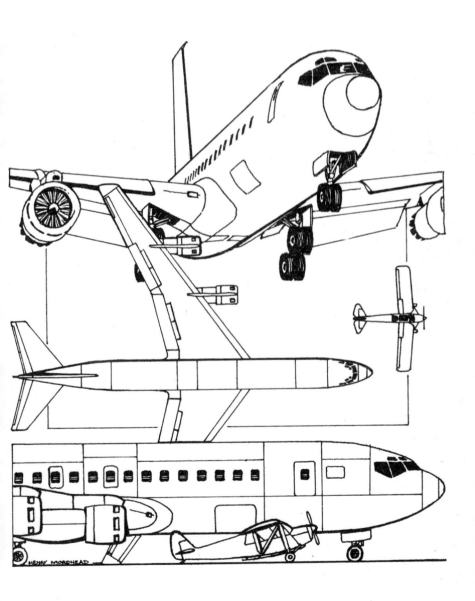

CHAPTER I

When The Skies Were Less Crowded

Early in this century, the US Government set aside 50,000 dollars to enable Samuel Langley to construct a craft that could fly. On 7 October 1903 Langley took off. Within seconds the machine plunged into the Potomac river.

Miles away in Dayton, North Carolina, two bicycle mechanics invested 1,000 dollars of their own to make a machine that could fly. Two months after Langley, Orville and Wilbur Wright were ready to test their machine on the windswept sand-dunes at Kitty Hawk. Their contraption was made of spruce wood and unbleached muslin.

Today, we take the shape of a plane for granted but they had to think of one. Orville lay down on the lower wing, his hips strapped in a padded cradle. The pilot's cradle was connected with cables to the wings and the rudder. The rudder was for turning and there was a wing that could warp up and down. In the years to come these two mechanisms would be the basis of design of every aircraft. The original plane of the Wright brothers is now displayed at the Aerospace Museum in Washington.

One December morning in 1903 the rather dignified Orville Wright lay in the strapped barrel. The engine started propelling the contraption on the ground and raised itself by its own power into the air in full flight. It landed at the same height as that from which it started. Orville's brother, Wilbur, was so excited he forgot to punch the stopwatch. But a North Carolina surfer did so. His timing was perfect and there was evidence that a man-made machine had actually flown. It was a small hop of forty yards that took about twelve seconds.

'Seven months after this event J.R.D. Tata was born. In 1909, Louis Bleriot was the first man to fly across the Channel. Until Lindbergh flew over the Atlantic in 1927, Bleriot's was the most celebrated name in aviation. As we saw earlier, it was Bleriot who built a house on the beach at Hardelot, close to R.D. Tata's home.

This accident of location was the beginning of J.R.D.'s interest in aviation. 'The early pilots have long been forgotten,' he says. 'Who remembers today that name of Adolphe Pegoud, Bleriot's chief pilot? To me Pegoud was one of the bravest and most foolhardy men that ever lived, for he was not only the first in the world to loop the loop but did so inverted, because as those early planes did not have enough power to climb into a loop in the normal way, Pegoud had to do so by diving beyond the vertical. Poor Pegoud must have been hanging on his flimsy belt the whole way round, with no parachute to save him in case the plane broke up, as he must have half expected it to do.

'After the early birds, came the legendary fighter pilots of World War I, such men as von Richtofen and Boelcke of Germany, Ball and McCudden of England, Guynemer and Fonck of France, Rickenbacker of the USA and Billy Bishop of Canada, to whose exploits my generation thrilled.'

J.R.D. says, 'Flying machines obviously had no future except in war-time and their pilots were a nuisance; the planes made a lot of noise and attracted large crowds that left a lot of litter to be carried away. Sometimes the aeroplanes and their pilots had to be carted away too! No one except those crazy people in aero clubs believed in a future for civil aviation other than as sport and spectacle. Officialdom only intervened, usually in the shape of a stern policeman with a peaked cap and handlebar moustache, when a farmer complained of the damage caused to his field or his cow by someone's aeroplane, or when crowds needed controlling at air shows.

'Then (in the mid-1920s) came the spate of long distance pioneers including Kingsford Smith and Lindbergh, whose flight across the Atlantic was perhaps the greatest individual feat ever performed in aviation. It brought a whole era of individual achievement to a climax, and, at the same time, opened an entirely new one in which commercial aviation reached maturity.'

When Lindbergh flew across the Atlantic, cooped up for thirty-six hours all alone in that little plane, the *New York Sun* wrote an editorial:

> Is he alone at whose right hand rides Courage, with Skill within the cockpit and Faith upon his left? Does solitude surround the brave when Adventure leads the way and Ambition reads the dials?

In J.R.D.'s red scrapbook is his own ink drawing of Lindbergh's plane "The Spirit of St. Louis" with a French poem by Maurice Rostand.

J.R.D.'s first flight was with a joy-riding pilot in Hardelot. Throughout the flight his father kept praying for his safety till he returned to *terra firma*. Ever since that flight, J.R.D. was determined to become a pilot but had to be patient. Almost ten years passed before a Flying Club opened in Bombay. Only twelve days later, on 3 February 1929, having collected three hours and forty-five minutes of dual flying experience, J.R.D. was let loose on his first solo flight and had within a week qualified for his 'A' Licence.

'No document has ever given me a greater thrill than the little blue and gold certificate delivered to me on 10 February 1929, by the Aero Club of India and Burma on behalf of the Federation Aeronautique Internationale (F.A.I.). The fact that it bore the Number I added to my pride in owning it, even though it meant nothing more than that I was the first one to have qualified in India. In the rest of the world flying had already become a fairly serious and humdrum business. The heroic days of pilots who constructed planes like the Wright brothers or Bleriot (there was a plane named after him), Farman, Caudron, Curtiss Latham, Breguet and others, and their strange looking aircraft made of wood, wires, canvas and bicycle wheels had long gone. Few even remember their names today. Luckily, I was born early enough to have seen some of them in action.'

J.R.D. nostalgically remembers those eventful and happy days 'when the skies were less crowded, the planes less reliable and men and women flew them because they loved to fly.'

J.R.D.'s initial flying experience was not without risk, for Cummings, who taught him to fly, had never been an instructor before. He was an ex-Navy pilot from an aircraft carrier. J.R.D. was not satisfied with merely having a pilot's licence. He felt that he wanted to be a good and safe pilot for which he believed he had to be an aerobatic pilot too. So he asked Cummings to teach him aerobatics but Cummings said that he had no time. He, however, gave J.R.D. a tip: 'Go high enough before you try anything.' J.R.D. had already read all that he could get hold of about flying and decided to teach himself aerobatics, which included the rather unpleasant need to know how to get in and out of a spin. As in a spin the plane would usually be pointing down, the natural instinct to pull out of a spin was to pull back on the joy stick, whereas the pilot had to do the reverse of releasing or even pushing the stick forward and applying the opposite rudder. J.R.D. had read the instructions, but momentarily dazed by being whirled around, he was slow in following them. This nearly cost him his life. Fortunately, the spin, which was unintentional and occurred while he was practising other aerobatic manoeuvres, started at 6,000 feet and this gave him the time to remember what he had read and to recover from the spin.

'Cummings had been watching me from the ground, leisurely leaning on the railings,' said JRD. 'After landing, I went up to him, somewhat shaken and angry, and demanded to know if he had seen what had happened!'

'Yes,' Cummings said casually, 'not too bad, but a bit untidy. Your loops were all right but your rolls were awful.'

'I hope you realize that you nearly got me killed. Because you never taught me how to get into, and out of a spin.'

Twenty years later J.R.D. had Cummings invited to a party given by the International Air Transport Association (IATA) Conference at Sydney. 'I reminded him that it was *not* thanks to him that I was attending the party myself!'

Within three months of obtaining his flying licence J.R.D. was in London in May 1929 to buy a plane as he had found that renting a plane from the club was proving expensive. A good light Gipsy Moth was available for £1200* and so he bought one.

In those early days young men indulged in their love of speed and flying by giving joy rides to people. Most of them landed their planes in the fields of farmers. Some had a companion, others a mechanic with them. Most of them had very little money and could not even afford the luxury of a small hotel in nearby villages or towns, so under the pretext that it was necessary to be near their plane, the pilot and his companion would induce the overwrought farmer, to allow them to use his barn as a means of shelter. That is the origin of the term "barnstorming". Whenever it landed, the Flying Bird, a novelty, stimulated great excitement.

When J.R.D. flew in France, he enjoyed the experience of landing in little fields and whenever possible visited various friends to give them joy rides.

J.R.D. had arranged with his brother Darab to land at his college near Rouen. 'I'd asked him to put a white strip on the college grounds to guide

---

* Around £30,000 in 1991.

my landing. Darab's brilliant sports teacher put the strip right in the middle of the football field, between the goal posts. So naturally I had to search for a more suitable one which I found in a field nearby. I managed to give Darab the promised joy ride.

'In those early days you really did have to fly, as they say, by the seat of your pants. All I had in my Moth was a compass, an altimeter, an engine RPM indicator and an airspeed indicator. That's all. No refinements like a radio, or brakes. Apart from landing in fields, I used to land on the beach in Hardelot. One had to taxi upto where the sand was hard, to take off. Small planes had no brakes then and the only way to steer on the ground was with a burst of engine and full rudder.'

J.R.D. remembers that this caused an accident that cost him some money, for, one day, as he was departing from Le Bourget, then Paris' airport, an unexpected blast from a parked plane blew him aside while taxiing and he found himself heading straight for an Imperial Airways plane parked on the tarmac. As he was taxiing slowly, he decided to let his plane drift into the side of the British plane rather than risk a more serious collision from a burst of engine accompanied by full rudder. Imperial Airways were not amused and made him pay £25* for repairs.

On his return to India he flew a couple of times to Jamshedpur. On one of these trips he had with him Kish Naoroji, grandson of Dadabhai Naoroji, as his passenger. When flying over the Aravalli Hills between Jamshedpur and Nagpur, he saw the compass between his legs swing violently from side to side. 'An hour later I was lost and landed in a field. Poor Kish was terrified.' J.R.D. learned from a farmer that they were near Ramtek, a hundred miles to the right of his intended course. When he finally landed at Juhu, Bombay, he asked the Flying Club engineer to test the compass which was obviously faulty.

Soon after this incident, J.R.D. saw in the 19 November 1929 issue of the London *Times*, an announcement headlined "Prize for England—India Flight". The news item read:

> The Aga Khan has offered through the Royal Aero Club, a prize of £500 for the first flight from England to India, or vice-versa, by a person of Indian nationality. It must be a solo flight completed within six weeks, from the date of starting.

---

* Around £600 in 1991.

The prize will remain open for one year from 1 January 1930.

A few hardy souls ventured to take up the challenge. One was Manmohan Singh, who announced with a flourish that he would take off from Croydon airfield, London. He did so, promptly lost his way over the Channel and soon returned to base. Manmohan Singh then announced he would make another attempt to get to India. This time he lost his way in fog over Europe and with difficulty returned to Croydon airfield. The Editor of *Aeroplane* magazine solemnly wrote:

> Mr. Manmohan Singh called his aeroplane "Miss India" and he is likely to!

Another competitor starting from England was Aspy Engineer, an eighteen-year-old lad. The third was J.R.D. Tata, from the Indian end.

J.R.D. got King, a ground engineer of the Flying Club, to streamline the plane so that he could get an additional five miles per hour. He asked King to double check the compass before the flight. On 3 May 1930 J.R.D. took off from Karachi, stopped at the port of Gwadar and flew on to Jask*. He spent the first night at the home of a Burmah Shell representative.

For strategic reasons Britain had dotted the route from Egypt to India with military airstrips. It was on these airstrips that J.R.D. landed and hence covered almost the same route as Sir Samuel Hoare. (In 1928, only a couple of years before J.R.D.'s flight, Sir Samuel Hoare—later Lord Templewood—as Secretary of Air for Britain covered the distance from London to Karachi for the first time in a civilian aircraft.) Writing about his experience in *India by Air*, Sir Samuel wrote:

> Jask is a corner of the world to which not even a Ford car has yet found access, and the only means of transport are the camels that proved as useful to us as they had been to Alexander the Great.[1]

Jask got two inches of rain annually and was in the Persia–Baluchistan area where,

---

* For the journey logbook, see appendix A.

The flies died of heat in summer and...(it was) not a subject of idle laughter that the cows ate the paper tape at the (telegraph) office.[2]

From Jask J.R.D. flew to Hormuz, Lingeh (when the annual rainfall was three-to-four inches), Bushire and Basra. For many miles between Bushire and Basra,

endless expanses of marsh and sand mark the dying efforts of the two rivers (the Euphrates and the Tigris) to free themselves from the desert. Bushire was an oasis on the Persian Gulf that juts out into the sea, a solitary circle of green in a landscape that is grey and brown.[3]

Sir Samuel continued:

Not a sign of human life in all this deserted country, not a spot where we should feel happy with a forced landing. Only the pink flamingoes, the white pelicans, an occasional school of turtles or porpoises, and the splashes of the sharks. Gradually the estuary gives place to a dryer land that, though of very different features, is equally desolate. Grey-brown mountains rise almost sheer from the sea, only to be broken by stretches of white desert that from time to time cut in and out of them. So savage and fantastic is the aspect of these crags that we likened them to the mountains of the moon; so arid is this country that for hour after hour we saw not a blade of grass.[4]

J.R.D. halted his plane at the Gaza airstrip (Palestine). 'While I was refuelling at Gaza, in came a new Gipsy Moth. It did a "split-arse" landing, with full power and turned to park alongside my aeroplane, just missed crashing into it by inches! It was Manmohan Singh. I must say he was certainly not short on enthusiasm. He even continued to hunt desert gazelles by flying low over them and shooting them with a revolver!'

Between Baghdad and the Mediterranean Sea, J.R.D., while on a compass course, ran into what seemed remarkably strong winds from the south which made him drift towards the north. He was to fly over the Dead

Sea, which is an elongated expanse of water, but found himself flying over a round sea which he definitely knew was not the Dead Sea. He realized he was, in fact, flying over the Sea of Galilee, a hundred miles north of where he should have been. Finally, he was convinced that it was his faulty compass which had forced him to drift in the wrong direction and had obviously not been repaired as per his orders in Bombay. However, he faced no trouble in hitting the Mediterranean Sea and landed on an old disused First World War airstrip which, to his discomfort, he found covered with high anthills. He walked to a road bordering the sea and waited till a peasant came trundling along in a cart. 'I asked him where I was. He didn't understand me but said "Haifa". Haifa was ten miles further on from where I should have hit the coast. What I'd flown over was not the Dead Sea but the Sea of Galilee.'

J.R.D. then flew into Cairo. He found an RAF aerodrome outside Cairo and they sent him on to Alexandria, where the Air Force ground engineer at the airstrip repaired the compass. 'Where did you say you came from?' he asked.

'Bombay.'

'You flew from Bombay with this compass?'

'Yes, I did drift a lot with it. What is wrong with it?'

'There is nothing wrong with it, it's only 25 degrees out!'

The engineer handed him a broken magnet. It had broken into two pieces over the Aravalli hills. J.R.D. was fortunately a very competent map reader.

'At Alexandria, at 7 a.m., I saw another Moth parked there and realised it must be Aspy Engineer, the third competitor in the Aga Khan Prize race. When he heard that I had landed, he came out to the aerodrome to meet me. I asked him what he was doing there. He told me he was waiting for some spare plugs since he had not taken an extra set of them. This was not very good planning! Since mine was a four cylinder aeroplane, and I had eight spare plugs, I gave him four of them. He was so pleased and grateful that he insisted I take something from him and he gave me his Mae West life-jacket. He had a Mae West, but no spare plugs!' The Mae West was a help if one had to fly over water. Aspy didn't need it any more and J.R.D. wore it when he crossed the Mediterranean.

With both Manmohan Singh and Aspy Engineer well on their way, J.R.D. had to get moving. He flew across the Mediterranean to Naples airport. He did not realize that it was a military airfield, which had strict instructions not to release planes before 6.00 a.m. He lost time there and at about the time he landed in Paris, Aspy Engineer landed at Karachi and captured the prize. 'I am glad he won, because it helped him to get into the Indian Air Force which was to start shortly.' (Engineer rose to be the second Indian to be Chief of Staff of the IAF.)

'Aviation and marriage came almost simultaneously because I started flying in 1929 and got married in December 1930.... In May 1930 I took on the Aga Khan challenge and came back somewhat as a hero. That may have decided her (in accepting me),' he casually observes.

A friend of the bride-to-be in Bombay says Thelly wrote to a friend excitedly, 'I am getting married to a famous pilot!'

*

Talking of his family's enthusiasm for aviation, J.R.D. says, 'My elder sister, Sylla Petit was the first Indian lady to get her flying licence in India. Later, also having learned at the Bombay Flying Club, my younger sister Rodabeh (Dabeh as she was nicknamed; later Mrs Sawhney) was the second Indian lady to get her flying licence in India. But unquestionably the best flyer and most naturally gifted airman among us all, was my youngest brother, Jamshed, or Jimmy. I also used to do a lot of formation flying together with him. Remember this was long before the days of specialised aerobatic planes like the Pitts. A born flyer, Jimmy was released solo after only four hours. I had the fullest confidence in him. On one occasion I overdid it and put my right wing between his left wing and tailplane. We were flying a little away from the Club so that we couldn't be seen, but old Vetch (J.R.D.'s new flying instructor) spotted us and was furious. Both Jimmy and I were grounded for a week.'

Indian Aviation under the head "Skilled Performance" reported in April 1933:

> Much skill was displayed in the balloon bursting event by Mr. J.R.D. Tata the well-known aviator and Mr. B.P. Dhargalkar, the Assistant Pilot Instructor of the Club. Owing to the squally wind the movements of the balloons were very tricky, some tried to take refuge near the earth and were burst as close as ten feet from the ground. All this from a plane in mid air.

Captain Vetch was an enthusiastic instructor. He inspired pupils and had enough confidence in J.R.D. to ask him to participate in instructing newcomers. They used to play games with planes, release balloons and prick them almost at ground level. They did formation flying with all three planes of the Flying Club. With Vetch in the lead, the planes tied with rubber bands and fluttering ribbons, they would fly and dive in formation over an excited crowd. Vetch was ready to fly at all hours of the day or dusk. There was only one fault in him and that was owing to a surfeit of enthusiasm—he had a propensity to crack up planes. One day when the planes were being lined up for the night, he suddenly rushed up to an aircraft, told J.R.D. and some others, 'Wait I'll show you.' He swung the propellers himself, as he always did, got in and demonstrated what he intended to do. However, by the time he landed it was almost dark and he collided with a parked plane. The Chairman of the Bombay Flying Club, a distinguished solicitor called Eastley, had had enough. Vetch was fired.

Of the qualities needed for a good pilot J.R.D. says, 'Fast reflexes are good but if you're high strung, you may take the wrong decision too quickly. To fly well you need two things, a good eye for distance and good hands, just as a good horse rider has. In addition, you want judgement and the capacity to keep cool. In an emergency, you must have coordination and rhythm. In addition to these natural and acquired gifts, when you are flying blind or you're in a spin, what is needed is not only coordination but the will to ignore impressions and trust only your instruments for one can be totally disoriented and misled by impressions which prove false.'

Some of these qualities related to flying were to stand him in good stead when handling an industrial empire.

When asked 'What has been the most satisfying experience of your life?' he replies instantly: 'The flying experience has dominated. No other can equal the excitement of the first solo flight.' Only next to this solo flying thrill does he place the opportunity to build up a great world airline, for which he opened a small door for himself the year after he qualified as a pilot; it was then that he planned the early creation of Tata Air Lines which ultimately became Air-India.

CHAPTER II

Getting Airborne

A bi-plane was flying over the Arabian desert in the First World War. Arab tribesmen were then fighting the British. The sun had set, and as darkness gathered over the desert, the plane had to force-land for the night. Hostile Arab tribesmen rushed towards the plane from the rear to capture its occupants, the pilot and the navigator. War planes of those days had a machine-gun mounted in a fixed position in front. The pilot asked the navigator to take charge of the gun while he quickly got out, physically lifted the rear of the plane, put it on his shoulder and slowly turned around to give the navigator scope for firing. He did all this at grave personal risk as he himself was exposed. The Arabs dispersed and the occupants of the plane spent a safe night in the desert before taking off the next morning. The man who accomplished this exacting feat was Nevill Vintcent, heavyweight boxing champion of the RAF in the Middle East.

A few years later a tall, strapping blond 'stranger with smiling blue eyes came to India in his De Havilland 9A and looked around. He had flown his aircraft all the way from England to come and see our land, and earn his living here. Aeroplanes were in his blood. It came to pass that this stranger met a young Indian, who was the first to get his A licence in our country...all that has since been achieved commenced at that meeting. The stranger was Nevill Vintcent, the Indian J.R.D. Tata.' So says a booklet on Tata Air Lines.

In the late 1920s Nevill Vintcent and Captain J.S. Newall went round the country offering joy rides. They landed in open spaces, collected a crowd of people, and offered them a ride. In due course they undertook aerial photography and survey work. As they traversed the great spaces of India, its great potential for commercial aviation dawned on them. When Nevill Vintcent heard that Imperial Airways were planning an air service from London to Karachi in April 1929, which would also carry mail, he realized that Peninsular India would be left out—the Imperial route would only link Karachi with Calcutta on to Australia and bypass the rest of the country. So even before the Imperial service could start, he worked out a proposal for a flight from Karachi to Bombay and on to South India and Colombo. Vintcent first contacted Russa Mehta, son of Sir Homi Mehta, a textile magnate, with his plan. Tatas had crossed his mind but he thought that the conglomerate was too big to bother with a small proposal like his own. Nevill made no headway with the Mehta family. Finally, someone suggested, 'Why don't you see J.R.D. Tata? He has just taken to flying.'

That is how the two met, in 1929, within about four weeks of J.R.D.'s getting his flying licence. They hit it off from the moment they met one another. 'He was a remarkably fine man,' says J.R.D. Vintcent's proposal and the exciting possibilities it opened up, struck a chord in J.R.D.'s heart. J.R.D. had read several books on aviation. The story of Aero-Postale, the French airline, that flew from France to Chile in South America, and across the towering Andes fascinated him. Saint Exupery had captured this saga in his book *Night Flight to Arras*. The French had dared in the early 1920s to fly the mail from Toulouse in France to Dakar in North Africa. A fast boat picked up the mail at Dakar and sped across the South Atlantic to Latin America. Then came the most challenging part—the flight across the Andes, the second highest mountain range in the world. The Andes were higher than the ceiling of those little planes yet skilled pilots negotiated between mountain passes to reach the west coast of South America. It was a daring journey. Aero-Postale had a motto that J.R.D. often quotes: 'Passengers may be delayed but must never be lost. Mail may be lost but must never be delayed.' The exploits of those pilots inspired J.R.D.

When Vintcent's proposal came to Tatas in 1929 they were recovering from the downswings of the 1920s and they had had to withdraw from some enterprises. Furthermore, the Chairman, Sir Dorab, was past his prime and was not likely to respond to J.R.D.

Once again it was John Peterson who came to J.R.D.'s rescue. He persuaded Sir Dorab to let J.R.D. have his way. 'Let the young man do it. It doesn't cost much.'

The initial investment needed was only Rs 2,00,000*. Finally, Sir Dorab agreed.

---

* Around six million rupees in 1991.

The first letter to the Member, Department for Industries, Government of India, Sir B.N.Mitra, was written by John Peterson as early as 20 March 1929—within six weeks of J.R.D. getting his licence. He was probing the government's interest in supporting the development of aviation in India, 'for military reasons, if for no other.' Peterson mentioned that two flying men, Newall and Vintcent, had approached Tatas with a proposal for a mail service. 'I do not myself know anything about them but they seem to have thought it out pretty carefully.' Sir B.N. Mitra indicated that the government was more keen on a link between Karachi and Delhi than the proposed link between Karachi and Bombay. Within days of the letter the first flight of Imperial Airways Limited to Karachi arrived. *Commerce* of 13 April 1929 reported:

> The first Indian mail landed in India at sunset on Saturday last... A large crowd cheered the arrival of the machine which made a perfect landing....Distinguished Air Chiefs travelled in the plane and every night they slept in a different country and in just over seven days after stepping on board at Croydon they disembarked on the hot and sandy aerodrome at Drigh Road, Karachi.

The original flight scheme of Tatas was to fly Karachi–Ahmedabad–Bombay. Before they could finalize their plan they had to resolve the following questions:

- Should Tatas use a normal plane or a seaplane as the flight was between two ports?
- Should it be a mail service or a passenger service or perhaps a passenger with a bag of mail tucked in?

Tatas first proposed seaplanes but abandoned the idea as both the capital and the operating costs were high.

Captain Newall was keen on a passenger service rather than a mail service. Peterson was sceptical about the prospects of such an enterprise. Newall wrote to Peterson: 'We do not agree with you when you say that you do not think that passengers will avail themselves of the service. At Rs. 100 per seat (Karachi–Bombay) we feel confident that they will do so...'

Tatas requested the government for a subsidy of Rs 125,000*. They were turned down.

From mid-1929, till the end of 1931, there was ceaseless correspondence between J.R.D., on behalf of Tatas, and the government in New Delhi. The period tested J.R.D.'s endurance and patience at the threshold of his career. A ray of hope appeared in May 1931. Vintcent, writing to J.R.D. from Cecil Hotel, Simla, on 20 May 1931 said: 'Yesterday, I lunched at the Viceregal Lodge and managed to get about ten minutes' conversation with H.E. (Lord Willingdon) and told him briefly the situation as far as I know it. Although, in such a short time it was not possible to discuss details I asked him whether in his opinion Indian firms should be encouraged to engage in air transport and he said most emphatically that he was in favour of it. His actual words on leaving were "Well, young man, I am glad we see eye to eye on that." I shall try to interest him further so that he may let his opinion be known.'

Vintcent continued: 'It is impossible to anticipate the verdict of the standing Finance Committee and I am inclined to agree with A.P. Herbert that any decision from so august a body is equivalent to an act of God and entirely beyond human control, but from information gleaned from various sources it does seem as if their decision will be satisfactory.'

Newall left for England as his wife was ill and Vintcent went on leave to get married. A little while before this, a rift had developed between Newall and Vintcent. When J.R.D. came to know about this he was quite disturbed. J.R.D. in his letter of 6 July 1931 to Newall said, 'Differences of opinion between you and Vintcent have been very disappointing to me.' He wrote a similar letter to Vintcent as well. His letter had some effect but the relationship between Newall and Vintcent did break down completely later. As the last proposal for the air mail service submitted to the government was prepared by Vintcent, Tatas preferred to continue their association with him. By this time J.R.D. and Vintcent had already come much closer to each other and were of one mind in their efforts to initiate an airline.

While in England Vintcent received a job offer from the Government of India, to serve as Deputy Director of Civil Aviation under Frederick Tymms. Vintcent accepted this position with the clear understanding with Tatas that he would leave the government and join them once Tatas' aviation operation began.

_____

* Around four million rupees in 1991.

Meanwhile, Vintcent advised J.R.D. to try for the Karachi–Calcutta route in preference to Karachi–Bombay. J.R.D. wrote back on 28 August 1931: 'Times are getting from bad to worse in India and with all my personal enthusiasm I would be reluctant to recommend to the firm the acceptance of proposals on a larger scale than our proposed West Coast line.'

By December 1931, J.R.D. was tired of the government's dilly-dallying. The strain was telling on him. He wrote on 9 December to Vintcent, who was in Delhi: 'I think that Government are treating us shabbily.... I hope that you will be able to save me this journey and to find out whether Government intend to say yes or no within the next 100 years...!'

At the end of December 1931, he wrote to Vintcent, 'The Chairman (Sir Dorab) is very disgusted and if we cannot make things turn for the better very soon, we may find that Tata Sons have definitely decided to give up trying to get an airmail contract from the Government. I really do not understand the position because every few days some sort of red herring is drawn across the trail. There is certainly more in it than meets the eye and I do wish you would try and find out what it is all about.'

Vintcent himself was very worried about the delay in the government's decision about Tatas' proposal. Writing to J.A.D. Naoroji—who looked after airline matters during J.R.D.'s absence—on 31 December 1931, Vintcent said, 'I hope you will use your influence to prevent the other Directors of the firm from losing interest on account of delay.'

On 11 January 1932, Vintcent had a word of cheer to pass on. Writing to Jal Naoroji he said: 'My wife and I lunched at the Viceregal Lodge last week and I managed to explain briefly the present situation of the West Coast route to Lady Willingdon who said that she personally was very anxious to see private enterprise develop in India. Whether this was polite conversation or not I do not know but there is no doubt that both she and the Viceroy are very pleased with their aeroplane and talk a great deal of possibilities of flying in this country.'

Finally, the government took the decision to approve their airmail proposal. According to J.R.D.: 'If Government could not, or would not, spend a lakh or half a lakh, or a cent for an air service would they see their way to accept one gratis, for nothing? Government would, and the Tata Air Mail came into being. Detailed negotiations followed, and (in 1932) a ten-year contract was signed between Government and ourselves which involved no subsidy of any kind but only payment of a certain rate per pound carried, based on a sliding scale according to the distance over which the mails were actually flown. It is worth mentioning here that income from the postal surcharge levied for carriage of mails by air in India covers the rates payable to us with the result that the service actually will cost nothing to Government and, therefore, nothing to the taxpayer.'

The contract was signed on 24 April 1932. Its main points were that government aerodromes could be used without any restrictions; the airline should employ only British subjects; the company should use British-built aircraft unless the government permitted it to use other aircraft; the company should extend its service to Ceylon (Sri Lanka) whenever the Government of Ceylon so desired.

The person deputed by Tatas to work out the details of the new airline was J.A.D. Naoroji. Like his grandfather, Dadabhai Naoroji, a pioneer of India's political struggle, it was in Jal Naoroji's blood to fight for the country's freedom. Just as the British conceded the airline, they arrested Naoroji for his political activities. Prior to his arrest Naoroji wrote to Nevill Vintcent on 7 January 1932, 'Things are getting very hot here. One of my sisters has been carried off to jail without any trial and another one looks like going tomorrow. What a Government? I do wish that those idiots in power would appreciate that Congress is not a few self-seeking banias as they suppose but that every decent Indian whether he is an avowed Congressman or not is behind the Congress and that the future of one-fifth of the human race is involved.'

Meanwhile, on 30 March 1932, from St. Moritz, J.R.D. wrote a detailed letter to Nevill Vintcent about working out the total quantities of petrol and oil required and other particulars he had earlier instructed Naoroji about. He was trying to estimate the operating expenses of the new airline.

'Naoroji was arrested under the Ordinances at about the time when he received my letter and I do not know whether he had time to handle the matter before his arrest,' wrote J.R.D. to Vintcent.

J.R.D. went on to tell Nevill Vintcent that he was thinking of hiring as a second pilot to Vintcent a man called Dastur, trained at Hamble, and highly spoken of. The only hesitation J.R.D. had was that when Dastur, 'was originally trained in Bombay he incidentally distinguished himself by indulging in a Don Quixote game with a steamroller on the aerodrome, with very little resulting harm to the roller!' He finally settled for a pilot called Homi D. Bharucha. Planes and pilots were still a novelty and Bharucha tried to impress people by wearing riding boots, lifting himself to the cockpit as if he were mounting a horse!

CHAPTER III

Song Of The Clouds

J.R.D. Tata said to the Rotary Club, Bombay, one hundred days after inaugurating the first air service in India: 'I want to express, perhaps unnecessarily, the unbounded confidence we have in the ultimate future of air transport in India. A few lean years will precede the great development that must come but that has always been so with any new enterprise and in the case of air transport, more perhaps than in any other, the difficulties are worth conquering. We look forward confidently to the day when none of you will think of travelling or sending your letters by any other way than by air, and when this time comes, if we have done our bit in helping India to make up for lost time, and to attain a position in aviation worthy of her, we shall have achieved our purpose and we shall be satisfied.'

Once government approval came, J.R.D. went to England to purchase two Puss Moths. He went to the factory of De Havilland and met the old man himself. Sir Geoffrey De Havilland was a towering figure in aviation but J.R.D. found him, in some ways, a sad person. He had lost three of his sons in the First World War. Of the two planes purchased, J.R.D. had hoped to fly one to India himself, accompanied by his wife. The Puss Moth could accommodate a pilot in front, two passengers in the rear or a few hundred pounds of mail. On the flight to India, they stopped at Naples where J.R.D. developed a high fever. He still took off, flew for ten minutes and returned to the airport. Flying on seemed out of the question.

A Lloyd-Triestino ship *Victoria* was to leave Naples soon. J.R.D. and Thelly booked their passage and towed the plane behind a car to Naples harbour. The Puss Moth was beautifully designed. All that needed to be done was to fold its small wings and then its width did not exceed that of a medium-sized car! The plane was hitched onto the deck and it came along with the Tatas as their personal baggage. Strapped to the deck it

97

arrived at Ballard Pier. From there, loaded onto a bullock cart, the plane trundled along to the Juhu mud-flats.

Karachi had an aerodrome to receive the Imperial Airways flight and they had even prepared a tower for the Zeppelin to land. But Bombay did not even have an aerodrome.

The only place suitable to fly from or land in at Bombay were the mud-flats at Juhu, a fishing village and a beach resort. Gulls flew overhead and cows looked for the odd tuft. The mud-flats were dotted with unending rows of coconut palms that stretched to the golden sands of the beach. Shaded by the trees were shacks owned by a few rich people for weekends of swimming and recreation.

At first 15 September 1932 was scheduled for the inauguration of the Tata Aviation Service. But the monsoon that year was heavy and the mud-flats at Juhu airfield were under water. So the date was postponed by a month. On the morning of 15 October 1932 there was a small crowd in attendance at Drigh Road airport, in Karachi. J.R.D. was clad in long white trousers and a short-sleeved white shirt. He was armed with only a pair of goggles and a slide rule that he always carried on flights. To give him a send-off came the Chief Officer of the Karachi Municipality, the Postmaster of Karachi and a few others. J.R.D. shook hands with all present and punctually at 6.30 a.m. the plane took off with the mail. The *Statesman* reported on 16 October 1932 that the small party at the aerodrome gave 'a cheering send-off to the handsome and nicely got up Moth plane and its pilot.' A couple of minutes earlier another small plane had taken off to give 'a fine salute to the new plane and its pilot.' Speaking about this episode years later J.R.D. was to say: 'On an exciting October dawn in 1932, a Puss Moth and I soared joyfully from Karachi with our first precious load of mail, on an inaugural flight to Bombay. As we hummed towards our destination at a "dazzling" hundred miles an hour, I breathed a silent prayer for the success of our venture and for the safety of those who would work for it. We were a small team in those days. We shared successes and failures, the joys and heartaches, as together we built up the enterprise which later was to blossom into Air-India and Air-India International.'

The flight was actually bumpy and hot. J.R.D. said on arrival, 'I had a delightful flight from Karachi to Bombay despite head winds all along the route, until I reached Ahmedabad. The result was that although my Puss Moth plane has a cruising speed of 100 miles an hour I was able to achieve maximum speed of only 90 to 92 miles per hour.' He touched down in Bombay at 1.50 p.m.

At Ahmedabad the plane was refuelled by Burmah Shell. A bullock cart trundled to the runway and four-gallon cans of petrol were poured into the tank of the little plane.

The only untoward incident in connection with the flight was that a bird flew into the cabin of his machine and J.R.D. had to kill it. On landing in Bombay that afternoon J.R.D. was asked to explain the future working of the service. He replied, 'I can say this much, that we shall try our best to keep up a regular service and hope the public will support us in our endeavour.' Among those to receive him at the airport were Thelly, Nevill Vintcent, Nusserwanji Guzder and Sir Phiroze Sethna, one of the senior business figures of Bombay.

The Postmaster of Bombay was of course present to receive the mail—fifty-five pounds of it for Bombay. J.R.D. had already deposited eight pounds in Ahmedabad. Within twenty minutes of J.R.D.'s landing, the mail was transferred to the second waiting Puss Moth. Nevill Vintcent climbed aboard the plane and took off with forty-seven pounds of mail for Madras and six pounds for Bellary—a refuelling halt en route. The *Statesman* of 16 October 1932 reported, that:

> Owing to the new service not having been sufficiently adver-
> tised in Europe many packets received by the airmail at
> Karachi, had to be left behind for being transported by ordi-
> nary routes owing to additional 2 anna stamps not having been
> affixed by senders in England and elsewhere. German mail
> also missed the connection and were not carried (by Imperial
> Airways). With these handicaps, today's post bag must be
> considered as a fair start.

The Imperial Airways plane arrived in Karachi every Friday evening, sat on the tarmac and left every Wednesday morning. The tiny Tata plane took off from Karachi Saturday morning, at or before dawn, and arrived at Bellary around sunset. The next morning the plane took off again at dawn and arrived in Madras at 9.15 a.m. The return mail left the next afternoon (Monday), spent the night at Bellary and left the next morning for Karachi, where it arrived the same evening and left the next morning by Imperial Airways for London. All this seemed very impressive but it was only a modest weekly service.

The main problem of the airline was that the Imperial Airways flight from London was often delayed, putting the whole schedule out of gear. Speaking at the Bombay Rotary Club, on 24 January 1933, J.R.D. said, 'We are today completing our fifteenth return flight and it's unfortunate, to say the least, that out of this number of flights the Imperial Air Mail (London-Karachi) was on time only on five occasions. This handicap is in fact becoming a serious matter to us, for every time the delay exceeds twenty-four hours we have to send a machine empty to Madras in order to be able to leave with the northbound mails on schedule. We sincerely hope that greater regularity will be achieved in future by Imperial Airways.'

'Those were adventurous days,' recalls J.R.D., 'we had no navigational or landing aids whatsoever on the ground or in the air, and no radio. In fact, we did not even have an aerodrome in Bombay. We used a mud-flat at Juhu and the sea was below what we called our airfield, and during the high tide of the monsoon the airfield was at the bottom of the sea! So we each had to pack up lock, stock and barrel—two planes, three pilots and three mechanics, and transfer ourselves to Poona where we were allowed to use a small field as an aerodrome, appropriately under the shadow of the Yeravada jail!'

The first year of functioning brought Tatas a worthy comment from the Directorate of Civil Aviation. Its report of the year 1933–34 said:

> As an example of how an airmail service should be run, we commend the efficiency of Tata Services who on October 10, 1933, arriving at Karachi as usual on time, completed a year's working with 100% punctuality...even during most of the difficult monsoon months when rainstorms increased the perils of the Western Ghats portion of the route no mail from Madras or Bombay missed connection at Karachi nor was the mail delivered late on a single occasion at Madras...our esteemed trans-Continental Airways, alias Imperial Airways, might send their staff on deputation to Tatas to see how it is done.

Speaking to the Rotarians in 1933, J.R.D. said: 'Ours is primarily and essentially a mail and freight service and passengers must take second place in our programme at least for some years to come....I'm personally convinced that air transport will not attain its true value until flying is done day and night but night flying is only possible with a very extensive, and experienced ground organisation in the shape of lighting equipment, emergency landing grounds, wireless communications, radio beacons, etc. The expenditure required for this could not possibly be undertaken by operators. It is clear that the provision of such a ground organisation must, and can, only be undertaken by Government. Considering the attitude of Government towards commercial aviation up to the present, the position is not very promising....So far as the immediate future is concerned, therefore, we shall probably have to fall back on faster machines as the only means of speeding up mails.'

Individual flying was still regarded as a rich man's privilege. At best most perceived the future of aviation as one of primarily carrying mail in peace time. It is in that setting that J.R.D.'s vision, quoted at the beginning of this chapter, is relevant.

Sir Frederick Tymms, the Director-General of Civil Aviation writing in the *Times of India* in October 1934 said:

> Scarcely anywhere else in the world was there an air service operating without support from the Government. It could only be done by throwing on the operator the financial risk. Tata Sons Ltd. were prepared to take that risk.

Earlier that year, in January 1934, Nevill Vintcent had declared to a Bombay audience 'Our unsubsidised service is making a profit.' The profit was only Rs 10,000 a year but the reward was far greater than the financial returns. They had the satisfaction of "conquering". In the second year of operation, too, the airline notched up a hundred per cent regularity while nine out of fifty-two flights of Imperial Airways were delayed. Meanwhile some British investors, mainly Govan Brothers, Delhi, set up another domestic airline called Indian National Airways (INA). Another domestic airline, the Indian Transcontinental Airways, had started about a year after Tatas. Its route was Delhi–Calcutta.

The Tata Air Lines kept expanding. In 1933 it started the Bombay–Calcutta route via Nagpur with Jamshedpur also as a halt. The route was bare of ground facilities or aids and the pilots had still to steer by a simple compass and a look at the ground. In 1935 the Nizam's territory opened up with a flight to Madras via Hyderabad. The same year Tatas linked Trivandrum via Goa and Cannanore.

'Flying single engined planes without radio,' says J.R.D., 'is pretty tough during the monsoon. On the other hand, as planes landed slowly in those days, forced landings in emergency in the small fields were possible. We had quite a few of them. I remember one forced landing in 1933 or 1934, between Bombay and Madras, when I spent a couple of hours on the ground, surrounded by wonderfully friendly villagers who had never seen a plane before, and then flying into the night and landing by a quarter moon at Bellary on a small 500-yard toy airfield. As I was very much overdue, and not expected at all that night, I spent the night under the wing of the plane. There wasn't even a watchman, leave alone a building. But my most exciting and cherished memory of those days was undoubtedly my inaugural flight from Karachi to Bombay, when a dream became a reality.

'We sat in the cockpit of those tiny planes, humming along at 80 or 100 mph, with little to distract—not even a radio or a human presence.'

'Did you feel then that there was a Creator?' I enquired.

'The fact that you found yourself totally alone in the immensity of space made you feel very humble and made you see of what little consequence you were. And you identify God with the immensity of nature. These are the only times I felt totally alone and was conscious of that loneliness.'

CHAPTER IV

Tata Air Lines 1933–1946

Speeding up the mail was the initial purpose for starting Tata Air Lines. As time went by and the planes became marginally bigger, a lone passenger was accommodated in an open seat behind the pilot. It took Tata pilots some time to get accustomed to a human being riding in the seat behind them. One day a skipper consuming a leg of chicken is reported to have flung the bone out of the cockpit. It was carried by the wind into the lap of his startled passenger.

At stops, Burmah Shell was always at hand, its bullock carts loaded with two-gallon tins of petrol poured into the tiny tank after being strained through chamois leather. The tins were passed from hand to hand and emptied into the tiny tank. There was a leisure to life. Those were the days when there were no runways, no radio facilities on ground or in the air, no aerodrome offices and no buildings.

Till 1936 all the planes had to be started by swinging the propeller by hand. There was a sign of relief in 1936 when the American 'Waco' single-engined aircraft of 225 horse power arrived with a self-starter. With it the gyro-compass also arrived. When the 'Rapides' joined the airline, the pilot sitting in the plane's nose had a perfect landing view straight ahead and on both sides. In some planes, the passenger cabin was enclosed while the pilot sat in an open cockpit. If the pilot wanted to communicate with the passengers, he would pass a chit through a little aperture in the cabin—not a good idea in an emergency.

Sir Phiroze Sethna used to say of those days of flying that the noise of the engine was so considerable that the passengers could never talk to each other and so you could read a book peacefully in flight. Later, when the planes were more advanced, passengers could talk but it was not 'an unmixed blessing.'

The 1930s were exciting days for aviation and women fliers. There was, for instance, the Mac Roberts Air Race of 1934, London to Australia, with Allahabad as one of the stops.

As the aeroplanes came in, Marshals were appointed to receive them. J.R.D. was one of them. De Havilland entered a Comet. J.R.D. received the KLM (Royal Dutch Airlines) entry, a DC2 with the chief pilot of KLM, which arrived second in Australia. The Comet came first. 'Also in that race were Amy Johnson and her husband, Mr Mollison,' says J.R.D. 'They were the first to get to Karachi but at Karachi they could not get high octane fuel and they had to put in ordinary fuel which damaged their engine. So by the time they came to Allahabad they were in trouble. I did not receive them—a friend of mine who did said they—Amy and her husband—were having a blazing row. She blaming him and he blaming her, when they landed. I only remember poor Mollison, who had come in a pin-striped suit from London, without even a spare shirt, on his way to Australia. He naturally had not shaved as he had left London only 18 hours ago. He stayed on in Allahabad, drinking for two or three days.'

J.R.D. recalls the story of an eccentric Duchess of the time, who was, he thought, the Duchess of Bedford. J.R.D., who spoke to her pilot, an ex-RAF officer, recalls: 'She took to aviation at about 75, bought a Fokker aircraft and with a pilot and engineer travelled to South Africa. The pilot related the story of one flight, where he landed somewhere in Africa. It was terribly hot—no air-conditioning. The pilot who was helping her brought her upto the tail of the plane when the poor lady collapsed. The pilot told the mechanic, "dead as mutton" at which point the old lady opened her eyes and said, "not as dead as you think."'

The inaugural flight of an air service from Willingdon aerodrome in Delhi to Juhu aerodrome in Bombay was a grand occasion. In 1937 two Tata planes inaugurated the service, one piloted by Nevill Vintcent. The mail plane carried 3,500 letters and one passenger. The accompanying machine carried three journalists. The planes were to halt at three princely states—Gwalior, Bhopal and Indore. In four silk bags of royal blue and gold, special messages from the Viceroy were despatched to the rulers of the three states and to the Governor of Bombay. The report of the whole ceremony was broadcast. The Viceroy, in his special message to the Governor of Bombay, welcomed, 'connecting the Imperial capital with the Gateway of India.'

In a brief inaugural speech, Communications Member, Sir Thomas Stewart, said: 'It is doubly appropriate that those responsible for this new venture should be the House of Tatas who have behind them so wonderful a record of commercial and industrial enterprise and achievement....The vision and the courage to follow that vision, which have brought into existence those great steel and electrical industries in this country, have also led the establishment of commercial flying in India.'

Kudos were fine but profits were slender. Captain K. Visvanath, who had joined in 1936, says they wondered why, for such little profit, 'this man (J.R.D.) was taking so much trouble. But he had a vision. In those days a pilot and three passengers could fly to Bhavnagar for Rs. 56 and Rs. 56 was for the return fare. Rich businessmen chartered our planes. Petrol was half-a-rupee per gallon and our charter rate a quarter rupee per mile. There was no commercial or traffic department. The year I joined we carried only 14 or 15 passengers. We had to carry the mail ourselves on the plane and offload it; including J.R.D.—we were postmen. An aircraft cost only Rs. 20,000.'

In 1938 the British Government decided on an "All-Up Empire Airmail Scheme". Both Tatas and Indian National Airways (INA) received generous mail carriage payments for their respective routes. The profit of the Tata airline jumped from Rs 66,000* in 1937 to six lakh rupees in 1938**. The number of passengers, thanks to bigger planes, shot up by six times during the year and the volume of mail grew four times.

It was J.R.D.'s hope that his brother Jimmy would carry on aviation activities along with him. When Jimmy died in an air crash in 1936, J.R.D. came closer to Nevill Vintcent. The following year something occurred which could have soured their relationship.

It so happened that in the early years, when the Aviation Department of Tatas was started, Nevill was promised a third of the profits. It was only in 1938 after the All-Up Empire Airmail Scheme, after the profits rose ten times, that solicitor J.D. Choksi advised a new arrangement be worked out with Nevill Vintcent. Profits once running into thousands for Vintcent, would now run into lakhs, noted Choksi. J.R.D.'s colleagues accepted the advice and so did J.R.D., rather unwillingly. J.R.D. observes, 'When Vintcent was presented with this suggestion he flatly said, "No, an agreement is an agreement", and said he would quit. So I went to another solicitor Dinshaw Daji who everyone respected. He was a little peculiar. He always used to wear his Parsee *feta* (a hard moulded black hat, with

---

* Around three million rupees in 1991.
** Around twenty-eight million rupees in 1991.

an elevated rim). He always wore the *feta* at an angle and kept moving it forward and backward over his head as he talked. His other peculiarity was he never sat down to work. He worked at an easel. He said he couldn't work sitting down.

'So I put the case to Dinshaw Daji. I told Mr. Daji, "I want to do the right thing." Mr. Daji replied, "Legally Vintcent is not entitled to the original share agreed upon, but morally he is." So I went back and told Vintcent, "Nevill forget it, you are not going."'

'But you must have had a job to persuade your colleagues.' I said. J.R.D. replied, 'No, I did not because I was very firm about it. I felt guilty. I said we should not go into the legal aspects of it. The moral aspect Tatas always respected. Nobody said "No". I just dictated a memo but in retrospect I asked myself, why did I allow a situation to be created where I made Nevill feel that I was not fair to him? It was one of those rare occasions when I regretted my inability to stand up immediately for what I believed.'

Within the first five years of operation Tatas' aviation service had flown a distance equivalent to sixty times the circumference of the world with a published punctuality average of 99.4 per cent. They owned fifteen planes, had fifteen pilots and thirty-four maintenance engineers. By 1938 Tatas invested in their first four-engined plane–a De Havilland 86. It carried mail and eight passengers. The first tragedy struck in January 1939 when a small Tata plane carrying mail and a pilot hit a hillside as it was flying in a cloud at Erpudu, South India. J.R.D. was heart broken.

By October 1940, eight years from its inception, the airline had covered a million-and-half miles and passengers could fly Bombay to Delhi at slightly less than the first class rail fare.

With the outbreak of war in late 1939, normal civil aviation activities were suspended and the government took over civil operations. Tatas' planes were put at the disposal of the government. The De Havilland 86 was commandeered for coastal operations. Tatas operated internal services for the government and for RAF Transport. Two aircraft, the Beechcraft Expeditor and that phenomenal plane, the Dakota (DC3), came on the scene.

The airlines were given many special assignments on behalf of the RAF. They had to transport priority mail, freight and civilian passengers besides carrying military personnel to various destinations in and out of India. Besides this, some other services which Tata Air Lines rendered to the RAF were:

- Surveying an alternate air route to the UK via Oman and Saudi Arabia;
- Transporting civilian refugees from Burma;
- Conducting flights in connection with the camouflage school at Pune;
- Transporting sick and wounded military personnel from the warfront;
- Overhauling and maintaining RAF equipment and flight testing of trainer aircraft.

J.R.D.'s personal contact with the British in wartime was primarily in connection with the manufacture of an aircraft for which a new company called Tata Aircraft was floated in March 1942. 'My devotion to aviation was not only to the Airline but to aviation and aeroplanes. So when the war came along, with the help of Nevill Vintcent, we decided to offer to build up an aircraft industry that would be useful after the war. So what should we build? Metal aeroplanes needed a lot of metallurgical materials and experience we couldn't import easily, so we decided to offer to build the Mosquito—a light, twin-engined fighter bomber. It was made of wood and could go extremely fast. The De Havilland company built it and it was designed to carry one big bomb. It was used to bomb Berlin. We would have to import the engine, of course. The body was a fairly simple structure, very successful. The British Government said, "yes," and we began to build a factory in Pune near the Aga Khan Palace.'

While preparations were still being made to establish the workshop and instal the machinery, the British Government cancelled the order. 'I think it was purely that they didn't want India to make a good aeroplane to compete with the British, so they killed it. Had we started with the Mosquito we would have gone on to other aeroplanes. Then they (the British Government in London) asked Tatas to switch to Invasion Gliders. At first Tatas said "yes" and plans were laid. Then I questioned the project. The more I thought of it, the less I thought it practical. So I wrote asking, "What are you going to do with the gliders? And how are you going to tow gliders from western India all the way to England or to India's North-East if the target is Burma?" The British realised it was totally impractical and cancelled that too.' J.R.D. always wanted to manufacture planes but apart from some help he gave to Hindustan Aircraft that dream was not to be fulfilled.

In spite of this disappointment, J.R.D. was excited with post-war prospects for aviation.

In November 1943, J.R.D. chose as the title of his talk to the Bombay Rotary Club, "The Dawn of the Air Age". To appreciate the time he is speaking of, one has to recall that till the late 1930s even an Imperial Airways flight from England to India flew only in the day and at night broke journey at airports. No stove was allowed in the planes. When passengers had breakfast a wide-mouthed thermos was opened and heated scrambled eggs were dished out. A speed of 250 mph was rated as high. It was about this time that J.R.D. envisioned the Air Age:

> It is evident that the aeroplane is still far from having come into its own as a means of transport in the same sense as the railway train or the motor car. If, therefore, this is not yet the Air Age, when can we expect it and what will it be like? I would define the Air Age as the time, some twenty* or thirty years hence, when facilities for air travel will be as widely available as railway and steamer facilities are today; when fares charged for various classes of travel will approach those of surface transport; when all long-distance passenger traffic will be carried by air; when the normal average speed of air transport will be around 400 mph, when no point on the earth's surface will be more than a day's travel from any other point.
>
> The first and most important consequence of the arrival of the Air Age will, therefore, be to make the world one neighbourhood and to bring its people closer together physically.
>
> The next will be a tremendous acceleration in the tempo of our life. With increased opportunities for business and cultural contact, the range and number of our activities will be much greater than they are today, and we may have to make room for these increased activities by cutting out unnecessary or unproductive occupations like listening to Rotary talks!

---

* In less than twenty years the Boeing 707 jet arrived and revolutionized air transport.

The strain on our hearts and nerves will probably increase proportionately, but this should be compensated for by the increased opportunities for sport, relaxation in better climates, and the greater health and longevity which advancing medical science will bring us. Travelling will have become so comfortable that one of the present causes of human wear and tear in our professional life will disappear. And, when the rush of our lives becomes intolerable, the aeroplane itself will furnish the ideal remedy. For what will be more restful and soothing at such times than to spend a few hours cruising in the upper air, far from the madding crowd.[1]

Some months before this speech, Nevill Vintcent had gone to Britain for discussions with the British Government on the manufacture of the Mosquito plane by Tatas. On the way back to India he took a lift in a Hudson bomber. When flying from Britain to North Africa, off the coast of France, the bomber was shot down and Nevill died. J.R.D. often quotes a French saying "When someone dear to us passes away, a part of us dies with him." Nevill took into the cold waters of the Atlantic, a part of J.R.D.'s life.

After Nevill Vintcent's sudden death, J.R.D. requested the British and Indian Governments for the loan of the services of Sir Frederick Tymms, Director-General of Civil Aviation. Tymms became a very good friend and practically joined Tatas as head of their Aviation department.

In 1943 the government commissioned Sir Frederick to prepare post-war plans for the development of civil aviation. In his report Tymms calculated that India would need forty Dakotas to meet its traffic needs and recommended a limited number of private airlines—not more than four to share the routes.

In June 1945, soon after Germany surrendered, the government published its own broad plan. It agreed with Tymms that, 'a limited number of sound and reliable commercial organizations be permitted' and that licences be granted by the Air Transport Licensing Board. Justice Din Mohammad was appointed Chairman of the board and Sir Frederick Tymms, Vice-Chairman. This is worth mentioning because a later Minister for Communications, Abdur Rab Nishtar, defied this recommendation and threw the airline industry into a deep nosedive by permitting about eleven private airlines to be licensed.

Meanwhile the compulsions of war resulted in the construction of forty-four new aerodromes well equipped (for that period of time) with meteorological services, radio communications and landing aids.

Tatas, since 1937, were training their pilots at Pune under F.W. Figgins, an instructor from the Air Services Training Establishment, Hamble, England. They later opened a ground training school in Juhu under Dr H.M. Wadia.

In 1946 Tata Air Lines carried about one out of every three passengers in the country. The rest were divided among the other airlines in the field. Of the forty-eight operating Dakotas and other aircraft, Tatas owned eighteen of the Dakotas, which were the workhorses of the industry.

J.R.D.'s dream was to launch his airline to the West and soon after the partition of India he made a proposal to the government to take India's wings abroad under the name of Air-India International—a joint enterprise between the government and Tatas.

J.R.D. kept up his personal interest in flying and flew his own single-engine Beechcraft often, though he was no longer personally piloting flights as in the early 1930s. In the 1940s he had as his private plane an Expeditor Beechcraft. Once, Captain Visvanath was standing by at Bangalore airport for J.R.D. J.R.D.'s meeting had ended late and he arrived at the airport at 6 p.m. The forecast was "thunderstorm". He told Visvanath, 'No, I must get back to Bombay.'

The thunderstorm hit them and it was dark. Captain Visvanath says: 'Usually pilots fight this force of nature. J.R.D. flies by his brains—a natural pilot who becomes a part of the machine. He let the small plane he flies go up and down like a piece of paper in the wind. He did not "fight the controls" or put the plane under the slightest stress for under stress a plane can break up.' He adds, 'There was no panic in J.R.D. at the controls.'

CHAPTER V

The Magic Carpet

You don't have to be crazy to be in the airline business, but it helps.

- Eddie Rickenbacker*

In the first five years of their air service even enthusiastic Tata pilots like Captain Visvanath wondered why J.R.D. who had a growing industrial empire to look after, spent so much time on aviation in which there was very little profit then. 'But,' says Visvanath, 'J.R.D. had a vision that was demonstrated time and again.' As we've seen in the previous chapter, it was only in 1938 that the government moved in to financially support the mail services by air and profits rose by ten times in one year to a respectable figure of Rs 600,000. Once the financial security was there, J.R.D. and Nevill Vintcent were ready to plan for the future.

The *Daily Gazette* of Karachi reported in July 1938 that Tatas had,

> ambitious and far reaching proposals for venturing beyond the pale of internal air services and capturing a share of the international air traffic by extension of their air services as far as Baghdad in the West and upto Singapore in the East.[1]

J.R.D. says this is not true. From the beginning he wanted to go West where the real traffic existed and not the East.

In the years of the Second World War, J.R.D. says, 'We built up our organisation and as we waited for peace, dreamed and planned for a bright and expanding future....'

---

* Pilot hero of the First World War; also President of Eastern Airlines.

J.R.D. recalls Vintcent and himself preparing, 'tentative post-war plans of development of our own which included in the last stage the operation of external services westwards and, if possible, all the way to England.' It was clear to them that if India were at all to enter the field of long-range international services she must do so quickly as, once foreign airlines were entrenched on all the world's best air routes, India's entry would become a difficult and financially risky enterprise. Apart from her own growing importance as a great trade and travel centre, India had a commanding strategic position astride the only practical air route from Europe to the Far East and Australia. She was thus in a strong bargaining position vis-a-vis other countries which operated services to or through India or intended to do so.

Soon after the war, KLM and Air-France resumed their normal services to India. British Overseas Airways Corporation (BOAC) continued to operate its service to Karachi, New Delhi and Calcutta. Big competition came with TWA and Pan Am both starting their services to India in 1947. If India had to move into the international airline business, it had to be done soon.

The year 1947 was one of rejoicing and mourning. At the attainment of independence a part of India was in a state of euphoria but there were other parts where communities locked in hate drew blood.

Air-India's New Delhi office was in charge of a Cambridge graduate called Maneck Dalal who had just joined Tatas. He had brought with him a young English wife, Kay, who did not see much of Dalal because he worked from 7.00 a.m. to 11 at night. 'We were ordered by the Government of India to transport Muslims out of Delhi, chiefly to Lahore,' Maneck Dalal says: 'We had eight Muslim servants, each of whom we had to carefully send out of Delhi to save their lives. One night, about 11.00, I received a telephone call at my house threatening to kill me. Since the house had a lot of glass around, and was very vulnerable, and my wife was expecting our first child, we decided to move into the Imperial Hotel in Delhi. There was a curfew and rather than leave my wife behind I had her drive my car. I sat beside her with a large clublike stick in case it was needed. There were corpses outside our house, in fact around the whole of Delhi. Pandit Nehru then decided to bring in troops from the South who were not able to make out whether North Indians were Hindus or Muslims, and they did a magnificent job in restoring order.'

Even forty-four years later Maneck Dalal recalls, 'It was an enormous relief to us when it was all over. One realises what a luxury it is to be allowed to sleep in peace.' The staff of Air-India, says Dalal, at risk to themselves, enabled hundreds of refugees to travel between the two newly born nations thereby saving many precious lives.

It was in these horrific circumstances that Air-India submitted its comprehensive proposal to create a new international airline. It was proposed to create a new company, to be called Air-India International Limited, the capital of which was to be subscribed by the Government of India, Air-India and the public. It would be managed and technically assisted by the domestic airline, Air-India Limited, and would initially operate regular services between India and the UK with modern, long-range pressurized aircraft. 'Government was at first somewhat cool towards this proposal,' says J.R.D., 'as it inclined to prefer a scheme for an airline wholly owned and managed by itself. The government realised that the Air-India scheme would save both money and valuable time because of the ready-made organisation and technical facilities placed from the start at the disposal of the project.'

When J.R.D. put his proposal to the government for an international airline he did not hope for an early reply. Yet within a matter of weeks it gave the green light. Years later, J.R.D. asked Jagjivan Ram, who had been a Cabinet Minister for more than three decades in independent India, 'Why is it that at a time of such a great crisis the government could give a decision in a matter of weeks when today it takes them years not to give a decision?'

Jagjivan Ram replied, 'We did not know any better then!'

This was not the whole truth because J.R.D. had pointed out that if the government started an airline of its own it would take at least two years, to get it off the ground. The fact that Tatas had, on their own initiative and without any guarantee, placed an order with Lockheed for three Constellation planes for early delivery, gave them a head-on lead. Furthermore, the expertise, infrastructure, marketing, staff training and maintenance facilities and the efficient organization of Air-India prompted the government to accept Tatas' proposal.

The government agreed to have forty-nine per cent of the equity holding, Tatas twenty-five per cent, with the public having the remaining twenty-six per cent. J.R.D. offered the government the option to take over another two per cent from Tatas to be the majority holder whenever it so desired.

This was the first joint venture proposal made by Tatas to the government. Its acceptance by the government gave J.R.D. hope that he could build an alliance for progress between industry and government.

As soon as J.R.D.'s proposal was accepted, Maneck Dalal was dispatched from Delhi to London. When he arrived in London, Dalal found the city recovering from the ravages of war. Accommodation was in short supply near the old airport in North London. Air-India's traffic department had to be accommodated initially in a caravan and after another six months a second caravan rolled up. Maneck Dalal remembers the winter of 1948: 'We had to trudge through slush and mud to get to the caravan and had oil heaters to keep us warm. It was a question of suffocating from the oil fumes or freezing of cold....London airport was a wide stretch of area with hardly any development—a large number of rabbits and hare could be seen jumping around. The only person who had the right to shoot them was the Commandant of the airport.'

As stated earlier, Air-India arranged for the training of pilots and other staff for the international airline. By a stroke of luck, delivery of the planes was advanced by six months because another purchaser cancelled an order. Thus it was that Air-India International was formally incorporated on 8 March 1948 and was able to inaugurate its Bombay–London service by 8 June the same year.

The little Maharajah, symbol and mascot of Air-India, rather cooped up within the subcontinent, donned a pair of pin-striped grey trousers, a black coat, a bowler hat and started swinging his umbrella like a London businessman. In time he was to dress as a prelate in Rome, a moustachioed Frenchman with a beret in Paris and an Alpine climber in Geneva.

On 8 June, a gleaming Constellation, the *Malabar Princess* was ready for take-off from Bombay to London. The passengers climbed aboard; J.R.D. and his wife, the Jamsaheb of Nawanagar and industrialist Neville Wadia were among them. A little distance from the crowd were two ladies in a pensive mood. They were J.R.D.'s sisters, Sylla and Rodabeh. Some years later Rodabeh, in a letter to J.R.D., wrote: 'When in 1929 you told me of your dream to start an airline, it seemed fantastic to me and it must have appeared as the most adventurous scheme to the old men at the helm of the company then—but when I waited for you in Juhu (in 1932) I knew that through your tenacity and faith you had made your dream come true and that *your* airline would have a great future—a few years later you spoke to me about starting an international airline. I knew that you would do it, but nevertheless when the first Air-India International plane took off one night from Santacruz, taking you on its first flight to London, Sylla and I held hands and shed a few tears of emotion and pride in our brother's achievements.'

At five minutes past midnight the *Malabar Princess* took off. 'I saw the stars over the Arabian Sea and dawn over the timeless Sphinx. I flew over the white mountains of the Alps and landed in old London Town. And as I rode down the runway at Santacruz in my giant 40-seater Lockheed Constellation I was full of great wonder for I was now an international airline,' wrote the Maharajah.

Years later J.R.D. was to recall: 'We flew across the seas with brand new Constellations, the queen of the skies of those days. I flew on the inaugural service and remember my anxiety throughout the flight as I watched and worried over our performance and the reaction of our passengers, most of whom were seasoned air travellers. I need not have worried as, all along the route, in the air and on the ground, our flight and ground personnel put up an impeccable performance and we landed at Cairo, Geneva and London dead on time.

'It was for me a great and stirring event which, as I reminisced on an earlier occasion, brought to life a dream first dreamed some ten years earlier by a few starry-eyed and slightly demented men, including myself. Seeing the Indian flag displayed on both sides of the *Malabar Princess* as she stood proudly on the apron at the airports of Cairo, Geneva and London filled me with joy and emotion.

'Many then thought that, as an Indian airline, we were foolish, to put it mildly, to invade the ferociously competitive field of international air transport in which great European and American carriers had established for themselves an impregnable position over the years. Who, I remember being asked, would want to entrust himself on long overseas journeys to a small Indian airline, its planes flown and maintained by Indians, in preference to such experienced giants as PA, TWA, KLM, BOAC or Air-France?

'Doubting Thomases could not have been proved more wrong.'

Maneck Dalal recalls the inaugural flight from the London end: 'The first person to step out of the plane into a battery of floodlights and cameras clicking was the lithe figure of our Chairman in a grey, double-breasted, pin-striped suit. As he came down the steps he called out to us

"Set your watches, boys! We are right on schedule.'" The Indian High Commissioner was there to receive the flight.

J.R.D. told the British Broadcasting Service (BBC) that this flight was the first by an Asian airline to link the East and the West by a regular service. The event was celebrated with dinner the following day at the Dorchester Hotel.

At the end of the year 1948, Air-India's load factor was seventy per cent. Its profit: Rs 77,000*. By 1968, seventy-five per cent of Air-India's passengers were foreigners who came from countries that had their own airlines.

In those early days Gianni Bertoli, who was married to Thelly's sister Kitty, started Air-India's office in Geneva and Fali Nariman the office at Cairo. S.K. (Bobby) Kooka was their immediate boss in Bombay. Kooka was to play a significant part in the growth of the airline.

J.R.D. knew that Air-India could not compete in size with well-established airlines—and he never tried. But he had the confidence that his airline's equipment, maintenance and decor were second to none and what is more Air-India could excel in the quality of its service. 'I want,' he told his employees, 'that the passengers who travel do not have occasion to complain. I want to establish that there is no airline which is better liked by passengers, that is safer and more punctual, where the food and service is better and which sets a better image than Air-India.'

The ever-bowing Maharajah was a symbol of that service.

---

* Around one million rupees in 1991.

CHAPTER VI

Setting Standards

'The joy and the feeling of accomplishment of the inaugural flight to London,' says J.R.D., 'was of the same order as on the inauguration of Tata Airlines sixteen years earlier.' Running an airline, however, was a task which required the coordinated efforts and cooperation of a varied, versatile group of experts. J.R.D. said:

> We had that in ample measure and with the excited enthusiasm of all concerned, the difficult and complex job of expanding our organisation, and adapting it to the needs of a highly competitive international operation was completed even faster than I had hoped and believed possible. It only shows what high morale and united effort can do. To me it was the final fulfilment of the dream which had begun in my early youth.[1]

The team he built began in a Juhu hangar which an old hand A.S. De Sa described as,

> far from the glamorous Air-India that we see today. It consisted of a small hangar with the smell of oil and grease mingling with the mud and slush around. It was not the riches we were after. We loved aviation and we were a happy family. Mr. Tata was very close to us then. We worked very hard.[2]

It was this "family" of the early years that was to carry the international airline to its heights. When the domestic airline started in 1932, J.R.D. recalls, 'there were no unions....We were about four employees in all. There was myself without pay, Vintcent, who was to be paid mainly out of profit, if any, MacWade who was paid in full, a chowkidar, and V.G. Gadgil, an apprentice from Nagpur.'

J.R.D. found that MacWade was upto his side business of repairing cars in office time, and worse, was getting young Gadgil to sweat for his private profit. 'We got rid of MacWade but not Gadgil. After nine months we decided to pay him a salary—a munificent Rs. 25 a month. After all, as I didn't get paid at all, it was 25 times more than I made.'

J.R.D. said:

> VG worked on all our aircraft—Puss Moth, Foxmoth, Miles Merlin, Q-6, Waco, Rapide, DH-86, Stinson Trimotor, Beechcraft, DC-2s, Dakotas, Vikings, Constellations and Boeings. All through these difficult years VG was always there and rose to be our Chief Engineer.

> VG is a real old-timer and personified the early birds who had come in with us right from the start. They lived and dreamed the same dreams that I dreamt, and built up this great airline with their sweat. Today when I see so many young men who have never done a day's real hard work in their lives, who think that the airline which they did nothing to build up owes them a good living for a minimum of work and an occasional strike, who have no understanding of what this airline stands for and means to people like VG and myself, is it surprising that I should feel particularly grateful to men like him?[3]

After J.R.D., Nevill Vintcent and Homi D. Bharucha, came B.K.N. Rao, who was soon followed by Captain Visvanath in 1936. Visvanath rose to be Director of Operations, Air-India, and retired in 1973. One evening in September 1989 I followed the busy Bangalore–Pune Road to a farm called Belmar Estate to meet B.K.N. Rao, M.B.E. In his eighties, he had had a stroke and sadly noted that he found his thoughts and memories 'a jumble, difficult to sort out.' But the one thing he said, with joy on his chubby face, was: 'Visvanath comes and sees me sometimes. He saw me the other day.' After fifty years, old friends still travelled long distances to meet each other. They were more than a team. They were, as De Sa said, part of 'a happy family.'

One of "the early birds", Visvanath, joined at the age of twenty-one as a pilot. Once J.R.D. became Chairman of Tata Sons in 1938, says Visvanath: 'There was an unwritten rule that he would not be alone in the air. Every 15 days of a month he would fly and a pilot would make preparations with the map and accompany him but J.R.D. did the flying. Everybody dodged being with him for they could not satisfy him. Some pilots said they had stomach pain and would opt out and I often got landed going with him. I have learnt more from him than from anybody. I remember once flying in his Beechcraft single engine, I think, to Delhi, and he asked me the ground speed which is different from air speed. I quickly worked it out and replied "145 miles per hour." He took out his slide rule and worked it out for there were few readings (on the dash board) those days. He turned to me politely and said, "145.5." From then on, I practised and insisted on the same standards of accuracy he taught me from the pilots I trained and checked.'

In 1938, Bobby Kooka, graduate in history from Oxford, came for an interview. He recalls how he crashed into aviation. 'I was warned that Mr. Tata was a terror. Heart in mouth, I went to his office. He asked me very searching questions, none of which I could answer. He was obviously impressed, so impressed, that within seconds, I was ushered out of the room and instructed to see Mr. Nevill Vintcent on a lower floor of Bombay House. After asking me more searching questions, Mr. Vintcent asked me how much money I expected. I said, "I leave it to you, sir." He replied, "That's very nice of you," and promptly offered me Rs. 100 per month. I was deeply moved! My close relations were irate and assured me that they could get me more money elsewhere, but they did not specify where.

'I asked my dear mother and she said, "Bobby, if you are offered Rs. 100, you'd better accept it, because that is what you must be worth to them." I have never regretted it.'

Till 1938 the sales of Tata Air Lines were handled by Thomas Cook. Then the airline opened its own sales office at Churchgate. Kooka showed his flair for marketing by deciding to decorate the show window of the office with a symbol of Eastern hospitality. He asked an artist to draw the figure of an Eastern potentate smoking a *hookah* on a flying carpet. 'The Maharajah was so good we never changed him,' says Kooka.

Jal Cawasji, the Art Director of Air-India, and his team were once responsible for the drawings. Kooka gave the words. Kooka's humour was a bit too robust for MPs and others and J.R.D. was kept busy for decades explaining and defending Kooka. But the Maharajah has brought light and humour into many lives and even solace to a few. Air-India's journal *Magic Carpet* of February 1988 published a letter from T.N.J. Raman, Personnel Manager, Jenson and Nicholson (India) Ltd., Calcutta. Mr Raman wrote to Bobby Kooka:

> Priya, the child of my niece, was suffering from cancer and this was detected when she was one year old in September 1966. The child was staying with her parents (Mr.& Mrs. G.C. Dore) at 32, Mistry Court, Dinshaw Vacha Road, Bombay-1. Your prominent hoarding of the Air-India "Maharaja" at the Marine Drive end was a favourite of this child. In fact the parents did not understand at first what it meant when it said that it wished to seek "Kulu"*. As the child's condition became worse it often used to cry to go out and see "Kulu". A beaming smile would usually follow even in the midst of great pain when the child would spot your "Maharaja". The child was brought to Calcutta in May 1967 in its last stages and in the locality in which it was staying it was not possible to show any of your Maharaja advertisements. A week before its final end, the child more than once cried to see "Kulu" and we then managed to get the miniature Maharaja which is kept in various Travel Agents' Offices. This was with the child until it passed away on June 9, 1967.

Behind every major enterprise there is a vision. 'Although he (J.R.D.) was one of our pioneer aviators,' says Kooka, 'he always looked on aviation through the eyes of an industrialist. He had the vision, back in the 1930s, to realise two things: first, that air travel would become economic only if it could be made into a mass-market, large-scale industry; and, second, that Air-India would participate effectively in air travel only if it could offer something unique to the air traveller. All that I have ever done is to dream up ways of putting these theories into practice.'

Profit was not the propelling factor in J.R.D.'s thinking. He was obsessed with the idea that Air-India must offer something special if it was to survive internationally.

---

* Possibly the Maharajah in that particular hoarding was extolling the attractions of the Kulu valley.

Next, J.R.D. insisted on operating the most attractive aircraft available to the consumer. 'It affects one's scheduling: unlike many airlines we'll always put a high priority on commercial requirements. It affects our budget,' says Kooka. J.R.D. persisted and his gamble paid off.

Air passenger traffic usually doubles every five years and cargo every three years; the newest planes usually paid off because they brought down the cost per mile.

Though he was not a full-time Chairman, J.R.D. was a full-hearted one, involved in every detail of Air-India. 'Some friends,' he observed, 'say one reason for my continued interest in Air-India is good-looking hostesses. I'd call this only a slight exaggeration.' J.R.D. notes that for him, and for some of his early birds working in Air-India, the airline business, 'was an addiction as some have to wine and women.'

During every flight—and he took many—J.R.D. kept copious notes. On arrival at his desk he would shoot off memos with his ideas, his suggestions, his criticisms and his compliments to whoever was the Managing Director at the time. One marvels at the energy and single-minded pursuit of a man who never undertook a flight without noting and communicating the plus and minus points he noticed.

On 8–9 April 1951 he travelled on Air-India's Bombay–Paris flight (VT-DAR) and on 17–18 April 1951 on the Geneva–Bombay flight (VT-DAS). On his return on 30 April, J.R.D. wrote a memo to the General Manager of the airline.

He began the memo by saying that his suggestions and comments were meant merely to help management. They were not to be treated as, 'complaints or instructions from the Chairman nor used for harassing the staff members.'

Excerpts from his memo:

Refreshments:
- A dark British beer is served on our planes. Few people, I think, prefer a heavy beer when flying and I suggest that a lighter brand of beer be stocked.
- The tea served on board from Geneva is, without exaggeration, indistinguishable in colour from coffee....I do not know whether the black colour of the tea is due to the quality used or due to excessive brewing. I suggest that the Station Manager at Geneva be asked to look into the matter.

121

Cabin:

- Chairs: I found on VT-DAR that some of the seats recline much more than others. As a result those seats are more comfortable....I suggest that all our seats be adjusted for a maximum reclining angle except, of course, the rear-most seats which are limited by bulkheads....

He next deals, at length, with the "shabby appearance" of the armrests and asks for free replacements from the suppliers.

On 29 April 1960 J.R.D. addressed a memo to a member of the Board of Air-India. He said, 'You may not be aware that I have throughout the history of Air-India made it my special responsibility to keep in touch with, and have the final say in, all matters directly concerned with passenger service. In fact, I have repeatedly given instructions which, incidentally, are often disregarded; that my specific approval must be taken in regard to any new proposal or change in arrangements for and supplies to passengers and also in respect of any cabin equipment. In other words, any change in anything that a passenger sees, listens to, eats or uses on board is to be subject to my approval.'

A decade later, on 6 January 1970, J.R.D. wrote to a senior Air-India official: 'A number of reports have come to me that cabin attendants are seen smoking in the galley while on duty. This practice creates a very poor impression and must be firmly put down. I suggest that a note be sent to all cabin attendants, under your signature, reminding them that smoking in the cabin or in the galley while on duty is strictly prohibited. Cabin attendants should be allowed to smoke only when sitting in the crew rest compartment, preferably with the curtain down.'

J.R.D further adds a small point about breakfast. 'I am told that while the scrambled eggs and omelettes we serve for breakfast in the first class are excellent, the bacon and tomatoes that go with them are often served stone cold....I suggest that appropriate action be taken to ensure that bacon and tomatoes are served hot along with the eggs.'

One day he received a note from Air-India to say, 'Sir, you've written a million words.' When the question of his meticulous attention to such matters comes up, he remarks: 'The only thing that can be said is that I as an airline Chairman took much more trouble and I attended to smaller things, more personal things than did those of the bigger airlines who were on the administrative side and couldn't possibly attend to details.'

Although it appeared from the decor of its offices and planes that Air-India was lavish, J.R.D. was ever conscious of expenses. After a year of Air-India International commencing operations, he wrote to Neville Wadia, head of Bombay Dyeing Mills: 'Regarding the few specific points you have raised, we *have* considered the question of purchasing an air-conditioning truck since A-I. I.'s inception, but as these costs landed about Rs. 50,000 and as ground air-conditioning would be required for only about six months in the year, we decided to use TWA's machine on hire during the hot season. Unfortunately it is unserviceable at present. We are again reviewing the possibility of buying one ourselves, but money is very tight and a number of more urgent needs have had to be postponed.'

When travelling on a foreign airline he noted its good points which he felt Air-India needed to practise and wrote about them too. He never tired. He never gave up. He never shut his eyes to what was wrong or amiss. Art Director Jal Cawasji recalls being woken up in the middle of the night. It was J.R.D. at the other end displeased with a hoarding he had just seen, probably on the way from the airport after a foreign trip. Very occasionally, while speaking to the staff of Air-India, he would also make some of his points.

Addressing the cabin crew of Air-India in 1970 he said:

> I want meals served during the day with overhead lights switched on. Wherever I travel I find that the overhead lights are not switched on. If you put the little overhead light on, the silver and crockery sparkles. It looks brighter and creates a better impression on the passenger. If he does not want the light he will put it off, but you must switch it on.

> The public judges you by the way you look, the way you dress and the condition of your clothes. With the sari you have a tremendous advantage over the other air-hostesses. It is also a disadvantage to you when serving. But it is one of the major attractions and I hope that when we get the churidar-kameez uniform right, it will be an attraction.

> I feel that our hostesses and pursers should have some liberty and freedom to do their own make-up, and drape their saris with an individuality of their own. You must not go too far.

We must know where to draw the line between the odd, the ridiculous and the attractive. Some of your pursers grow sideburns right into their collars! Some have grown drooping moustaches, that make them indistinguishable from Fu Manchu. Some hostesses have their buns bigger than the head. You may dress as you like. But at the same time remember that you are being looked at by hundreds of passengers. Don't go over the edge of what seems and is a little ridiculous. I have seen hostesses with white lipstick. Now white lipstick may be smart and chic nowadays, but from a distance it makes you look like a corpse, and the passenger is not very interested in being served by one who looks like a corpse. So please do pay special attention to make-up and appearance.[4]

Such attention to details can be infectious. The staff too acquired the ability to care for others. In 1968 Air-India topped the list of airlines surveyed by Julian Holland of the *Daily Mail*, London. She noted:

I left some chocolates on my seat while the plane was delayed in Rome for an hour. When I returned, they had not moved but the blinds had been drawn at the window to keep the sun off them and prevent them from melting.

Perfection, it is said, is in trifles; but perfection is no trifle.

CHAPTER VII

Night Air Mail

For J.R.D. 1948 was a year of spectacular success in aviation. The first scheduled airline of Asia to land in London had standards of food, service, punctuality equal to any in the West. Little did J.R.D. dream that within months of its success—and Prime Minister Jawaharlal Nehru's warm appreciation of it—Nehru's closest Cabinet colleague, Rafi Ahmed Kidwai and he would be set on a collision course.

Michael Brecher describes Kidwai 'as the second ranking nationalist Muslim in India,' the first place belonging to Maulana Abul Kalam Azad. In some ways Kidwai scored over the Maulana. As Brecher says:

> A superb administrator, Kidwai was indefatigable and very successful as Minister of Food and Agriculture. He was, too, a nimble and shrewd political tactician with a brilliant sense of timing, qualities which he used with great skill in engineering the overthrow of Kashmir Premier Sheikh Abdullah in 1953....As a close personal and family friend, an old colleague from the U.P. and one of the few senior Congressmen who shared Nehru's basic ideology, he had ready access to the Prime Minister; he was also one of the few Cabinet Ministers who spoke his mind fearlessly, and Nehru listened.[1]

Kidwai was a man of strong opinions and a will of iron. As Minister for Communications he felt that the air fares charged were too high. The government, under his predecessor in office, had sanctioned so many airlines that each could fly for only a few hours a day thereby raising their operating costs. J.R.D. observes that, 'Either Kidwai or somebody for him thought out a remarkably imaginative and clever scheme. Under the Night Air Mail Scheme aeroplanes carrying mail would leave the four corners of India, Bombay, Calcutta, Madras and Delhi and meet at Nagpur in the centre of India, exchange their mail and go back to where they came from. Letters posted at the four metropolitan cities would reach their destination the next day. Now that involved night flying. I had done night flying in India in little planes and knew that in bad weather and under certain conditions you certainly needed ground aids—both in flight and at either end. You needed approach lights, you needed proper radar. You needed all those things that were not there and therefore I felt there was a safety problem. I knew of how an American President had once cancelled all mail contracts with various airlines and handed the mail to the U.S. Airforce pilots for night flying and there were many crashes.

'This happened in America not because of lack of facilities on the ground but because the weather was very unpredictable—much worse than ours. We've got beautiful weather for flying, even during the monsoons, compared to America or Europe. In America you've got tremendous thunderstorms and even hurricanes. Very unpredictable kind of weather. So I was always thinking of safety and I remembered the words "Mail may be lost but never delayed; passengers may be delayed but must never be lost." The Government proposal was that in addition to mail, some passengers should also be carried and at a lower rate than that of day flights. I thought it would be safer to add more flights during the day.'

The Night Air Mail was inaugurated on 31 January 1949. The operator, Indian Overseas Airlines, ran into financial difficulties, having accepted government terms. The service was soon suspended. Undeterred, Kidwai started, on 1 April 1949, an All-Up Scheme, carrying all letters by air and the Night Air Mail was a part of it. This benefited all the airlines. J.R.D. advised on safety grounds that the night service should be suspended during the monsoons. Kidwai heeded the advice. After the rains the government wanted to resume the Night Air Mail and asked the operators to join on the same terms that drove the earlier air operator to the ground.

This was not acceptable to Air-India and four other airlines—Air Services of India, Airways (India), Indian National Airways (INA) and Deccan Airways—who were jointly discussing this issue with the government.

Rafi Ahmed Kidwai did not like the idea of the airlines coming together on this issue.

When the airlines and the government did arrive at a settlement, the government withdrew the subsidy it had offered. Tatas next suggested that instead of starting an independent Night Air Mail, the objective could be achieved by converting some of the day services into night services. The government ignored this suggestion and announced the reintroduction of Night Air Mail.

Kidwai did his best to get Air-India to take on the Night Air Mail from mid-October. In a letter to J.R.D. on 29 September 1949, he said,' I've been anxious for night service not only because it will provide facilities for quicker exchange of mail between cities but also to attract more traffic to air transport. I think our high rates of airfare confine the air travel to a limited circle....The air operation companies are reluctant to do so (reduce fare). I, therefore, propose to suggest a lower schedule of charges for the night service....As you're aware I was very keen on Air-India taking up this service. I'm sure they are in the best position to offer such a service efficiently.'

Meanwhile J.R.D. was getting restive. He wrote to the Director-General of Civil Aviation on 23 September 1949: 'I am forced to believe that the Honourable Member and yourself have already made up your minds to force the scheme through, come what may. Government are, of course, free to pursue any policy however damaging it may be to the air transport industry, but I wish they would spare us, at the same time, obviously unrealisable assurances that they do not wish airlines to suffer losses.'

He wrote to Kidwai that he thought it was wasteful to have more aircraft for night flying and more fuel mainly for mail and reiterated the fact that some of the day-time services could be converted into night-time services of the airlines. What also concerned J.R.D. was the financial condition of the airlines. He had foreseen the crisis the moment several airlines were licensed—over ten (when the Tymms report had advised not more than four.)

Undeterred, Kidwai reintroduced the Night Air Mail with Himalayan Airways as the licencee and issued on 15 October a press statement painting a rosy picture of the condition of the airline industry. J.R.D. in a rejoinder on 17 October said that the Honourable Minister for Communications had, 'made a number of statements which, I regret to say, are incorrect.' Quoting some further points of Kidwai, he said,' Nothing could be further from the truth.' He concluded his vigorous rejoinder to the press with a personal attack against the Director-General of Civil Aviation—so unlike J.R.D.: 'Mr. Bhalla,' he said, 'whose experience of civil aviation was, up to a year or so ago, limited to that of an air passenger, summarily rejected these proposals as impracticable, without taking the trouble of discussing them with their sponsors who have had 17 years of successful air transport experience.'

On October 25, Khurshed Lal, Kidwai's Deputy Minister, called P.A. Narielwala, Tatas' Delhi representative, and discussed the Kidwai–Tata controversy with him. In a confidential note to J.R.D., Narielwala said: 'He (Khurshed Lal) said that this long-term bombardment between the two of us should cease as it was not doing good to either of us.' Air-India had already issued a pamphlet presenting its case.

One has to recognize that this clash between a high ranking Congress leader and a private individual was taking place just a couple of years after independence when the prestige of the Congress and its leaders was still high. The two-way bombardment did not cease, and Nehru intervened and advised a ceasefire. Heeding the Prime Minister, there was a short pause but within days the same issue came up before parliament.

Kidwai loved the cut and thrust of politics and, in a loudly cheered speech, replied in the Dominion Parliament to the charges reproduced against government in the pamphlet of Air-India. He branded them as "lies". 'Since the introduction of the Night Air Mail,' thundered Kidwai, 'all services except Air-India have agreed to lower their rates.'

To add to the merriment, one MP said that maybe this was because Tatas had to spend too much money on hostesses. It may be mentioned that at the time Tatas were the only airline to have hostesses. Others had male pursers. Kidwai dragged J.R.D.'s name more than once into the debate and accused Air-India of forming a combine with some other operators.

Not unlike Mark Anthony who said 'And Brutus is an honourable man', so Kidwai after piercing J.R.D. with his words said, 'I have already confessed my weakness—I do not want to injure Air-India.'

To the surprise of everybody, Minoo Masani, MP, who had worked for J.R.D. Tata earlier, came out in favour of night flying, subject to the industry being relieved of the disadvantages he complained about. Nehru's intervention somewhat redressed the balance. Nehru said: 'I have travelled half a dozen times by Air-India International. There is a feeling of surprise among the people of international companies in Britain and other places that an Indian international air service has come to succeed because they expected that in view of the highly technical and complicated difficulties which they themselves have experienced any attempt by an Indian company would fail.' Nehru said he would refer the case of the air transport industry to an impartial committee.

The next day Masani wrote to J.R.D. about the implications of the Prime Minister's speech: 'One view is that apart from paying a compliment to Air-India International, which had nothing to do with the debate, the Prime Minister backed the Minister on all essentials. In fact it is urged that he came to the Minister's rescue by putting his case in a more sound and dignified way than the Minister had known how to. The other, more widely held view, is that this was Nehru's way of restoring the balance and that his opening the door to the possibility of an investigation through the channel of the scientific coordination committee is in a way a triumph for those of us who asked for an independent enquiry.'

In an editorial, the *Times of India* of 3 December 1949 flayed Kidwai. The debate on the subject, it said,

> bade fair to go on record as a deplorable exhibition. The contribution of Mr. Kidwai to the debate was unworthy of a responsible Minister. Nothing is easier than for a politician, by jibes, prevarication and misrepresentation, to persuade susceptible ears that he is a champion of cheap travel boldly refusing to be thwarted by potential profiteers and monopolists.

The *Times of India* concluded:

> There is no acceptable alternative to the present party governments or Air-India leadership, but surely there are alternatives to Ministers such as this.

The *Indian Express* took the view on 2 December 1949 that the Night Air Mail Service,

> has proved both popular and profitable to the public. Popular because their letters reach them early morning instead of late in the evening, profitable because it has had the effect of an immediate reduction in the passenger fares demanded. From the people's point of view therefore the Government's decision not to give up is welcome.

Soon after the parliamentary debate J.R.D. felt that his commitment to the Prime Minister ceased when he and Air-India were attacked in parliament. On 8 December 1949 he addressed a letter to the Prime Minister* in which J.R.D. said, if Tatas did not reply to the attack in parliament, permanent damage would be done to Air-India.

Two months later, on 8 February 1950, the Air Transport Enquiry Committee was appointed under Justice G.S. Rajadhyaksha of the Bombay High Court.

It was assigned the task of advising the government on the measures required 'to ensure that the operation of air services is placed on a firm economic footing and that the future development of air transport proceeds on sound and healthy lines.'

In its report, submitted in September 1950, the committee criticized the government's licensing system which, it said, was mainly responsible for the sad plight of the industry. Thereby, it vindicated J.R.D.'s stand. It further expressed itself in favour of licensing only four operators, as suggested earlier by the Tymms report. It further recommended the merger of a few airlines and the delicensing of two on account of their poor performance in the past. The committee also recommended the maintaining of the status quo in respect of the airline industry and the introduction of a system of subsidies, similar in concept to that proposed by the Tymms report. However, this report was totally disregarded by the government, such was Kidwai's power.

Even as this battle between two of Nehru's friends was going on, Rafi Ahmed Kidwai invited himself one day to Tatas' head office to consult the Tata physician, Surgeon-Commander Dr Jal Patel, at the medical clinic on the ground floor of Bombay House. As luck would have it, J.R.D. rang Patel from the fourth floor just then, and Dr Patel said, 'Guess who I've got with me?' Dr Patel paused for effect and added: 'Mr. Rafi Ahmed Kidwai!'

'What is he doing here?' asked J.R.D.

'He has come for my medical advice,' said Dr Patel.

Instantly, J.R.D. shot back with Churchill's words, (used when he heard of the illness of his labour opponent Nye Bevan) 'Nothing trivial, I hope.'

---

* For the complete text of J.R.D. Tata's letter of 8 December 1949 to Pandit Nehru, see Appendix B.

Recalling the incident J.R.D. says, 'That idiot Patel repeated my comment to Kidwai within my hearing and roared with laughter.'

Kidwai, says Minoo Masani, had a sense of humour. Even so, it would be stretching it a bit far to say Kidwai would have appreciated this one.

At the time of this controversy, J.R.D. called on Sardar Vallabhbhai Patel and told him 'Kidwai knows nothing about aviation.' With his left eyelid closed as usual, his lips turned up at one corner, the Sardar gave an understanding smile to J.R.D. Little did J.R.D. know then that within a year a serious disagreement was to develop between Sardar Patel and Nehru over Kidwai.

Patel, as Deputy Prime Minister, had complained to Nehru that Kidwai was collecting money from businessmen for the *National Herald* newspaper in Delhi in return for official favours. Nehru stoutly defended Kidwai before Patel. If Nehru was willing to risk his relationship with Patel over Kidwai, J.R.D. at a lower level of influence had hardly any chance against Nehru's favourite.

Looking back over this episode forty years later, J.R.D. said that maybe he was too cautious regarding the Night Air Mail Scheme and that he may have needed to be bolder in some of his decisions. I reminded him of something he had once said: 'You told me once you may have been wrong opposing the Night Air Mail Scheme.'

J.R.D. replied: 'Oh! definitely by hindsight. I considered it a dangerous or not a safe operation. It is true that I had done unsafe operations myself, I had flown during the monsoon (without aids), I had landed in fields, but once you operate an airline you have to be careful. I told them that they must instal equipment, put radar beams, take a year or two... I thought it was premature....In actual fact it worked well.'

Kidwai brought to bear all the resources of the state and made the Night Air Mail Service a success. Passenger fares were reduced and mercifully the accidents J.R.D. feared, due to ill-equipped airports, did not take place.

In the conflict between the Maharajah of aviation and the government—it was the Minister who won. It was a harbinger of the fate that was to await hereditary maharajas, twenty years later.

CHAPTER VIII

Nationalization Of Airlines

At the end of the war hundreds of Dakotas were left behind by the American Air Force based in India. Tata Aircraft had the contract to buy and sell them. The planes, mostly DC3s, lay in aerodromes all over India. 'Some of them were literally sabotaged; some of them had to be sold as scrap....I remember one airport I went to where the Americans had destroyed instruments with pickaxes because they did not want these thousands of planes to be sold as no one would then buy new planes from the manufacturer,' recalls J.R.D.

Some planes in reasonably good condition were bought by the then seven existing airlines, refurbished for civilian use, and put into operation. The refurbishing usually cost more than the price of the plane. At the same time the government decided to issue licences indiscriminately. Birlas, Dalmias and some others jumped into the act. Abdur Rab Nishtar, the Communications Minister in the Interim Government, confirmed issuing licences despite the warnings of J.R.D. that the government was launching on a dangerous course. 'The large number of Dakota aircraft available from American disposals and hopes of securing route licences under Government's scheme tempted promoters, wholly ignorant of the facts and requirements of air transport, into forming far too many airline companies and purchasing far too many aircraft.

'The Tymms plan (for post-war development of aviation, 1943) which was accepted by the Government, contemplated the licensing of only three or four airlines in all, each with adequate route mileage and scope for developing traffic on the routes allotted to it. As two long-established airlines already existed, only two additional companies at the most should have been licensed. Instead, for reasons into which I need not go here, the then Minister for Communications, Mr. Abdur Rab Nishtar, rejected the fundamental requirement of a small number of strong and efficient operators and encouraged a ruinous scramble for route licences by a large number of concerns created almost overnight. The scene was set for the troubles which plagued the industry from then on. Capital issues far in excess of requirements were sanctioned and in spite of our protests and appeals, no less than fourteen companies were initially licensed.'[1] Looking back J.R.D. felt that Nishtar—a Muslim League man—deliberately took this disastrous course because he knew that Pakistan would come into being and that India could stew in its own juice.

Not only had J.R.D. warned the government but he alerted a fellow industrialist, G.D.Birla, when Birlas entered the industry. As early as 23 August 1946 he wrote to G.D.Birla: 'There is already, as you know, a considerable body of public opinion in the country favouring nationalising air transport, along with other public utilities and key industries. The formation of innumerable new companies will create a chaotic situation which will undoubtedly strengthen the case for nationalisation.'

G.D.Birla replied that his son wanted to enter the field and he was opposed to the idea. But he eventually withdrew his objection, 'in spite of the fact that air transport has no attraction for money making.'[2] G.D.Birla claimed he decided, 'to enter the field...purely on patriotic grounds. Civil Aviation' he said, 'is closely allied to the defence of the country and India's size and distance demanded a much wider interest in air transport.'

J.R.D. was right about nationalization. For, in the very next session of the Central Assembly in November 1946, Sardar Mangal Singh moved a private member's resolution that Civil Aviation be run as a State Department—in other words, nationalized.

Sardar Vallabhbhai Patel, then Minister for Home spoke in the animated debate. He said: 'Government have to plant their feet firmly on hard soil before venturing into such projects as nationalisation of any big industry like aviation without considering pros and cons of the business would mean not nationalisation but liquidation of the Government.'[3]

On the final day of the debate Abdur Rab Nishtar said the government needed more time to take a policy decision.

During the parliamentary debate J.R.D. gave the Associated Press of India (API) his views on nationalization:

> If by nationalising a particular industry or service, or any section of it, the interests of the public will be served better

than by leaving it to state controlled private enterprise, I would be in favour of it. If, on the other hand, a careful and impartial analysis of all the factors involved shows that the national and public interests can best be served in any particular case by private enterprise closely controlled by the state, if necessary, I would be against nationalisation. In the present instance, there can be no doubt that if the subject is considered purely on its merit, on the basis of a scientific study of the facts and of the requirements of the industry and of the country, there is an overwhelming case against the nationalisation of Indian airlines.

One of the general factors to be taken into consideration in this matter, J.R.D. said, was 'nationalised airlines would be subject to political influence and pressure in their management with disastrous results in the conduct of a highly specialised industry in which technical progress, safety and business drive and initiative are of paramount importance.'

On 21 November 1946, J.R.D. forwarded this API interview to the Prime Minister with a letter. Perhaps at Nehru's prompting, Nishtar summoned a Civil Aviation Conference in Delhi, three months after the Assembly debate.

At the conference J.R.D. noted that civil aviation was thirty-five times as costly as transport by ship and fifteen times the cost of rail transport. And its human and commercial aspects were very different. Government should not, he said, take financial risks in developing it, and 'would lack the ability to take quick decisions. What was required, on the other hand, was a management which would give personal service and courtesy to the user.' He advocated 'strictly controlled private enterprise (as)....the desirable mean between the two extremes of nationalisation and unrestricted private enterprise.' G.D. Birla also favoured government control but said healthy private competition should not be stopped. The Minister assured that despite his preference for nationalization, no doctrinaire decision would be made.

For the first couple of years most of the airlines had a bonanza and it looked as if J.R.D. was wrong after all. Then profits fell sharply, many lost heavily and two airlines went out of business. J.R.D. struggled to get the government to understand that even at this stage it was possible to cancel some licences, reallocate air routes and save the industry. But the government was not interested.

In a review of the state of the Indian air transport industry, issued before nationalization on 24 November 1949 and marked, "Not for Publication" Air-India says, 'Although 30 to 40 Dakota aircraft, operated efficiently, would have been sufficient to fly all the services licensed by the Board, about five times that number were purchased from American war surpluses. As a result, whereas efficient and economical operation of aircraft requires that they be flown 2,500 to 3,000 hours per year each, the average rate of utilisation in India fell to the absurdly low level of about 500 hours per year, and the incidence of depreciation, insurance and other overheads was four or five times what it should have been. Even those companies the size of whose fleet was kept at a reasonable figure, were unable to achieve an adequate rate of utilisation owing to the restricted scale of their operations.

'By the middle of 1948, the situation had become critical for a number of companies and by the following winter, two operators namely, Jupiter Airlines and Ambica Airlines, were driven to liquidation.'

Between 1949 and 1952 the airlines drama unfolded like a Greek tragedy. Much earlier, in a second personal letter to G.D. Birla on 22 October 1946, J.R.D. had tried to explain that there were only two alternatives to run the airline industry efficiently. Either to licence a small number of companies or take over all airlines. 'The Government will be driven by the chaotic situation which I foresee as a result of the creation of all these companies.' He tried to persuade Birla not to go into the airline industry.

G.D. Birla, unconvinced, replied on 8 November 1946: 'We must face the fact that there are a large number of new entrants and yours or my firm should be the last people to oppose them for more than one reason. I am sure it would be a bad policy to do so. In the long run only the efficient firms would continue to function. But that apart, I am sure the traffic is going to expand so enormously that all may have sufficient load to carry.'

Seeing that the situation was getting out of hand, the government, as noted in the previous chapter, appointed a committee under Justice G.S. Rajadhyaksha in February 1950, to investigate the state of the industry. The committee recommended in September the same year: first, cutting down the number of operators; second, introducing subsidies and, finally, maintaining the status quo on the issue of nationalization. The report, as we've seen earlier was a vindication of J.R.D.'s stand.

J.R.D. was absolutely clear about the impending crisis. Later he admitted to his shareholders that, 'we expanded our organisation and fleet faster and to a greater extent than subsequent events justified....From being a high-efficiency, low-cost operator Air-India became a high-cost operator.'

By 1952 the situation had deteriorated so much that the Planning Commission recommended the merger of all scheduled airlines into a single corporation with the government having a controlling share. J.R.D. submitted an alternative proposal. He recommended two corporations for domestic and international flights, in which the government would have a controlling share, viz., a joint enterprise like in Air-India International.

J.R.D. was greatly concerned that any attempt to form a single government corporation for aviation, which would lump Air-India International with other less-efficient and poorly-equipped airlines, would endanger the reputation of Air-India International, which he, along with his colleagues, had built up into a force to reckon with in world aviation.

However, instead of opting for a controlling share, the Union Government decided to nationalize the whole industry and put this issue beyond consultation or discussion with the airlines. J.R.D. was upset. The government was willing to discuss two issues: first, whether to set up two corporations—one for domestic and another for international service—or only one; and, second, the basis of compensation. Tatas urged the government to accept world market prices as the basis for compensation. The government rejected the plea. Tatas urged the government that, as it was an interested party, it should leave the decision on compensation to an impartial tribunal. The government rejected this proposal as well and decided upon cost minus depreciation on evaluating the assets. The compensation offered was twenty-eight million rupees* for Air-India and thirty million** to be divided among the other eight domestic airlines.

Looking back, J.R.D. feels that things had come to such a pass that nationalization was inevitable but years later he spoke with feeling about the approach of government to an industry pioneered by private enterprise.

It raised in J.R.D.'s mind larger issues, viz., whether it was worth struggling to expand business elsewhere if this was the government's conduct. J.R.D. led the struggle for adequate compensation for sharehold-

---

* Around 350 million rupees in 1991.
** Around 375 million rupees in 1991.

ers. He wrote to G.D. Birla on 24 March 1953: 'Were it not for the hard fight put up by Tatas the basis adopted by Government (for compensation) would have been even more onerous than it is.'

By November 1952 the government was clear on nationalization. On 5 November 1952, J.R.D., accompanied by Sir Homi Mody, met the Communications Minister, Jagjivan Ram, at Parliament House, New Delhi. Ram had replaced Kidwai by then. When the meeting began, both J.R.D. and Sir Homi were surprised and distressed that the Minister's purpose in arranging the meeting was not to get their views or suggestions but to tell them that the Minister's ministry had decided to nationalize the industry. In the course of a somewhat heated discussion, J.R.D. mentioned that he had an alternative scheme to suggest and had prepared a memorandum on the subject. Jagjivan Ram was obviously not interested. He did not even ask to see the note on the subject.

The only thing Ram wanted from J.R.D. was a suggestion regarding, 'the best and fairest method of fixing the compensation payable to the acquired companies for their assets.'

J.R.D. and Sir Homi expressed their resentment 'in extremely blunt terms' and their indignation at the manner in which nationalization had been brought in by the back door at the cost of heavy losses to the shareholders.

In the course of the discussion J.R.D. asked Jagjivan Ram whether he thought a bureaucratically run airline system was in the interest of the country. The Minister replied that it would not be run as a government department and hoped that J.R.D. would help the government to run it.

Citing this incident, J.R.D. told one of his colleagues, 'I told Mr. Jagjivan Ram, that I was so indignant at the manner in which the Government had treated the air transport industry during the last three or four years and had deliberately brought it to its knees, in order to acquire it for a song, that to ask me to be the head of the nationalised company was to add insult to injury and that I had no intention to preside at my own execution, my duty being only to protect the interests of the shareholders and the employees.'

Subsequently, when J.R.D. met Nehru at lunch he expressed his distress about the outcome of his discussion with Jagjivan Ram. As J.R.D. described it: 'I told him that I had prepared an alternative scheme. The P.M. listened patiently and, it seemed to me, sympathetically, but did not express any views. When I asked him whether the decision to nationalise had actually come upto the Cabinet, he replied somewhat cryptically: "In

137

the last three years this proposal has come before the Cabinet a dozen times." My impression was that, while he did not wish to commit himself in any way, he would go further into the matter and discuss it again with those concerned.'

Within days of his return to Bombay, J.R.D. received a letter from the Prime Minister.* Nehru began, 'My dear Jehangir, I was very sorry to notice your distress of mind when you came to lunch with me the other day. You told me that you felt strongly that you or the Tatas, or at any rate your air companies, had been treated shabbily by the Government of India. Indeed you appeared to think that all this was part of a set policy, pursued through the years, just to do injury to your services in order to bring them to such a pass that Government could acquire them cheaply.

'You were in such evident distress at the time that I did not think it proper to discuss this matter with you. Nor indeed am I writing to you today with any intention to carry on an argument. But I feel I must write to you and try, in so far as I can, to remove an impression from your mind which I think is totally wrong and is unjust to Government, to me as well as to you...

'...the charge you made the other day which amounted to a planned conspiracy to suppress private civil aviation and, more particularly, Tatas' air services, astounded me. I could not conceive of it and I am sure that nobody here could do it.'

He concluded the two-page letter by saying, 'I do not want you to carry in your mind the impression you gave me when you came here. We want your help in this and other matters and it is a bad thing to suspect motives and nurse resentment. Coming from an old friend like you, this distresses me greatly.'

J.R.D. replied on 12 November: 'My only idea was to convey to you frankly my view of the policies and actions of Government which have brought about the present situation in the air transport industry.' J.R.D. noted that Sir Homi Mody and he had brought to Jagjivan Ram, 'an alternative scheme which in my humble judgement was better calculated to achieve Government's objective. The Minister sought our advice only on questions of compensation and the like.' J.R.D. even offered to Nehru to come again to Delhi to present the scheme to him and he ended by

---

* For the complete text of Jawarharlal Nehru's letter of 10 November and J.R.D.'s reply of 12 November 1952, see Appendix C.

stating, 'My only anxiety is to see a strong and efficient Indian air transport system built up and at the time to see justice done to investors and staff who have suffered heavily.'

Then followed laborious negotiations on compensation. In a letter on 5 March 1953 to Jagjivan Ram, J.R.D. pointed out that the Draft Bill regarding civil aviation which government had been good enough to show to Tatas made it clear that all differences regarding evaluation were to be decided by the tribunal set up for the purpose although Tatas' proposals for personnel for the tribunal were not acceptable to government. When the Draft Bill was put before the Cabinet it came as a painful surprise to J.R.D. that the underlying principle of the tribunal being in charge of evaluation and compensation was nullified. Instead this power was given to the Director-General of Civil Aviation, a government official, who was to be the sole judge of whether the aircraft or plants were of approved standards.

'If the bill is proceeded with on this footing, we shall have completely failed in our duty to our shareholders,' concluded J.R.D.

It was not so much whether the airlines were nationalized or not that bothered J.R.D. by then. What bothered him was the dishonesty. He told the author: 'My charge was later of sheer dishonesty on the part of the Government of India. First, they were going to take over...to pay for what they took—and then did an absolutely dirty thing....The Government banned their (the planes') export. We said these planes have a value of Rs. 500,000 in the world. They said exports were banned and they banned them at that time just before negotiation. So what they paid us was the original cost minus the depreciation.'

'At what price were the planes bought?' I inquired.

'Actually the cost of the planes in those days was very little and we'd used them for 5 or 6 years. But we had to spend more money in refurbishing and converting them from airforce planes and troop carriers and buy spare engines.... It was the same thing when they nationalised the banks and the insurance companies. With identical dishonesty they said "When we can save money, why not?" So they didn't care a damn, including Jawaharlal!' J.R.D. noted that the petrol tax alone, paid by the airlines in seven years, exceeded the total amount paid for the assets by the government.

The government, as we've seen, was thinking of having one corporation for domestic and foreign airlines. Nehru was keen to have the

continuing advice of J.R.D. and wanted him as Chairman of the nationalized airline. J.R.D. was thus in a strong position to have his advice heeded
on the future set-up.

'I protested on the 11 domestic airlines and Air-India International
being lumped together (in one Corporation). Because Air-India
International at least had created the same standard of flying, of food, of
hostesses—the whole thing—on international standards. Except for Air-
India Ltd. (the domestic Tata airline) the other airlines in India had a very
low standard and in some cases appallingly low. Also the ground services
of these airlines were very very poor. So I said, "Look this is what's going
to happen—a man having flown on Pan Am or BOAC will then fly on
this domestic Indian airline. His only memory of having flown on India's
airline is of a third class airline. He would never fly Air-India
International" (thinking it to be of the same class as the domestic line)—so
I said, "For goodness sake keep them separate." To that they agreed—to
separate the two corporations.

'They invited me to be Chairman of both the domestic and the international airlines. "To be the Chairman of both airlines will involve too
much work," I said. "I would have to give up almost everything else. But
if you want me to, I'll consider staying on as Chairman, as long as you
want, of the international airline, to ensure that it will maintain the
standards that I introduced from the start." These standards it did maintain
until they got rid of me in 1978. From then onwards everything has
changed.'

Whether or not he should accept the Chairmanship of the international
airline was also a question mark in his mind and in that of his colleagues.
If he accepted, would the government be emboldened to grab another
industry, pay minimum compensation, and then get the top industrialist
to preside as a government nominee?* Feelings ran high on the issue. A
democrat, he launched on an exercise in consensus.

---

*This fear was not unjustified. Twenty-five years later, when George Fernandes
and Biju Patnaik, Cabinet Ministers in the Janata government, suggested the
nationalization of Tata Steel, Fernandes thought he was conferring an honour on
J.R.D. when he said the government would make him 'The czar of the steel
industry', putting him in charge of all the steel plants in India *including* all the
government steel plants that were making heavy losses.

CHAPTER IX

To Be Or Not To Be

Though the airline was dear to J.R.D. he always saw it in the context of his larger responsibility to Tatas. On 10 December 1952, J.R.D. called in his eighteen heads of departments at Bombay House. He reviewed the situation briefly and expressed the dilemma he was facing— whether to accept or to refuse the Chairmanship of Air-India International. He requested all those present to give their views individually. There was a lobby that felt acceptance of the Chairmanship was a surrender to the roughshod ways of the government and that it would encourage the government to repeat its performance in other industries—especially those in the Tata group. Many gave a qualified "aye". Even a vigorous opponent of socialism, M.R. Masani (who had joined Tatas as head of Public Relations) said he was not opposed if J.R.D. felt he had enough of a free hand and autonomy to function effectively. J.J. Bhabha gave a five-point case for acceptance: '...fourthly, because Mr. Tata will have a chance of completing the task pioneered by him of establishing civil aviation in India with success; finally, because the national interest demands that Indian civil aviation be saved from failure, and it will be a catastrophe for Tatas if at this juncture they go counter to Jamsetji Tata's ideal of national service above all other considerations.' The "ayes" won the day.

J.R.D.'s own thinking is reflected in a letter he wrote in December 1952 to Sir Frederic James, head of Tata Limited, London. 'I have been and am still much exercised,' he wrote, 'over the question of Chairmanship; my first inclination was to turn down the offer. I consulted my colleagues, the senior members of the staff of our various companies in Bombay House, and also some of our outside directors and friends. I came to the conclusion that if I was satisfied on the basis of the valuation in respect of compensation to the shareholders, on the fate of the personnel and staff and on the most important point of the structure and form of management of the new corporation and the scope I would be given for independent judgement and action, I should not shirk the opportunity of discharging a duty to the country and to Indian aviation. I am particularly anxious that the present high standards of Air-India International should not be adversely affected by nationalisation. My final decision is still in the melting-pot as the main issue of compensation to shareholders is still under discussion with Government.'

The day after his meeting with heads of departments of Tatas, he sent a letter to his friend Sir Miles Thomas, Chairman of BOAC, confirming an earlier cable which read:

GOVERNMENT OF INDIA HAVE DECIDED IMMEDIATE NATION-ALISATION OF ALL AIRLINES IN INDIA INCLUDING AIR-INDIA INTERNATIONAL STOP THEY HAVE INVITED ME (TO) BECOME CHAIRMAN OF NEW STATE CORPORATION AND TO MAKE REC-OMMENDATIONS ON ITS CORPORATE STRUCTURE MANAGE-MENT AND TECHNICAL ORGANISATION STOP FOR THIS PURPOSE INFORMATION ON INTERNAL ORGANISATION OF YOUR COM-PANY WOULD BE MOST HELPFUL TO ME STOP AM DEPUTING OUR MANAGING DIRECTOR K.C. BAKHLE TO EUROPE TO COL-LECT SUCH INFORMATION REGARDING BOAC AND SOME OTHER NATIONAL CORPORATIONS IN EUROPE STOP HE WILL BE IN LONDON FOR THREE DAYS FROM MONDAY EIGHTH INSTANT AND WILL CALL ON YOU STOP I WILL BE VERY GRATEFUL IF YOU WILL EXTEND TO HIM NECESSARY FACILITIES STOP RE-GARDS-J.R.D. TATA CHAIRMAN AIR-INDIA INTERNATIONAL

Although J.R.D. had the various documents on reorganization of Britain's nationalized airlines (which earlier had been split between British European Airways and BOAC), he felt the need for discussion in view of Britain's experience in the field.

By December 1952 J.R.D. was laying the lines for the future, spelling out the new set-up. 'My tentative view,' he wrote to the Communications Secretary, 'is that all long range international operations be based in Bombay.' He made out a case that two corporations would not cost more than one.

Even as he bowed to the inevitable, he tried a few last salvoes. At the final meeting of Air-India Limited, before nationalization on 21 June 1953, he said: 'As the dismal chapter closes forever, I hope that some of its obvious lessons will be appreciated by all concerned and kept in mind for the future; for instance, that private enterprise should not rush into a new field without making sure that the total productive capacity so created will not exceed the available demand and that the industry will be allowed to function with a reasonable measure of freedom; that a detailed economic plan, once accepted by Government, should not be changed arbitrarily without careful consideration at the highest level and without consultation with those concerned...'

He concluded his address to the final Annual General Meeting of Air-India Limited by observing: 'I must utter a word of warning that unless the greatest attention continues to be paid to the maintenance of high standards of training and discipline amongst flying and ground crews, the resulting deterioration might destroy the good name of Indian civil aviation.

'And now my task is done. If I have dwelt at length on past events and misfortunes and if I have been forthright in directing criticism where I felt it was due, it is in no spirit of bitterness or rancour that I have done so. In laying down the stewardship of this Company's affairs I considered it a duty to place the correct facts before you and to endeavour to draw from them lessons that may be of value in the future.

'Grieved as I am that this venture to which I devoted so much thought, energy and time and for the success of which I have held such high hopes, has come to this end, my sorrow is tempered by the thought that our twenty-one years' endeavour has not been in vain, that the work we pioneered is being carried on and that, in laying the foundations of Indian air transport and building upon them the greater part of the present edifice, your Company and its predecessors, Tata Sons Limited, served the country well.' After the shareholders' interests were as protected as possible under the circumstances, J.R.D. decided to accept the offer of the Chairmanship of the international airline.

On 31 July 1953, a day before the takeover, he sent a message to the staff of Air-India and of Air-India International. 'So as this last day of July marks the passing of an enterprise born twenty-one years ago my thoughts turn to happier days gone by and in particular to an exciting October dawn, when a Puss Moth and I soared joyfully from Karachi....We were a small team in those days; I knew them all personally and each that went took with them a little of me.

143

'My thoughts turn to Nevill Vintcent, that gallant and immensely able man, who conceived the project and managed it with zest and efficiency until he was shot down over the Atlantic ten years later on a dangerous flight back to India. I think of Homi Bharucha who, besides Vintcent and myself as part-time pilot, constituted the whole of the original pilot strength of Tata Airlines. I think of N.G. Gadgil who took his "B" licence with me in England; of my brother Jamshed who lost his life while training to become one of our pilots. I think of other fine men who are no more, and I grieve.

'But more happily my thoughts also turn to friends who are still with us...towards them all....I have a deep sense of comradeship. We shared the successes and the failures, the joys and the heartaches of those early days as, together, we built up the enterprise which later was to blossom into Air-India International. I thank them for their loyal support which I shall ever remember.

'And now the time has come to say goodbye. As we turn the last page and put away the book, regret or bitterness has no place in our hearts. Instead, we may find content in the thought that what we did was worth doing, that we set our standards high and would not lower them, that we never need part with our memories.'

A nostalgic farewell (illustrated by Umesh Rao) was written by Bobby Kooka, Traffic Manager of Air-India International, the person credited with the creation of 'The Maharajah.'

"Little man, you have not long to stay"
with softness in their eyes, they gently ask
"what will become of thee?" I do not know,
for like the dog that knows two masters in one life,
how can my heart be aught but full of sadness! They gave me birth,
they nurtured me, this very Hou... was where I saw
the light of day these fourteen years ago.
Though small at first, strong did I gr... who would not thrive on love and
kindliness!
When I had roamed the length and breadth ... our great land,
and famous was our name from north to south
and east to west, another world became my beat,
my wings I spread to peoples and to countries new.

Our country's flag I bore to them, as strong of limb
I flew across the seas our precious freight.
Toil never dies; no pioneering feat sinks to the earth in vain.
If seeds are sown that are honest and true,
winter can but be on the wing.
Our name may change, our footprints stand.
Bowing low, I take leave of thee,
and if the world's a stage, I've played my part.
And if the crowded house
in chorus thunders out SHABASH
humbly shall I stand...
and to those who take my place,
GODSPEED, shall I say.

The message stirs J.R.D. whenever he reads it. Twice has he read it aloud to me and each time his voice broke and he collected himself. 'What a gift Bobby had of writing!' he said.

Once J.R.D. had accepted the Chairmanship of Air-India and the Directorship of Indian Airlines, he was asked by Jagjivan Ram as to who, from a small list, would be the most suitable as Chairman of Indian Airlines. J.R.D. suggested H.M. Patel, ICS, (who later rose to be Finance and Home Minister in the Janata Government, 1977–79) as the first choice and S. Lall as the second. He advised the Minister not to pack the Boards of the two corporations with government officials, especially Air-India International, as they would face intense competition and required people with a "business outlook". He, however, advised that a representative of the Indian Air Force be on the Board of Air-India International.

I asked J.R.D.: 'You naturally were upset about nationalisation but were you more cut up about Air-India International than the domestic Air-India because the domestic airlines were already in difficulties?'

J.R.D. replied: 'I was cut up naturally also about Air-India Ltd. for it ran Air-India International. The two were literally inseparable and it led to a very difficult problem for me when Air-India was nationalised. All the management of Air-India International had been done by Air-India. All the files had to be sorted out. We had to transfer equipment. All the engineering, all the buying, maintenance was done by Air-India. The operations were of Air-India International. The only thing Air-India International had was aeroplanes, and this whole operation had to be done in the face of a man who was hostile to us.'

"Hostile," because the Minister did not accept J.R.D.'s recommendation and instead of Patel or Lall gave the Chairman's job to an official called B.C. Mukharji. Mukharji behaved in a most peculiar fashion. In a letter to a colleague, A.D. Shroff, on 27 July, four days before nationalization, J.R.D. wrote: 'You will be interested to know that the Chairman of the domestic corporation took up such a hostile attitude and created such a difficult situation by cancelling successive appointments with me without notice or explanation.... I even sent for him and told him, "Why, why is it that everything I suggest to you, you find an objection to? I'm trying to be helpful. Between the two of us, we've got to satisfy the Government. How are we going to do it if we can't agree?"'

As nothing worked, finally he had a showdown with Mukharji in the presence of senior government officials. Then J.R.D. went to the Minister of Civil Aviation, Jagjivan Ram. 'To my pleasant surprise,' J.R.D. wrote on 27 July 1953, 'I found the Minister, backed me completely and said he would speak to B.C. Mukharji. I do not know what he told him but the next day we (Mukharji and J.R.D.) had a long meeting at which we did everything but kiss each other and at which he agreed to everything asked for.'

Then a day came when B.C. Mukharji needed J.R.D.'s help. He wanted to appoint a General Manager for Indian Airlines. The Ministry refused. They expected him—as full time Chairman—to manage the airline. (J.R.D. was full-time Chairman of the international airline and drew no remuneration.)

Mukharji turned to J.R.D. for help. 'I am going not only to defend you but I am going with you to Jagjivan Ram.' J.R.D. recalls: 'I took him and I told Mr. Ram, "Look, your Ministry is entirely unfair to Mr. Mukharji. What he is arguing on this issue is absolutely right." After that this fellow realised I was not an enemy but a friend and then he opened up to me. "I am very sorry. Yes, you are quite right. I have been hostile. But I was told that you would keep interfering (in the domestic airline) because you have been Chairman for so long—that you will boss over me all the time. I am wrong. I am sorry."'

'There was a certain largeness of heart on your part,' I observed. 'You were not holding the past against him.'

J.R.D. says: 'There was no largeness of heart on my part because it was obvious that he was misinformed and had got a totally wrong idea about me. Don't forget it's easy to have largeness of heart when you are indestructible yourself in this sense. When a man like him was fighting me, it was not a question that he could destroy me. He could be destroyed. He was the weak one, he knew all the time that he was vulnerable. So it's easy for me whenever I have a thing like that happen to me—what am I afraid of? All that they could do, was get rid of me. Largeness of heart is required when *you're* vulnerable.'

When J.R.D. took Mukharji to Jagjivan Ram, the Minister was polite to J.R.D. but J.R.D. sensed that he did not take kindly to his intervention in support of a colleague in the other corporation. J.R.D. wrote to Mukharji on 24 December 1953: 'I sincerely hope that in my anxiety to be helpful to you....I have not done more harm than good. I would be deeply distressed if I thought I had.' He also wrote to Jagjivan Ram that coming to him was his (J.R.D.'s) idea and not Mukharji's. Mukharji did not last long. J.R.D. said later, 'Apparently Mr. Jagjivan Ram was annoyed at him but not at me, for having brought me over with him.'

Not long afterwards Mukharji suffered from ill-health and it appears the strain was too great for him. He resigned from Indian Airlines and wrote to J.R.D. on 12 March 1954: 'Believe me I am as sorry as you are that we shall not ever have the opportunity of working together. I do hope that wherever I will be I shall continue to enjoy your friendship which I have come to value very greatly. I shall never forget all the help and cooperation I received from you in such ample measure, particularly during the past few months.'

The months following nationalization, August and September, were months of intense activity. So many decisions had to be made of "unscrambling" Air-India International from the Indian Airlines Corporation. B.K. Patel, an ICS official, was appointed Managing Director of Air-India International.

Letters poured in expressing relief that J.R.D. was still at the helm in Air-India International. Many letters were from the staff of Air-India International or their family members. Peter Menezes, one of the early starters with Tatas, wrote to J.R.D. J.R.D. who knew him personally replied: 'It is, more than anything else, the friendliness, support and loyalty of men like you that has compensated for the many disappointments and worries that we have had in our company in recent years.'[1]

'One of the main reasons for me to accept the Chairmanship of

147

Air-India International was my desire not to lose touch with all the fine men and women who worked for the airline.'

Sometime after nationalization the word 'International' was dropped and 'Air-India' became the international carrier. About a year after nationalization J.R.D. wrote to a friend, Floyd Blair, about his other friend—The Maharajah—mascot of Air-India. 'Fortunately he has survived in the new Air-India International where he retains his popularity with the clientele. I am not abandoning aviation for good yet and I hope to serve India's overseas air transport industry for a few years more provided that, as at present, I am given adequate freedom of action.'[2]

Little did J.R.D. imagine that the "few years" would stretch to a quarter century.

CHAPTER X

Triumph And Tragedy
Air-India 1948–1964

Once the Air-India International airline started J.R.D. began to devote more and more time to this absorbing venture. On many days half his time went in running Air-India International and some Tata Directors remarked they wished J.R.D. would give more time to the rest of Tatas. As an aeroplane's initial ascent needs far more engine power than when it is cruising, so did Air-India International, when it took off, require a great deal more attention. Then only forty-four, J.R.D.'s remarkable physical energy was at its zenith. Not only was his reputation at stake in aviation but that of India, for the Maharajah represented India to the world.

Punctuality was one of J.R.D.'s obsessions. Nari Dastur, who for twenty-two years worked in different offices in Europe and later became the Regional Director of Europe, relates how in the 1950s, Air-India flights used to arrive in Geneva at 11 a.m. He overheard one Swiss ask another 'What's the exact time?' His companion looked out of the window and said, '11 o' clock.' 'How do you know? You haven't looked at your watch.' The other replied in a matter-of-fact way, 'Air-India has just landed.'

'These are the sort of things that made J.R.D. proud of Air-India and this is what he expected of us,' says Dastur.

J.R.D. knew he could not do the job alone. As a team builder he was firm in demanding standards, and generous in giving credit to others which brought out the best in them. Occasionally, he even tended to be over-generous in boosting the egos of his colleagues. Most of the important men he had to work with, he had chosen himself over the years. They were a varied group. Bobby Kooka, the Traffic and Commercial Chief, was brilliant in advertising and promotion. Witty and sarcastic, he was capable of crossing the borderline between humour and obscenity. Some pilots say they found him difficult. But J.R.D. gave him tremendous leeway and got the best out of him for Air-India.

There were the old faithfuls of the domestic airline, now in Air-India, pilots of the 1930s like A.D. Guzder, K.Visvanath, V.N. Shirodkar, D.K. Jatar and Captain Screwvala. 'Screwvala was among the best,' says J.R.D. and recalls that (unfortunately) his flights were introduced on the intercom: 'Good morning, ladies and gentlemen—Captain Screwvala and his (s)crew welcome you on board....' Some passengers would burst into laughter. J.R.D. offered Screwvala a chance to get into management but he declined and instead went on to change his surname to Spencer!

The year following the inaugural flight to London, Air-India took off for the African continent. Nairobi with its considerable Indian population, was the first destination there. Ultimately Air-India stretched eastwards upto Japan and Fiji, and westward to the USA across the Atlantic.

Air-India's West German link opened in the 1950s before the post-war German airline, Lufthansa, started flying. Nari Dastur was the Manager.

In-between flights J.R.D. once spent thirty-five minutes with Dastur at the Dusseldorf airport. Dastur suggested that Dusseldorf was not the correct centre for Air-India and it should be shifted to Frankfurt. 'He went at me hammer and tongs,' says Dastur. J.R.D. said the suggestion was ridiculous as ninety-nine per cent of Air-India's passenger traffic originated in the Dusseldorf area. 'I tried to explain that there was only one other airline by which the Germans could fly directly to India and they had to come to Dusseldorf (to take it).

'This was on a Thursday. On Friday I got a phone call from the traffic manager in Bombay: "I don't know what you've done but the Chairman has ordered we move from Dusseldorf to Frankfurt."'

Dastur goes on to relate that at Frankfurt, the Government of India summoned a conference on how to encourage European tourist traffic to India. The Minister for Tourism, Dr Karan Singh, J.R.D., the Director-General of Tourism as well as all the European station heads of Air-India were present. Everyone was enthusiastic and talked a lot about India's biggest attractions which they felt were Goa and Kovalam beach in Kerala. Only one person sat silent. The effervescence of the others passed with the conference but J.R.D. came home and suggested to the Taj Mahal Hotel to start a complex in Goa, says Dastur, which resulted in the Fort Aguada Hotel and the Taj Village. 'He listens,' notes Dastur.

J.R.D. was proud of his Boeing 707s and even more so that Air-India was one of the first airlines in the world to place an order for Boeing 707s with Rolls-Royce Conway Engines, while other airlines preferred the American Pratt and Whitney engines. This combination proved formidable and was perhaps a precursor of the Rolls-Royce engines being demanded for Boeings by other airlines—even in the United States.

Even though things had begun looking up, J.R.D. had to continue his battles with the government. In July 1954, Air-India was taking a proving flight to Tokyo. It was not an inaugural flight and J.R.D. had declined requests from newspapers and others for a flight to Tokyo. J.R.D. felt that guests should be invited only for the inaugural flights. He mentioned that to the Secretary of the Communication Ministry, then handling Civil Aviation. In spite of that journalists and youth activists of the Congress were invited. Later when he took the matter up again the Ministry pleaded that it was too late to de-invite the guests. J.R.D. would not be moved. 'If government (insisted) on our carrying these persons on our flight I will be compelled to withdraw from the trip.' J.R.D. explained to the government that inviting journalists and members of the youth organization who had no connection whatsoever with the airline would result in unfavourable publicity and loss of goodwill, as he had already rejected requests from regular newspapers and others.

He spoke on the telephone to Jagjivan Ram and followed it up with two telegrams and insisted that the, 'government having decided that the national airline will be managed not departmentally but by corporations such a matter as arrangements for a proving flight should have been left to myself and Mr. Patel as Chief Executives entrusted with the management of the airline. I have devoted time and energy without reserve to the affairs of the corporation even to the extent of heavy strain on my health and neglect of my duties to the firm (Tatas). I've done so willingly in a spirit of service to the national airline and to the country. In return I only expect that my advice should not summarily be disregarded as it has been in this case. I am sorry to disappoint you. The Prime Minister has been informed of this development and if you have no objection I will be grateful if this and my previous telegram be shown to him.'

J.R.D. followed the telegram up with a six-page, closely typed letter to Jagjivan Ram on 27 July explaining his stand and referring to his earlier phone calls to the Minister. He concluded his letter with the words: 'So long as Government require my services and allow me the necessary scope for the efficient discharge of my duties, I shall cheerfully continue to carry that extra load. If, however, Government were at any time to feel that the corporation would operate better under departmental management or supervision and that I was in the way of such an objective, I hope you would not hesitate to say so. In that event, I would retire from the scene readily and without fuss or bitterness.'

It is typical of the man that once he realized nationalization was inevitable, he could say in 1961 in a speech at the Royal Aeronautical Society of London: 'Nationalisation opened a new and, thank goodness, a more orderly chapter in the story of Indian Air Transport.'[1]

On one occasion Air-India was pulled up by the Union Government's Secretary for Civil Aviation when there were some adverse remarks in parliament on Air-India's publicity (courtesy Bobby Kooka.) J.R.D. shot back an answer: 'At the risk of sounding boastful, I have no doubt in my mind that Air-India is today by far the most efficient and successful State owned enterprise in the Public Sector although it has to operate in a fiercely competitive and difficult field which none of the others have to face.'[2]

*

Even at its best the airline business can be one of triumph as well as danger. After he had ceased to be Chairman of Air-India, when the *Emperor Kanishka* was sabotaged and crashed in the Atlantic in 1985, he confided to his old friend Sir Frederick Tymms: 'I allowed myself to be so emotionally involved with Air-India that after guiding its destinies for 46 years, I still find it difficult, subconsciously, to accept the role of an outsider....Throughout that period the fear of learning about serious accidents was always present and affected me deeply when they did occur.'[3]

J.R.D. had to live through four accidents during his Chairmanship. The first was in 1949 when a Constellation carrying forty seamen hit Mont Blanc. The second, in 1955, was one of political sabotage and hit the world's headlines on the eve of the Bandung Conference of non-aligned nations. In the third, again over the Alps, his dear friend Dr Homi Bhabha and his brother-in-law, Gianni Bertoli, died. Their photos are still displayed prominently in the bookshelves near his writing desk at home. In the fourth, a Jumbo soon after take-off from Bombay, dived into the sea in 1978.

Gianni Bertoli was an extraordinary man who had led a most interesting life. An Italian, he knew the English language and had British friends. J.R.D. recalls that as an Italian Naval officer during the Second World War Bertoli and his co-pilot were involved 'in friendly bombing of the British fleet in Alexandria.' When their plane was hit, he forcelanded in the desert, set fire to the plane and started walking with his co-pilot in the direction of the Italian Army. In the desert instead of meeting the Italian Army he ran into a British platoon and much to his surprise the officer said: 'Gianni, what are you doing here?' Recognizing the British officer, Bertoli replied 'Going for a walk.' 'In that case,' said the British officer, 'you better come for a walk with me.' Bertoli landed in India as a Prisoner-Of-War of the British. After release he met Thelly's sister Kitty in Bombay and married her. He was an highly intelligent administrator. He joined Air-India and was in charge of its Geneva office and later rose to be the Regional Director for the whole of Europe (excluding Britain.) J.R.D. was devoted to him. Bertoli had come to India for consultations with J.R.D. and a couple of days before he left, Minoo Masani invited him for dinner. Gianni replied that he had made his booking on the day of the proposed dinner. Masani said 'You can easily postpone your flight by a day.' Gianni declined. He had a tryst with his destiny and stepped on the ill-fated flight.

In the mid-1950s, India was way ahead of China in civil aviation. When Chou En-Lai came to confer with Nehru in May–June 1954, he took a chartered flight of Air-India. J.R.D. wrote to Nehru: 'I was glad that the special charter flight for the Prime Minister of China went off smoothly and that he expressed himself in complimentary terms on the service. With affectionate wishes–Jeh.'[4]

Since the Communist takeover of China, Nehru had worked hard to bring India's giant neighbour into the comity of nations. China had been treated as an outcaste by the United States, a policy that was adopted by many other nations. At the first conference of the Non-Aligned Movement at Bandung, Indonesia, in April 1955 the heads of states were looking forward to meeting the Chinese Prime Minister. China did not have the requisite long distance aircraft to carry their Prime Minister and his party, so an Air-India flight was chartered. It was to collect Chou En-Lai from Hong Kong at noon. The *Kashmir Princess* was ready for take-off at Hong Kong at the due time but neither Chou En-Lai nor his party turned up, only a handful of junior officials with their typewriters did. After an uneventful flight of five hours when the aeroplane was cruising at 18,000 feet above the sea, an explosion occurred on board. It was caused, "by an infernal machine" says Maintenance Engineer, A.S. Karnik who was on the flight. 'I still feel as though the crash had just taken place, so vivid is my memory of everyone aboard. I can see Gloria (Berry, the air-hostess) with untold courage emerging out of the dense black smoke with a handful of life-jackets moving towards the cockpit with unfaltering steps; Captain (D.K. Jatar) holding the control wheel with all his might, up to the bitter end, in a heroic attempt to keep the airplane flying level—courageously cool, not a line of fear across his pleasant face, giving orders in unperturbed tone....'[5]

There were only three survivors out of the eleven passengers and eight crew members. Communications was very difficult and J.R.D. was totally absorbed in the rescue operations at the Bombay end and in keeping touch with the families of the crew members, especially Mrs Jatar. Captain D.K. Jatar and air-hostess Gloria Berry were decorated posthumously by the government with the Ashok Chakra Class I and II respectively.

The hand of fate had intervened to claim Captain Jatar. Not he but Captain Visvanath was to pilot the ill-fated flight. 'I was to be on that flight leaving Bombay at 6 p.m.,' says Captain Visvanath, 'but at 3 p.m. Captain Jatar came and said, "I want to do some shopping in Hong Kong. Please let me take the flight." So I rang Delhi and requested a change so Jatar could go. It was a high security flight so they took two hours to clear it, and Jatar went.'

On receipt of news of the crash Captain Visvanath was rushed to Singapore for the rescue operations and there he met Chou En-Lai who had just flown in on a DC 4 aircraft of Indian Airlines, also chartered by China.

'Chou En-Lai asked me, "Were you not warned?" I said, "No." Chou En-Lai replied, "But the New China News Agency warned us!"' Captain Visvanath told the author he was too stunned to inquire: 'But why did you not inform us? And why did you risk the lives of the crew and eight of your own citizens?'

A couple of years later, on 15 October 1957, India celebrated the silver jubilee of Indian aviation and the spotlight was on J.R.D. On this occasion *Current* magazine brought out a supplement to which the President of India contributed a message. In a letter to President Rajendra Prasad, J.R.D. said, 'I felt particularly honoured and moved by your generous gesture. While I feel I have been given far more credit than I deserve for the part I have played in the development of Indian Civil Aviation, it is deeply gratifying to know that you, as the distinguished Head of our country, place some value on whatever services I have been able to render in this field. May I assure you that I shall continue to do my best to help in keeping high the name and prestige, particularly abroad, of Indian air transport.'

India's second highest civilian award the Padma Vibhushan was pinned by President Prasad on J.R.D. What stirred J.R.D. was a nostalgic letter from the young boy of eighteen he had met at Aboukir Bay on the solo flight to London in 1930—Aspy Engineer.

Aspy Engineer had become second-in-command in the Indian Air-Force. In reply to Aspy's letter, J.R.D. recalled: 'Those days were fun, weren't they? Although you were only seventeen or eighteen at the time, I, at least, did not underestimate you in the Aga Khan competition!...I took you so seriously as a competitor that I spent at least a day more than I need have in checking everything on the plane and everything else connected with the trip! Actually, I was glad I lost to you....Our friendship ever since has been much more worthwhile than winning the competition would have been.

'I must say I enjoyed every moment of that adventure as I am sure you did too. Incidentally, one of the highlights that remains imprinted on my memory was my arrival at Karachi by Imperial Airways on my return to India when, to my embarrassment, you met me with a platoon of scouts and presented me with a medal. That was terribly nice of you and undeserved.'[6]

In a speech a few weeks before the silver anniversary, J.R.D. said,: 'I have been mad about aviation since childhood. I consider myself fortunate in having been associated with the airline business for twenty-five years by now. Romance and glamour have gone out of flying now for it is now a big and highly organised business. Individuals don't count much any more, not even the pilots, who made girlish hearts beat faster in my younger days. In fact, confidentially, that is how I impressed the one who later inevitably proposed to me!'[7]

Nari Dastur says of J.R.D.: 'He loved Air-India and it would have hurt him if there was anything which was not of a high standard and he did that by example. Whatever he wanted of others, he was doing much more himself. Of course, we had our conflicts yet all worked together for a common purpose. If your chief lives, dreams, talks only of one thing, it is infectious.'

'Could it not be also boring?'

'It was never boring—always fresh,' replied Dastur, 'what J.R.D. enjoyed most was to meet the junior-most along with the senior staff at evening parties in Europe. All that stopped after he left. Now it is "we" and "they".' During the twenty-two years Dastur was in Europe, 'No senior officers left Air-India. Even after J.R.D. left they still referred to him as Chairman.'

On 15 October 1962, the thirtieth anniversary of his inaugural flight, J.R.D. decided to re-enact it in a single-engine plane of the same vintage. J.R.D. at fifty-eight was as fit as ever. While no Puss Moth was available, a De Havilland Leopard Moth of the same vintage was found in Calcutta and refurbished. A couple of days before the commemorative flight— Karachi–Ahmedabad–Bombay—the Leopard Moth flew to Karachi with J.R.D. at the controls and Visvanath accompanying him in the uncomfortable seat in the rear compartment which was originally meant to carry the mail.

Visvanath relates: 'After Ahmedabad we ran into head winds as well as a dust storm. We were 5,000 ft. high. No visibility in front. We descended...we were literally crawling. We could see the road crossing below at a distance but after a few minutes it was still at a distance. The head winds had reduced the actual speed to 55 miles per hour. J.R.D. asked if we had enough fuel. We had; however as a measure of abundant precaution, exhibiting his mastery, he manipulated the altitude control and throttle in such a manner as to get the maximum speed with the minimum of fuel consumption. We finally landed after a flight of five hours and twenty minutes.' There was enough for at the most a six-hour flight.

Visvanath was Director of Operations, Air-India then. Re-enacting the original flight Karachi–Bombay, this time by himself, in the fragile plane he landed at Juhu alone and observed, 'It was an uneventful flight. The only difference was that a radio was added this time which mercifully packed up at the end of the Karachi runway so the re-enaction was "a little closer to the original than I had intended."'

Why did J.R.D. undertake the commemorative flight? In a message to the Air-India staff he said: 'I re-enacted the inaugural flight, not for sentimental reasons alone, or for the pleasure of doing it. I hoped that particularly to those of you who had never even seen a Leopard Moth it would bring home the fact that the great airline which we all serve today, could be, and was actually built from the smallest beginnings, with little more to sustain it at first than the love, the sweat and the devotion of those who worked for it. If my flight helped to bring this message to you its purpose has been fulfilled.'[8]

Replying to a well-known trade unionist, N.S. Kajrolkar, who wanted to felicitate him on the occasion, J.R.D. wrote: 'Throughout my life I have endeavoured to avoid personal publicity of any sort and when circumstances have made it impossible for me to avoid it, I have found myself put to considerable embarrassment. The credit for much of the progress attributed to me really lies with the pilots, engineers, staff and workers who, by their devoted work, have built up Indian air transport over these thirty years.'[9]

J.R.D. also observed that he decided to re-enact the flight on the thirtieth anniversary, as he feared he might not be around on the fiftieth!

How wrong he was! He did re-enact it on 15 October 1982 when he was seventy-eight years old.

CHAPTER XI

President Of IATA

As the first airline from Asia to have a scheduled service to the West, Air-India International was "up against the old timers", the giants of aviation—KLM, BOAC, TWA and Pan Am.

In the words of Bobby Kooka, 'For us it (was) like singing at the Metropolitan realising that Caruso once sang there.' Nevertheless, Air-India made its voice heard.

Within weeks of its inaugural flight to London, the International Air Transport Association (IATA) invited J.R.D., 'to take a more personal and prominent interest in the affairs of IATA,' by joining its Executive Committee.

Founded in 1919 by Dr Albert Plesman, President of KLM, the organization was started to make air transport safer and more efficient by enforcing technical standards and standardizing air fares. For almost two decades Sir William Hildred's powerful personality as Director-General (he was earlier head of Belgian Airlines) was to dominate IATA.

To IATA's invitation extended by Sir William, J.R.D. explained on 3 July 1948 that he had travelled extensively that year and his future plans included being an Indian delegate to the United Nations. J.R.D. said, 'I hope you will, therefore, kindly excuse me for this year.'

Sir William replied, 'I am personally disappointed...however, there is always another year.' On 4 March 1952 Hildred came back to say that IATA wanted 'someone from your part of the world, specially at this interesting and growing point in the history and development of your own country. I would very much like to ask you to give us your help for a spell and if you think I am selfish in asking you, believe me I have the best of reasons. Will you please consider this and not turn me down lightly.' To take up Sir William's offer meant at least a three-year term on the Executive Committee of IATA.

J.R.D. replied on 27 March 1952: 'Although the reasons for which I regretfully declined your previous invitations apply with equal or even greater force today, I have not the courage to say no this time and you may therefore take it that, if elected, I shall be willing and shall consider it a privilege to work on the Executive Committee.'

Sir William replied: 'Individually and geographically this would be a splendid appointment.' It was also the beginning of a splendid friendship.

A couple of years later Sir William wrote asking for a photograph of J.R.D. to hang on the walls of his IATA office.[1] J.R.D. obliged with a covering letter, 'As requested, I send herewith a photograph of myself to join the "Rogues' Gallery" on the walls of your room.'

Sir William promptly wrote back: 'Thank you very much indeed for complying so quickly with my request and for sending so dangerous looking a photograph....As I look at my pictures some are benevolent, some contemplative, some wise, some complacent, some interesting, some handsome, some happy, but only one or two come boldly within my category of dangerous—not, of course, to the exclusion of many of the other adjectives too.'[2]

J.R.D.'s term ended in 1955 when he tried to step down by writing to Sir William in March that year: '...I find even these relatively minor obligations somewhat of a burden, hard-pressed as I am in other directions and would like, if possible, to be relieved of them. At the same time, I know that you have found it difficult in the past to obtain suitable representation on the committee for Asia and if that position still continues, the area might no longer be represented.'[3]

In response Sir William replied: 'You have done that stint and done it, if I may be allowed to say so, very well indeed. The Committee has benefited at every meeting from your questing mind, your knowledge and experience and commonsense. I know how busy you are and do not feel I can ask you outright to take on another three years, but I think without doubt there is no one geographically who could dream of replacing you....'[4]

Following this letter J.R.D. agreed to continue.

It is just as well that J.R.D. continued for he was elected President of IATA for the year 1958–59. In that capacity he invited the first IATA conference outside Europe or America to come to New Delhi, where thanks to the newly constructed Ashoka Hotel, conference facilities were available. He invited Jawaharlal Nehru to inaugurate. Nehru, equally hard-pressed, wrote back: 'Ever since my return from Europe two weeks ago, I have been naturally full of work and worry. Also, there are a fairly large number of VIPs and dignitaries, Heads of States, Prime Ministers, etc. coming here during the cold weather, and I did not wish to add to my engagements....Nevertheless, I cannot say no to you and I am making a note of your IATA meeting. If all goes well, I shall present myself on the 27th October (1958) at the Vigyan Bhavan to inaugurate the opening session.'[5]

The General Meeting of IATA was preceded by a meeting of the Executive Committee some days earlier. The wives of the committee members normally accompanied them. Two hundred delegates flew in at their own expense, though Air-India looked after them in India.

This was the one enterprise that Thelly Tata could join her husband in organizing. Whilst the men were attending the conference, the women had to be kept occupied. It was an exciting time for a young Polish lady, Tina Khote, married to an Air-India officer. She recalls 'We stayed at the Ashoka. We had the Jaipur House and we had a big banquet and dance recital by Yamini Krishnamurthy, at the beginning of her career then. We had a trip to Jaipur for the Executive Committee, the usual practice, and that was lovely, and a whole day out in Agra, where we had tents, shamianas, conjurers, entertainers, a Moghul lunch and everything else, beautifully done, with Jeh also in conjunction. I have good memories of that. Mrs. Tata could not have been a better hostess to all of us. We had fashion shows and all kinds of things. Thelly was involved with it all. Intense care was taken on the entire programme.'

Tina Khote comments, 'The International Conference set the standards and for many years afterwards we would always hear, "That was the best IATA Conference to happen." Presidents of airlines, now retired, who came to the conference still talk about it many years later and say that the standards that were set were very difficult to meet after that.'

More than the proceedings of the meeting his entertaining of the IATA members and their wives in Jaipur, lingers in J.R.D.'s memory. He had arranged with the Maharaja and Maharani of Jaipur to be the hosts of the meeting. 'They (the guests) were so excited to be entertained by a genuine Maharaja and a genuine Maharani—though they were not dressed as such. We arranged for them to be entertained at the palace of Jaipur where in order to enter the main big courtyard you had to go through thick walls. You had to walk a fairly deep passage and you burst out into the open courtyard. As visitors emerged, on one side there were two magnificent horses all dressed up and on the other side an elephant which saluted every single one of the guests. The guests were in their element, particularly the wives. I was there, receiving them as they came out of the passage, when a woman started sobbing, so I asked myself, "My God, what has the elephant done to her." I went up to her and said anxiously to the American lady, "What's happened?" She said "I'm so happy! I have never been greeted by an elephant before!" They were so happy. Whilst in Jaipur I asked the Maharaja if we could use his hunting lodge for dinner.

'All the Executive Committee members and their wives normally got a souvenir. I decided to do something that had never been done before for the ladies...for the men it's easy, you give them a cigarette box or something...but for the ladies I said, look, let's give them a choice. So we had a room in the hotel which had all the different kinds of presents they could choose from, like sarees, scarves, shawls, some simple jewellery, so they were given a choice. The ladies wrote to say, "What a wonderful idea."'

Twenty years later, when J.R.D. was relieved of the Chairmanship of Air-India by Prime Minister Morarji Desai, Sir William Hildred wrote to his old friend with some anguish and concluded, 'I want to thank you for your brilliant and witty years in IATA, your presidential year and the marvellous and unforgettable Annual General Meeting which you gave us.'[6]

J.R.D. was President of IATA in the fortieth year of commercial aviation. At the end of his term, when handing over charge in Tokyo to the head of Japan Airlines, Mr Yanagita, J.R.D. said: 'I do not know the position in other similar trade associations, but I doubt that any of them takes greater care to ensure that its President is incapable of doing any mischief other than that contained in his introductory and outgoing speeches. I have had quite a fruitful twelve months as President. I started my year of office with quite a rush of work: I had to acknowledge no less than four messages of congratulations on my assumption of the Presidentship. Then I waited breathlessly for an opportunity to serve. I did not have long to wait. Within two months, I received from the Director-General a large envelope containing a charming Christmas card. Ready as I was to deal with any emergency, it took me no time at all to send him a New Year card in return. After that, thank goodness, things were fairly quiet for a couple of months. Then came a snappy exchange of birthday greetings with the Director-General, after which the pressure died down again for a while only to return in a crescendo with a cryptic cable from headquarters informing me of a legal action in which I had, it seems, been named in person. Unfortunately, this was followed by another cable which made it clear there was nothing whatever for me to do in the matter. I must confess I was disappointed as no one had ever taken the trouble to involve me in a multi-million-dollar suit before and I had visions of basking happily in the limelight of a *cause celebre*. You will appreciate how relieved I feel today, after such a harassing year, at being able to transfer such a heavy load on to the shoulders of my friend, Mr. Yanagita. Joking apart, and even allowing for the fact that we are blessed with such an outstanding Director-General who makes a President somewhat superfluous, I wonder whether we would not have been the gainers if such Presidents of IATA as Albert Plesman, Juan Trippe, Gilbert Perrier, Warren Lee Pierson, Sholto Douglas, Max Hymans and others had been given greater opportunities during their terms of office as President, to contribute personally to the welfare and development of this great Association and to the cause of international air transport as a whole.'

IATA, says J.R.D., put up a 'very fine technical organisation which was used by all the airlines and represented the view of the industry so that the manufacturers could consult them. One of its most important functions was to be an international clearing house so that you could have all ticket vouchers sorted out for different airlines. Finally, they had a very pronounced safety section to ensure airworthiness of planes. In the '50s they were a very powerful organisation though today they are a little less.'

'In those days,' recalls J.R.D., 'IATA could enforce its approved fares which the Americans initially resented and ultimately the American Government prohibited as part of their anti-monopoly policy. For years they objected to the fact that if American airlines were members of IATA they had to fix fares, because in America you cannot (fix prices in conjunction with other manufacturers.) In America, even Chairmen of two companies are afraid to be seen together on a golf course or in a restaurant for they may be accused of conspiring not to compete or to increase their prices, thereby preventing real competition. So ultimately the Americans decided they would not allow the American airlines to accept any regulations regarding fares.'

This considerably weakened IATA. Now everybody cuts fares. 'I tried to keep Air-India from the start from doing it and doing the complaining rather than being complained of. But then IATA became a completely different organisation when it could no longer enforce fares.'

The Director-General Sir William Hildred, according to J.R.D., was a 'somewhat dictatorial man who had tremendous executive power and you needed that to be the Director-General. None of the members on the committee had any authority except the Director-General himself, and the administrators of IATA. So we were all equal but there were some who were more equal than others. I think they treated me as one "more equal than others".' At IATA meetings J.R.D. talked freely, utilizing the added advantage and prestige attached to his being a pioneer aviator. The Executive Committee found that in cases of disagreement J.R.D. often spoke up for the passengers. 'I felt they used to turn to me more than I really deserved because I was running a small airline compared to theirs.'

'Who were the famous personalities of aviation you came in touch with?'

'Perhaps the most towering was Juan Trippe, founder of Pan Am. Trippe established Pan Am in the year following Lindbergh's flight across the Atlantic. In those days planes had a limited range in the air and it was very difficult to cross the oceans. Flying boats were therefore developed (specially) for Pan Am and called "Clipper Ships". The old sailing ships (Clippers) became the symbol and were used in all the Pan Am publicity. It is he who developed the concept of flying across the Pacific with passengers having a break at night at Pacific Islands like the Midway Islands. Trippe helped Pan Am to circle the globe—the first airline to do so. He was a man of vision.'

Then there was Eddie Rickenbacker, the First World War hero and fighter pilot, who was the head of Eastern Airlines in America.

IATA gave J.R.D. a chance to meet these men. 'It was great fun—of greater interest, something new was always happening.'

J.R.D. was keen that there should be a retiring age for the IATA Executive Committee members and in pursuance of that decided to call it a day, after seventeen years, in 1970. J.R.D. wrote to Sir William Hildred: 'My years as member of Executive Committee of IATA were perhaps the most pleasant ones of my life, during which I met so many wonderful people.'[7]

When he laid down the IATA Presidentship in 1959 he had said that he was looking ahead to an era which was yet to dawn: 'Our Jeremiahs have already looked upon the prospects of the 2,000 miles-an-hour civil transport as an unmitigated disaster which would doom forever any chance of achieving a mass market in air travel. I don't share that view at all. The prospects of flying over the Atlantic in an hour-and-a-half and from Europe to Japan in six or seven hours are so dazzling and would bring such immense benefits to mankind in promoting trade, travel and communications that, in my view, nothing should be done to discourage or retard such a phenomenal step forward.'

This is a dream he cherishes.

CHAPTER XII

The Jet Age

On a cruise ship off Seattle were a few hundred people. All of them were leading lights in the world of aviation: there were Air-Force Chiefs, the heads of the world's leading airlines, suppliers of aeroplane parts and other VIPs. They were the guests of Bill Allen, Chairman of Boeing Company, which was about to demonstrate the first commercial jet liner—the 707. There was mild excitement aboard as the first 707 approached the ship. It was announced that Tex Johnson was the test pilot.

Johnson brought the aircraft in pretty low at 1,000 feet and as it came near the ship slowly the plane rolled over and appeared to go out of control. This was not on the programme. Bill Allen's heart sank as he thought his company's masterpiece was about to crash and with it a multi-million dollar programme. Mercifully it did not. After the slow roll, Tex Johnson went on and came back over the ship.

Bill Allen told J.R.D. he was so furious with Tex he waited for two days before summoning him. He questioned his Chief Test Pilot: 'Tex, why did you do it? Do you realise our whole fortune was involved. The whole world was watching.'

Tex replied, 'Mr. Allen, I'm sorry. I just couldn't help it.'

Talking of pilots J.R.D. says: 'There are two types of pilots, the engineering type, probably the best, who knows all about flying and the natural pilot who flies by the seat of his pants.' Though J.R.D. himself is quick-tempered, and appears highly strung, you will seldom find him making a decision in a hurry—unless he has to. He has amazingly strong nerves to bear pressure, what some call "a high threshold of endurance".

J.R.D.'s knowledge of flying is based on not only his practical experience but also on his wide reading.

'I don't get many books on aviation any more, because now they are mainly about new aeroplanes. In the old pioneering days, you had the

French aviator Antoine de Saint Exupery with his classics, *Wind, Sand and Stars, Night Flight to Arras* and that wonderful *The Little Prince.* Then books about World War I. There was an American pilot, Elliot White Springs, who later owned a good textile mill in America. He was a member of the famous Lafayette Squadron of World War I, with the great flying ace Eddie Rickenbacker, who as you know was founder and Chairman of Eastern Airlines. There used to be a saying that whatever the weather, even if the seagulls walked, Eastern would fly. Springs wrote what I thought were some outstanding books about the early flying days, *The Bright Blue Sky, War Birds* and so on. You naturally don't get those kinds of books anymore. Then, of course, you had Anne Morrow Lindbergh and the great Lindbergh himself with *The Spirit of St. Louis.* Later you had modern novelists like Nevile Shute with his remarkable, prophetic *No Highway* about the early commercial jet operations and also Ernest Gann who is very good. Today flying has become so obviously an accepted, even a rather prosaic way of life that there are no longer such books written.'

A day came in 1956 when Air-India ordered the Boeing 707. It marked a quantum jump in the operation of the airline. It was not only twice as fast as planes in the existing world fleet but the capacity and size was three times that of the Super-Constellation. These seated about sixty passengers; a Boeing 707 could accommodate nearly 150. The induction of jet planes meant a complete re-orientation of both the flying and the service department, as well as of the galley service of every airline taking to jets.

'The crew had to shift from aircraft-related functions to passenger-related functions.' says Bakul Khote, who was one of those chosen to start the customer service department in flight and on the ground. He had to bring samples of everything—crockery, cutlery, and so on for the approval of J.R.D. even though J.R.D. was supposed to be a non-executive Chairman. A Boeing 707 demanded 5,000 individual items to run an efficient service. 'We had to teach the crew a new way of life.

'The lasting impression we have is that at all times we were pushed towards excellence. He (J.R.D.) had a great capacity for infinitely small details which made for excellence,' says Khote. 'When your excellence comes from the top it becomes the ethos of the organisation.

'J.R.D. personified many areas of excellence—maintenance, high utilisation and skilled pilots,' Khote continues. 'He was meticulous where the passengers interfaced with the airline, and displayed a real concern for the travelling public. Every time Jeh travelled by Air-India or any other airline he kept long notes on everything he had noticed. If well done he was quick to praise. When he went through a group of people even the humblest felt he was noticed.'

'Was J.R.D. decisive?' I ask.

Khote replies, 'Whenever he was knowledgeable he was decisive and never went back. He made a mistake when he jumped into negotiations in staff matters too early. As Chairman that was not his role, getting involved in the union and commercial staff negotiations!'

J.R.D. himself touched on this point when I questioned him on the mistakes he had made in his aviation career. Summing up his three decades in Air-India with J.R.D., Khote says, 'Jeh projected the personality and image of India which, though a developing country, was considered on par with the Western standards. As a pioneer in aviation he had a very considerable personal standing.'

K.G. Appuswamy joined Air-India in 1946 as an engineer and rose steadily to become the Managing Director during J.R.D.'s last year as Chairman (1977–78). Appuswamy was put by J.R.D. on flights when Jawaharlal Nehru was flying and he was fully responsible for the maintenance of the Prime Minister's flights. Three years before jets were introduced in 1960, Appuswamy was commissioned to organize the technical side and the maintenance of the jets.

Appuswamy says J.R.D. was enthusiastic about the Boeing 707 coming in and although at the start 'we ordered only three Boeing 707s we decided to develop 100 per cent of the maintenance facilities—our own engine overhaul system, instrument overhaul etc. We also arranged with the manufacturers the training of all technical personnel. Pilot instructors and operating pilots were sent to America one-and-a-half years ahead of time. Others were trained in India.'

When J.R.D. started his aviation career in the early 1930s, even the twin-propeller Dakota (DC3) did not exist and the jet was a quarter century away. In these twenty-five years several aircraft were created, some of which were purchased by Tata Air Lines and Air-India. Quite a few of these J.R.D. flew himself.

As we've seen, one of the innovations of Air-India International's Boeing order (as J.R.D. notes) was 'That it was the first airline in the world to specify Rolls-Royce Conway engines. The exceptional range and fuel economy of this combination of aircraft and engine rendered possible the first non-stop flight ever made between London and Bombay by a transport aircraft. This was achieved in 1960 in the course of a delivery flight, when the distance of 4,850 miles was covered at an average speed of six hundred miles an hour in exactly eight hours and five minutes. As we landed at Bombay I recalled, somewhat nostalgically, that it had taken me that many days and hours to cover the same distance thirty years earlier when I flew solo from Bombay to London in eight days and five hours.'

Air-India's first Boeing arrived in January 1960 and in that decade not only did most established airlines switch over to jets but more and more new airlines came into being as newly independent countries started airlines, often as a prestige symbol. These countries, commonly, were very small and did not even have the necessary home traffic. So the competition became more ferocious between the airlines.

Air-India was somewhere in between the traditional giants like KLM, Pan Am and British Airways and the newly established airlines. Air-India had to compete for passengers coming from Western countries, with established airlines, with the added handicap that Indians then had severe restrictions on travelling, with a P-form to be cleared by the Reserve Bank of India. Even so Air-India managed to show profits and in the 1960s managed to get seventy-five per cent of its traffic from non-Indians. Twenty years later the scenario had reversed because so many Indians had settled abroad that they and their families provided the main business of Air-India. Meanwhile the P-form was also lifted. J.R.D. notes: 'Even though commercial aviation has now become pretty much of a routine industry like any other, from which most of the individual enterprise and leadership and the sense of adventure has disappeared, there is still something about aviation which one finds in no other activity I know of today. Perhaps it is the dazzling tempo of technological progress in aviation, the size, speed and beauty of its equipment, the wonders of its conquest of space and time, the feeling it gives you of doing something worthwhile, of helping to bring peoples and nations together. Whatever it is, it seems to satisfy the urges and the inner needs of restless souls like myself.'

By December 1967 something much bigger was at hand. The 707 was almost three times the size of the propeller airliners. Now the newest plane, the Jumbo Jet 747, was three times the size of the 707. J.R.D. saw the jet age through the eyes of not only an aviator but one who was alert to the potential for tourism and what it would mean to have the ground facilities when the age of mass tourism began. Just about this time J.R.D. discovered the drive and talent of a man called Ajit Kerkar, a senior but new officer of the Taj Group of Hotels, and with him began planning for the expansion of the hotel arm of Tatas, which grew from one hotel to a chain of hotels, the largest in India. Twenty years later the Taj Group had forty hotels in India and abroad.

Just before the first Boeing 707 arrived Air-India celebrated its twentieth anniversary. J.R.D. noted then, 'At one time I remember we were the thirteenth largest airline in IATA. Now we are the nineteenth. So, it is obvious, that six airlines have grown ahead of us. Some of them by mergers, many by getting deeply into debt by expanding too fast. Throughout this period, we have always made a profit for Government, and we paid them a dividend. For the last five years we have not asked Government for a single penny to finance our expansion. We don't lack plans for the future. We are anxiously waiting to see an aircraft three times the size of the 707 parked on our aprons. Our two Jumbos will come in 1971. We will get two more and from then on we hope that we shall probably get one every year. Then come the supersonics. We have booked options for two Concordes and two American SSTs. I have no idea as yet whether we will take the Concorde, nor do we know when the American SST will be coming. But we are all set and we have our eyes on the stars.'

Wisely, J.R.D. did not rush into the acquisition of the Concorde. He foresaw that they were meant to fly long distances only over the oceans and not over land because of the sonic boom this created in flight. And he proved to be correct.

In April 1969, after two years of work, J.R.D. completed his report to the government on International Airports in India—a study in depth of present and future requirements.

He favoured an independent airport authority in order to give it the ability to act quickly and not be subject to regulations and inevitable delays. Though a separate authority was established for the four metropolitan airports, his hope that there would be little interference from New Delhi was not fulfilled.

In a career of forty-six years, of which thirty were with the international airline industry, a criticism I heard was that J.R.D. was more sympathetic to fellow pilots than others. I asked two people who worked closely with him about this—Bakul Khote and K.G. Appuswamy.

Khote replied, 'They (the pilots) were heard, they had greater access to Jeh.' When asked again whether J.R.D. had any favourites, he said, 'Whoever was seen in a favourite light was always efficient. There was never any pandering to lack of performance.'

K.G. Appuswamy replied: 'It is a tricky question. In the early days everybody felt he had a soft spot for the cabin crew. A Captain is hardly seen. People on the ground felt he was devoting more time to the cabin crew. After I clubbed with him and he came with the whole Air-India Board to see the workshop for Boeing at Santa Cruz, people felt that he was partial to engineers. He would take a lot of interest in the commercial side. In the out stations of India the staff thought he was partial to offices abroad. He did that more to maintain our image. The degree (of his attention) varied from time to time. If he yielded to the cabin crew (who were agitating) people said he was supporting the cabin crew. But he had a soft corner for the flying crew.'

'Was his interest in Air-India as intense after nationalization in 1953 as before?' I asked.

'His interest was as intense. We always thought that his first love was aviation and the Tata business came second.'

I asked Appuswamy how he conducted the Air-India Board meetings. 'Did he speak a lot?'

'Oh yes, he spoke most of the time. But such was his stature that very few could oppose him. If you call that a weakness I wouldn't know.'

In reply to another question Appuswamy said that people could always give their ideas even at such meetings.

'But was he open to other people's ideas?'

'Oh yes,' replied Appuswamy.

When Appuswamy was asked what it was like working for him, he singled out his own experience. 'I will give you an instance of his loyalty. I was on the Board of Air-India with Nari Dastur as a full time Director. During the Emergency, 1975–76, Sanjay Gandhi had me removed from the Board not because he had anything against me but because he wanted to get rid of another Director of Indian Airlines and he wanted it to appear that one each was removed from the international and internal airline. When my name was dropped I told Jeh I was resigning but he said, "No." He wrote three letters to Indira Gandhi and when he didn't hear back he telephoned her in my presence and spoke to her about my case. She claimed she didn't even know about it. Then J.R.D. took an appointment and went to see her and the Prime Minister pleaded she had nothing to do with my removal from the Board. Jeh told Indira "Appuswamy is going to leave and Air-India will suffer and I too may have to consider (indicating that if his people were treated like that he wouldn't stand for it.)"

'The P.M. said to J.R.D., "It will take some time now. It has just been done and it looks bad (to put him on the Board again). I'll do it in a month's time."

'J.R.D. said to her: "Anyway you will now have to convince him to stay on."'

J.R.D. took an appointment for Appuswamy to meet Indira Gandhi and rang Appuswamy from Delhi and instructed him 'Go and see her but don't be difficult with her.'

When Appuswamy called on Indira Gandhi she was very pleasant and promised that she would reinstate him on the Board and she did. Sometime after the Emergency, when Indira Gandhi was no longer Prime Minister and Appuswamy resigned, she had the courtesy to write a letter to him to say that she had nothing to do with the sad episode that happened and that she was under considerable pressure at that time. She was deeply disturbed to hear that Appuswamy had to leave Air-India finally.

Looking back at his forty-year career in aviation, J.R.D. says: 'I am, of course, highly overrated as a daring flyer and I've become a bit of a legend—a father figure and all that but I've not done all that much....I've got others to do things and, yes, you can say Air-India was one thing I did on my own for 46 years....'

CHAPTER XIII

The Emperors Arrive

In 1967, at Boeing's headquarters in Seattle, J.R.D. saw a model cabin of a 747, three times the size of a 707. 'You may be amused to know,' J.R.D. wrote to Sir Frederick Tymms 'that the 747 is referred to in the industry as "the Savior" for the reason that when visitors enter the enormous mock up cabin their automatic reaction is to say "Christ".' J.R.D. added, 'I'm looking forward to a better year in which we shall have to make a very important decision in regard to the Boeing 747.'[1]

The man who had once carefully inspected the 2,000 pound wood and fabric Puss Moth at the De Havilland factory, made bold to place an order for the 750,000 pound metal giant from Seattle. The Puss Moth had a power of a hundred horses. The Jumbo had a thrust of 80,000 pounds. Some years later *The Magic Carpet*, Air-India's magazine reported:

> *Emperor Ashoka*, Air-India's first Boeing 747, the world's largest commercial airliner, swept over Bombay in regal splendour on April 18, 1971. Shimmering in early morning sunlight and resplendent in its new red and white livery the 747 glided over Bombay airport precisely at 8.20 a.m., escorted by two Air Force Mig-21 fighters. *Emperor Ashoka* was given a standing ovation by a large and distinguished gathering and thousands of people who had converged on Santa Cruz to welcome the new super jet.... As the 747 came overhead, the two Migs screamed into a near vertical climb with their after-burners on to disappear into the wide blue yonder.

Speaking before a distinguished gathering in a *shamiana* at Santa Cruz airport, J.R.D. said, 'For me and for the few old timers still with us, who

172

started it all nearly 40 years ago, a couple of miles away from this spot, it is a tremendous event which brings back a flood of nostalgic memories. All our memories are not lighthearted or happy ones for on our journey to this happy day the road has often been a rough and rocky one, and we have had our frustrations and our heartaches.' He traced Air-India's history from its modest beginning in 1932 to 1970. Though only sixty-six years old, J.R.D. spoke of himself 'as an old man who has never quite recovered from the blinding view he had of the stars some 40 years ago.'

The first 747's arrival needed intense preparations for two years. Then a steady stream began. *Emperor Shahjehan* followed *Emperor Ashoka* and not long after came *Emperor Rajaraja* and *Emperor Vikramaditya*. The Constellations were named after princesses, the Super-Constellations after the ranis of India, the 707s after Himalayan mountain peaks, and the 747s after the emperors of India.

The humble Puss Moth, said J.R.D., had,

> no radio to go wrong. *Emperor Ashoka* carries so much electronic and automatic equipment that it literally flies and navigates itself.... The Puss Moth carried no passengers to feed and pamper but only a few mail bags behind its pilot who, oddly enough, occasionally bore my name and looked like me. *Emperor Ashoka* accommodates in luxurious comfort, 340 passengers, wined and dined from four galleys, their every need met by fifteen hostesses and pursers. We have indeed come a long way in these thirty-nine years, and if I told you I was not a proud man today I would not be telling the truth.

J.R.D. began with a handful of people and by 1970 Air-India had a staff of over 10,000, spread over fifty-four countries, not to speak of a first rate reputation.

When asked in 1967 how Air-India could afford Jumbos costing twenty crore rupees apiece, J.R.D. replied that, 'while the new generation of huge subsonic and supersonic aircraft were very costly, it is not so much the capital costs of a plane that count as its operating costs.' J.R.D. made sure this expenditure was not a drain on the Indian exchequer but got loans from thirteen US commercial banks and the Export Import Bank and arranged to pay off the loan from Air-India's earnings.

The arrival of the Jumbo and the challenge to prepare for it gave added zest to J.R.D.'s life. He wrote to Rose Benas, an American publisher and friend: 'I am certainly continuing to be active in the airline business and have no intention of quitting the Chairmanship of Air-India until removed or incapacitated.'[2]

Looking at his own role in Air-India, he observed to Sir William Hildred: 'The best thing I have done actually has been to build up such a good team and also not to allow any of us to take ourselves too seriously.'[3]

As the airline grew, the technical side of Air-India did not worry him, staff relations did. In a message to the staff on 15 October 1967, he said: 'My only regret is that the growth of the airline has made it more and more difficult for me to keep in touch with any but a few of you. This, I suppose, is inevitable in large organizations such as ours, but even so, I have increasingly felt that communications between members of the top management, including myself, and the bulk of our staff have become too tenuous, and in some places actually non-existent. This inevitably results in a lack of understanding of each other's problems, motives and feelings, and leads to misunderstanding and mistrust.

'I am anxious, therefore, to find ways and means of improving communications between top management and all categories of our officers, staff and workers.' He invited their suggestions. Soon afterwards he invited the Indian Institute of Management (IIM) to make an independent appraisal of management-staff relations.

On 25 March 1971, in a letter to Mohan Kumaramangalam, Union Minister for Steel and Heavy Engineering he said: 'I enclose a copy of a letter I have just written to Karan Singh (Minister for Civil Aviation) and hope you will support my request that even at this stage Government should agree to convert the two corporations into joint stock companies. If you and Karan Singh both agree, and you speak to the P.M. on the matter, I am sure that this most desirable objective can still be achieved.'

In a prompt reply, on 31 March, Kumaramangalam (who at one time was a Communist) said, 'Regarding the question of the two corporations being transformed into a joint stock company, I will try my best to see whether I can help in that direction.'

The idea was not accepted. J.R.D. was a decade ahead of economists calling for privatization.

How did he manage to run both Air-India and a number of Tata companies? J.R.D. says, 'I put copies of all important reports (of

Air-India) in my weekend box and if it was of any importance I would phone or see the Managing Director.'

Managing Director, K.G. Appuswamy, recalls that when he walked into J.R.D.'s Bombay House office, however important the Tata Directors were, they would say, 'Here is Air-India.'

While J.R.D. continued to give as much priority as he could to Air-India, this compact airline which began with four people slowly grew into scores, hundreds and thousands and the original camaraderie was little in evidence.

A couple of years after the Emperors arrived, Air-India celebrated its fortieth anniversary on 15 October 1972. It was time for introspection: 'I ask myself whether what we have achieved over these four decades has been commensurate with the time, the love, the heartaches, the sacrifices. the sweat and the tears devoted to it.'

As the founder of the Air-India family he asked his staff whether they were, 'a closely knit family to which we all are happy to belong? How far have we succeeded in achieving this happy state of affairs? Do we all care enough? Are we united? Does a high morale prevail throughout our ranks? Is there full confidence and cooperation between management and staff and within the various categories of our employees?

'As things stand today, the answer would, I fear, be no, so far as the situation in India is concerned, and this failure in human relations, for which, as head of the airline, I must take my full share of responsibility, is, I am convinced, the only reason for our failing to become what we all want to be—the best and most sought after airline in the world.

'This is a problem which is not only ours, but which, unfortunately, permeates most of Indian industry and commerce. Yet there is no reason why we should not be a shining exception, for we have all the brains, the skills and the experience to be the best. We all mean well towards one another, and deep down in our hearts we are proud of what we and our predecessors, some of whom have sacrificed their lives in the process, have achieved in these 40 years, and I am sure all of us are anxious to ensure that their sacrifice was not in vain.

'As I stand alone, the last surviving member of that small band of men who started it all 40 years ago, I realise I may not be there much longer to serve Air-India, but in whatever time is left to me I shall do everything in my power to remove causes of misunderstanding and mistrust, and to promote goodwill and cooperation throughout the organisation. I hope you will all join me in this endeavour which will bring immense benefits to all and serve the airline's and the country's interests.'

Some years later, in 1977, he wrote to B.W. Figgins: 'Air-India is indeed doing well. In fact, remarkably so compared to many other airlines in the world. Apart from using the best available equipment and giving what is now recognised as first class ground and inflight service, our progress is largely due, I think, to our consistent policy of slow but steady growth....'

His frequent memos which pointed out lapses were balanced by his kudos to the staff. On 4 April 1972, he wrote to the then Director of Engineering K.G. Appuswamy: 'I travelled back from Geneva last week on *Everest*, which I believe is by now one of our oldest ships and was most impressed by its spotless condition. I do not know whether it had recently come out of overhaul, but it looked as if it had just been delivered brand new from the factory.'

About a year after the Jumbo arrived it was time for Bobby Kooka to retire in 1971. At a function in his honour, he recalled his first flight by Tata Air Lines in 1938. First it was weighed, then unleashed on the apron. Kooka mounted the Bombay–Goa flight. There was no radio communication so the procedure was that on landing the pilot had to send a telegram declaring his safe arrival. The system worked well in the British Raj. But the Portuguese in Goa worked at greater leisure and the telegram from Bobby's pilot about their safe arrival did not reach for three days. There was pandemonium in Bombay House—not at the thought of losing him but the Waco plane, says Kooka.

At the function for Kooka, J.R.D. began with the opening words: 'First of all, let me remind you all that you haven't been invited here to enjoy yourself. This is a farewell function. As I have often remarked before, a French poet has said, "To depart is to die a little". The appropriate behaviour at a farewell party is therefore like at a funeral.'

J.R.D. went on to say that prior to nationalization, when Air-India came under the wing of Tatas, Bobby Kooka was given the run of the Tata house magazine. 'Mr. Kooka proceeded to play havoc with the whole Tata organisation by demolishing the ego and assassinating the character of every Tata Director and senior official. They were so happy about it that I was, one day, peremptorily faced with the choice of either silencing Mr. Kooka or getting a new set of Directors and executives. Well, as I was one of the victims myself, having been depicted in one of the cartoons taking dictation from my secretary sitting on a throne, I decided it would be easier to muzzle Mr. Kooka than to get a new set of Directors, and I ordered him to reserve his mischief-making propensities and talents for Air-India. That was damn silly of me because it gave him the opportunity for the next 25 years of taunting not only the Directors of Tatas, but Ministers, Members of Parliament and any VIP he could think of. Every time he created an uproar I had to face the music. Through these (Air-India) hoardings he demolished and punctured innumerable egos, which placed me at the receiving end of endless complaints from MPs and Ministers, including Mr. Morarji Desai, and Mr. Krishna Menon who was depicted in red pants running a track race with Mr. Kripalani. I forgive him all the apologies I had to tender on his behalf. I forgive him all the scars that I have borne because of the pleasure, the laughter and the relief from frustration and boredom that he provided to thousands, and perhaps millions, of people.'

*

If Kooka provided "the laughter and the relief" to people J.R.D. demonstrated the joy of achievement. He did that through two commemorative flights in a plane of the same vintage as the Puss Moth—the Leopard Moth. A series of planes, the Gipsy Moth 1925, the Puss Moth 1930, the Tiger Moth 1931, the Fox Moth 1932, and the Leopard Moth 1933, all came from the stable of Geoffrey De Havilland of England. (The De Havilland company is today owned by Boeing of Canada.) He named these magnificent little planes Moth because his own hobby was the collecting of moths.

The first operating flight of Tata Air Lines, as we've seen earlier, took place in 1932 and its first commemorative flight, also piloted by J.R.D., on the airline's thirtieth anniversary in 1962.

In 1982, although he was no longer Chairman of Air-India, J.R.D. thought it a pity not to repeat the performance on the golden jubilee of the first flight. He wrote to the then Chairman of Air-India, Raghu Raj, that he would like to undertake another commemorative flight on the golden anniversary. While the Air-India Board approved of the proposal of the flight, Raghu Raj replied that the Board had grave apprehensions about J.R.D. doing the flight solo at the age of seventy-eight. J.R.D. was quick to reply: 'I did this flight solo in 1932, and again in 1962, and the whole purpose would be lost if I did not do it in exactly the same way as on the previous two occasions. Otherwise I might as well do the flight as a passenger, which would be pointless. I would prefer in that case to forget the whole thing.'

He underlined the fact that he not only had renewed his Private Pilot's Licence, but had practised on the Bombay Flying Club's Piper planes and Tata Steel's Bonanza plane. He also told Raghu Raj, 'I have recently got myself fully checked up medically, including hearing and eyesight, and have been declared fit in every way.' In full measure he added, 'I continue to ski every winter and this year even did some hang-gliding on skis in the mountains.'

Raghu Raj wrote back in August 1982 that 'under the circumstances it is but only right that "J.R.D should fly again"... we are solidly behind you. I would earnestly appeal to you to reconsider our suggestion about a co-pilot with you on your flight.' Come October, J.R.D. carried his old friend Captain Visvanath as a co-pilot to Karachi from Bombay but he flew the final lap of the return to Bombay alone. He was due to land at 4 p.m. and awaiting him in a huge *shamiana* were the Governor and the Chief Minister, personalities of the aviation world and the public. First, a dot appeared in the clear blue sky and with it excitement on the ground. As the plane neared, two airforce helicopters escorted the little plane for a distance and then gracefully parted company. As the plane approached the *shamiana* at Juhu it came in pretty low. It did not land but flew over the *shamiana* at a height of hardly 200 feet, took a circle and landed flawlessly on the dot of 4 p.m. If J.R.D. landed sooner he would have been five minutes early!

After seven hours in the air he listened patiently to the speeches of the dignitaries. Then he spoke extempore to the waiting audience: 'I am a little disappointed that I have not been asked (about the commemorative flight) on "Why the hell did you do it if it was so simple?"'

He answered his own question, saying he wanted "to celebrate and commemorate the occasion". 'I wanted to dedicate a gesture to those, at first in handfuls, then in hundreds, and finally in thousands, the men and women who, over a period of 46 years, have helped me to build up Air-India and Indian air transport. I wanted to express in some way my gratitude and pay tribute to them and I did not know of any other way of dramatising this event...the other reason...was to re-live a memorable occasion of the past, something one often wants to do—for instance, one's engagement or marriage, some people do it by marrying more than once but nowadays with taxes as they are, very few people can afford more than one wife...I also had another reason. As I get older I feel distressed that in recent times there is a growing sense of disenchantment in our land, that the hopes, the aspirations, the enthusiasm, the zest, the joy with which freedom was received in our country some years ago, the achievements that we participated in, including the creation of Air-India had faded. So I thought that perhaps this flight would rekindle a spark of enthusiasm, a desire to do something for the country and for the good name. This flight of mine today was intended to inspire a little hope and enthusiasm in the younger people of our country that despite all the difficulties, all the frustrations, there is a joy in having done something as well as you could and better than others thought you could.'

CHAPTER XIV

Dismissed

———

Thou dost not know my son, with how little wisdom the world is governed.

- Count Axel Oxenstierna, in a letter to his son, 1648*

The relationship between Prime Minister Morarji Desai (1977–79), and J.R.D. was an uneasy one throughout their thirty years of association. J.R.D. vaguely recalls their first meeting when Desai accompanied Sardar Vallabhbhai Patel to a meeting with him at the Taj Mahal Hotel. The purpose of the visit was perhaps to ask him some election funds for Congress.

The first significant meeting of Desai and J.R.D. was in the mid-1950s, when Desai was the Chief Minister of the large undivided State of Bombay, which then included the states of Gujarat and Maharashtra. J.R.D. called on him with Sir Homi Mody, then head of the Tata Electric companies which supplied power to Bombay city. The meeting commenced with J.R.D. telling Desai that Tatas had made a projection of electrical power demand for the coming years and estimated that there would be a power shortage. Morarji Desai replied, 'No, there will not be. I have seen to that.'

J.R.D. rose from his seat.

'Where are you going?' demanded Desai.

J.R.D. replied, 'We have worked out, sir, the demand for electrical power in the coming years. We say it will lead to a power shortage if additional generating capacity is not created. You say it won't. We do not want to waste your time, leave alone ours.'

———

* Chancellor of Sweden at the height of that country's territorial greatness.

180

Taken aback, the Chief Minister commanded, 'Sit down.' J.R.D. resumed his seat.

Ultimately the Government of Bombay accepted Tatas' assessment and sanctioned the expansion Tatas asked for. But the unfortunate incident was not lost on either.

When Morarji Desai went to Delhi, first as Minister of Commerce and then as Finance Minister and as second-in-command in Nehru's Cabinet, there was occasional friction with J.R.D. on policy matters. But no major confrontation took place. In January 1977, just before the Emergency was lifted and elections ordered by Prime Minister Indira Gandhi, J.R.D. in an interview with the *Hindustan Times* on 2 January 1977 said that,

> dramatic changes and improvements that have taken place as
> a result of the bold and sustained steps taken by the Govern-
> ment since the proclamation of the Emergency. Among the
> achievements he mentioned the "unique" one of "not only
> arresting but reversing inflation"*.

Two months later, constituents of the Janata Party swept into power and in the midst of conflicting claims for Prime Ministership, Jayaprakash Narayan and Acharya Kripalani, adjudicated and selected Morarji Desai as Prime Minister.

In March 1977, one of the first acts of the new government was to reconstitute the Atomic Energy Commission (AEC) and J.R.D. (who, with Dr Homi Bhabha, was a member from its inception in 1948) was removed. It was a straw in the wind whose full impact was not evaluated at the time.

On 1 February 1978, the Boards of Air-India and Indian Airlines came up for reconstitution. At a Cabinet meeting it was decided that the name of J.R.D. Tata be dropped from the reconstituted Board of both the airlines. As he was the Chairman and the moving spirit behind Air-India, it virtually meant a dismissal from the Chairmanship. What was worse was that the founder of the airline and the man who had brought it up to international standards, and who had worked for the government without salary for twenty-five years, was not shown the elementary courtesy of being told what the government was going to do. 'I came to know from a third party,' says J.R.D., and the third party happened to be the man who

---

* See chapter "The Indira Gandhi Era" in Part III.

was appointed to succeed him, Air Chief Marshal P.C. Lal, Chairman of Indian Airlines. What made things worse was that P.C. Lal, was then also serving Tatas as the head of the Indian Tube Company. In a telephone call from Delhi, P.C. Lal told J.R.D., on 3 February, that the Prime Minister had called and told him that the government had decided to appoint him (P.C. Lal) as the Chairman of both Indian Airlines and Air-India. 'What did you say?' J.R.D. enquired. Lal replied, 'I said, "What about Mr. J.R.D. Tata?" Mr. Desai answered, "He has been there long enough."' The same evening the government released the news item on All-India Radio. There was a stir next day. The *Indian Express* headlined:

TATA OUT OF AIR-INDIA IN HUSH-HUSH MOVE

Following the virtual dismissal of their Chairman, two Air-India officials submitted their resignations—Managing Director K.G. Appuswamy* and the Deputy Managing Director, Nari H. Dastur. 'The way the top set-up has been changed has left a bitter taste in the mouth of the top executives in the airline,' said the Air-India Cabin Crew Association. The Officers Association also protested.

On 7 February, a letter dated 4 February arrived in J.R.D.'s office from Morarji Desai. The Prime Minister wrote:

My dear J.R.D.,

We have had under consideration the question of reconstitution of the Boards of Air-India and Indian Airlines. After taking into account various considerations which were relevant to the importance of securing efficient working of the

---

* When Appuswamy resigned immediately in protest at the way J.R.D. was treated, J.R.D. advised him, 'It is not good for the airline (for you to resign). Anyway it is upto you. But I would recommend you don't do it.' Even so Appuswamy forwarded his letter via the new Chairman to the government but they would not relieve him and a committee, including the Cabinet Secretary, tried to persuade him to continue. 'I came back to Tata,' says Appuswamy,'and again he advised "Take it (their offer) you've got some time to go."' This time Appuswamy took J.R.D's advice and spent another year and rose to be Vice-Chairman of Air-India and Indian Airlines. A year after that he got disillusioned and asked for voluntary retirement.

two Airlines we came to the conclusion that there should be one Chairman of both the Undertakings. We have, therefore, decided to appoint Air Chief Marshal P.C. Lal who is already part-time Chairman of Indian Airlines and who we thought could be spared by you,* as the Chairman of both the Airlines. You know P.C. Lal very well and I do hope you will agree with our choice.

Let me assure you that we are very sorry to part with you as Chairman of Air-India. We are fully appreciative of the distinguished services you have rendered to Air-India during your long association with it and the great contribution you have made to its build-up. I have no doubt that it is your association which is responsible for its being able to hold its own among the airlines of the world despite certain disadvantages and drawbacks with which you had to contend. I am expressing these views particularly because I do not wish you to entertain any impression that we have in any way made this change because of any lack of appreciation of your conspicuous work.

With best wishes and regards,

Yours sincerely

sd/-
Morarji Desai

On the original letter J.R.D. has put four question marks in the margin including line one of para two where the PM says "very sorry to part with you."

J.R.D. replied to the letter on his return to Bombay from Jamshedpur on 13 February 1978. He wrote: 'As I told you in 1953, when Government, having nationalised Air-India, invited me to remain at its helm as Chairman, I informed the Government as Chairman that my services

---

*As stated earlier, P.C. Lal was employed by Tatas as Chairman and Managing Director of the Indian Tube Company, Jamshedpur.

would always be at their disposal for as long as they felt I could serve a useful purpose. In the light of the work I performed for Government as a labour of love during the past twenty-five years, I hoped you will not consider it presumptuous of me to have expected that when Government decided to terminate my services and my 45 years' association with Indian civil aviation, I would be informed of their decision directly and, if possible, in advance of the public, instead of the news being communicated to me by my successor who was good enough to telephone me after you had informed him of his appointment and he had accepted it.

'What pained me more, in view of our long personal relationship over the past three decades, was the fact that when I paid a courtesy call on you in Delhi on the 24th January, during the half hour or so of which you were particularly cordial and friendly towards me, you did not give me the slightest hint of your decision to bring my appointment to an end.

'May I also express some surprise at the statement in your letter that the reason for my removal and the appointment of a common Chairman, was to secure efficient working of the two airlines, clearly implying that they had not been working as efficiently as Government would have wished in the past. Perhaps, busy as you are with more important national matters, you were not aware that at no time in its history has Air-India been so successful and profitable as in the current year 1977–78, with the whole of its organisation at a peak of morale and enthusiasm, gearing itself for doubling the magnitude of its operations in the next five years, under a capital expenditure programme of about Rs. 500 crores which I presented to the Board only a few weeks ago. Our sister corporation, Indian Airlines, has also done extremely well and is showing better results than ever before.

'In closing, may I thank you for your kind words of appreciation of my services to Air-India and contribution to its development.'

On 26 February 1978, Morarji Desai replied that J.R.D.'s letter of 13 February had caused him "some surprise and some distress". He then gave a long explanation* in the course of which he said, 'Frankly speaking I thought that after so many years of useful services to the corporation and the manner in which you had brought it up to its present stage of

---

* For the complete text of Morarji Desai's letter of 26 February 1978 to J.R.D. Tata, see Appendix D.

remarkable distinction, you would yourself offer to place the responsibilities on younger shoulders. You did not offer to do so which is why I did not broach the subject.'

J.R.D. replied on 17 March 1978: 'I do not wish to prolong this correspondence which I am sure is as distasteful to you as it is to me. If I did feel somewhat hurt, it was on the two grounds mentioned in my letter of February 13, to which I note you have avoided making any reference in your reply, namely the fact that you chose to give me no hint whatsoever of the impending change during our half hour's conversation on the 24th of January, only a week before it became effective, and that the first news of my removal was communicated to me not by Government but by my successor and through the Press.

'My greatest regret is that the manner in which this matter has been handled has aroused so much dismay and heart-burning within Air-India's fine organisation whose high morale and dedication have been so largely responsible for the airline's outstanding progress and international reputation.'

One person who was watching the episode at a distance, with some interest, was Indira Gandhi. In a personal note written on a flight from Gauhati to Calcutta on 14 February, Mrs Gandhi wrote:

Dear Jeh,

I am sorry that you are no longer with Air-India. Air-India must be as sad at the parting as you yourself. You were not merely Chairman but the founder and nurturer who felt deep personal concern. It was this and the meticulous care you gave to the smallest detail, including the decor and the saris of the hostesses, which raised Air-India to the international level and indeed to the top of the list.

We were proud of you and of the Airline. No one can take this satisfaction from you nor belittle Government's debt to you in this respect.

There was some misunderstanding between us but it was not possible for me to let you know of the pressures under which

I had to function and the rivalries within the Ministry of Civil Aviation. I would not like to say more.

With all good wishes,

Sincerely

sd/-
Indira

Sir William Hildred, former Director-General, IATA, wrote on 27 February, 'I was shocked and angry when I learned that political manoeuvering had led to your leaving the air industry which you have served with supreme distinction for so long. When the British public dropped Churchill, after what he had done for them, I was shocked and I have the same feelings now.'

In his reply J.R.D. said, 'Consideration and appreciation of work done are not qualities one expects to find nowadays in politicians, even in our country whose people are somewhat soft-hearted and sentimental. My total severance from the airline I served for forty-six years and from aviation generally which filled so much of my life, was indeed a wrench, but at my age, I shall be seventy-four this year, one learns to be philosophical and accept knocks as they come with equanimity. In any case, my job was done, the airline and its organisations well established. I am glad that I left it at a peak of financial success.'

The *Daily Telegraph*, London on 27 February 1978 headlined its story:

UNPAID AIR-INDIA CHIEF IS SACKED BY DESAI

The Air Chief credited with making Air-India one of the world's most successful airlines, basing its appeal on the beaming "little Maharajah" emblem, has been fired by Mr. Desai, the Prime-Minister. His abrupt removal, apparently for political reasons, after 30 unpaid years at the helm of the national flag carrier, has left the business community in uproar and brought Mr. Desai some of the worst publicity since he took office. Mr. J.R.D. Tata, 73, is a legendary figure, known to legions of executives around the world and envied by most for his success...with a fleet of only 14 aircraft, Air-India last year flew more than one million passengers and made a profit of £11 million. But despite this, and without warning, Mr. Tata was replaced.

The reconstituted Board of Air-India expressed their regret on the departure of J.R.D. Tata, crediting him with the building up of one of the finest airlines in the world and making it into a shining example of public sector enterprise. It went on to state: 'He gave Air-India a personality and style which has become the envy of others. Despite his many other preoccupations, Air-India remained his first love and he devoted much of his time to its affairs. There is not a single aspect of Air-India's operations which Shri Tata did not know about, or which does not bear the stamp of his personality, whether it was purchase of new aircraft, or interior decor of the Corporations' offices abroad. The departure of this pioneer marks the end of an era in India's aviation history and his achievements will continue to be remembered by many generations to come.'

The demoralization of Air-India began with the dismissal of J.R.D. As the airline steadily declined in service, popularity and profitability, even a decade later it was evident that all this was due to the petty-minded folly of the government, and worse the way it was done. The sacking of J.R.D. was the first major unpopular action of the Janata government that made its own supporters think again. Eighteen months after J.R.D.'s dismissal, contradictions and quarrels within the ruling Janata party broke it. They afterwards treated each other no better than they had treated J.R.D.

If the Prime Minister of India was ungrateful, Air-India's staff was not. They came to his office and presented him, in a simple ceremony, a silver model of the Puss Moth which he flew when he inaugurated the airline in October 1932. They did it on the thirtieth anniversary of the inauguration of Air-India International on 8 June 1978. By a strange coincidence just as the ceremony was going on, J.R.D. received a message from the US informing him that he had been selected for the Tony Jannus Award instituted in Florida in honour of the man who started the first scheduled airline service in the world in 1914. Among those who had previously been given the award were the founders of Pan Am, Eastern Airlines, Douglas Aircraft and the inventor of the jet engine, Sir Frank Whittle.

The staff of Air-India in a message to their founder, on the first page of their journal, *The Magic Carpet*, said:

It is with a heavy heart that we in Air-India send you this message on your relinquishing the Chairmanship of the great airline which you founded almost half a century ago. Only a handful of us were privileged to share in the joys and sorrows of those early days, in your sense of pride, your achievements, and your excitement when you created the airline which we today are proud to call "AIR-INDIA".....You, Mr. Tata, signify everything to us that is dedication, loyalty and single-minded devotion to a cause; involved in every phase of Air-India's operations, staunch defender of the ubiquitous Maharajah (even when the little fellow was on many an occasion criticised) constant in your endeavour to have the latest and most sophisticated equipment for the airline, invariably a stylist—be it the interior decor of a new 747 or the length of a hostess's hair, you were unswerving in your attention to detail, in your pursuit of excellence. And from those humble beginnings, as Air-India's fleet grew, as our country's flag-carrier expanded and developed to operate over five continents, you nurtured and cherished it over the years with imagination and foresight.

As you yourself said a little over fifteen years ago: "The international airline business is still the love of my life. Because I was present at the birth of Air-India, I feel a little like a mother who can't realise her baby has grown up." Yes, Mr. Tata, babies grow up and one day leave the family fold, but the indelible stamp of your personality, your love and your affection will forever remain.

And so the time has come to say "au revoir". As we turn the last page of the Tata Saga we will undoubtedly find some consolation in the thought that what we in Air-India did under your stewardship was worth doing, that we set our standards high, that we held our heads up erect both in

success and in adversity—and that no one can ever compel
us to part with our memories of you.

*

Within a month of Indira Gandhi assuming power, she re-appointed J.R.D.
on the Boards of Air-India and Indian Airlines in April 1980. The Minister
for Tourism and Civil Aviation, J.B. Patnaik, said to J.R.D. that the Prime
Minister and he wished to consult him on the reconstitution of the Boards
of the two air corporations. In June 1980, when the Board was reconstituted, J.R.D. was on it. When the Board was reconstituted in 1982 the
government appointed only officials to the reconstituted Board, dropping
all the non-officials including J.R.D. and Field Marshal Sam
Maneckshaw. J.R.D. was then planning his commemorative fiftieth anniversary flight in October 1982, through Air-India, and the resultant
publicity would have enhanced Air-India's prestige. But the founder of
the airline was not reinstated on the Board. The date of the flight was fast
arriving. On 7 October 1982, about a week before his flight, he said to the
Press Trust of India (PTI) that since he had not been told he was on the
new Board he presumed he was not on it.

Perhaps this statement, appearing along with the press publicity on the
impending flight, stirred the government to appoint J.R.D. to the Board
again. On 14 October, one day before the flight from Karachi to Bombay,
the Joint Secretary of Civil Aviation, C.M. Chaturvedi, sent an official
letter congratulating him on his appointment to the Board of Air-India.
He added a handwritten P.S.: 'I personally look forward to the successful
accomplishment of the super adventurous task undertaken by you to
re-enact your first flight after 50 years.' The letter was received after the
flight on 19 October 1982.

On both occasions after Morarji dropped him from the Board of
Air-India, and again when he was dropped in June 1982, J.R.D. was
apparently restive. He wanted to get back on the Board.

I asked him: 'Why?'

He replied: 'For the sake of the morale of the staff of Air-India.' But
it was too late to affect the morale. Once he was no longer Chairman
powers he had exercised previously were taken over by the Managing
Director of the airline.

There is little doubt that he took it to heart when he was dropped from the Board of the baby he had created and nursed for forty-six years. He would have been less than human had it not hurt him. J.R.D. was reappointed to the Board in 1984. In 1986, when the Board was reconstituted, J.R.D.'s name (he was by then eighty-two) did not feature but it was some satisfaction to him that the new Chairman of the airline had the magic of his surname. The new chief was his cousin Ratan Tata.

*

In 1985 the Federation Aeronautique Internationale's gold Air Medal came to J.R.D. In 1986, the International Civil Aviation Organization gave him the Edward Warner Award. The climax came in 1988 when he was given the Daniel Guggenheim Medal for distinguished services to aviation. Its earlier award-winners were Orville Wright, Charles Lindbergh, General Doolittle, Juan Trippe of Pan Am and Bill Allen of Boeing. The award was given at the unusual setting of the Great Hall of the Pacific Museum of Flight in Seattle where vintage planes hung from the ceiling, including a Puss Moth. Against this backdrop a glowing citation was read out. J.R.D. began his reply: 'Ladies and gentlemen, I would suggest you do not take too seriously what the Chairman has so kindly said about me... When I first got the message requesting me to accept the 1988 Guggenheim Medal Award for services to aviation, my reaction was "My God, they must be running out of names!" After all by the time I played my very little part in aviation history by launching India into commercial aviation, the real pioneering work had been done.'

He did say that what he was proud of was the fact that 'for 47 years I remained the head of an airline though I never got a dollar for it!' He told the audience, 'We are still historically at the beginning of air travel and I personally feel very lucky that I was born just about when aviation started and I am still around 85 years later. So much so that I have decided to come back in the future, hopefully in the next century... If and when I do come back I shall certainly visit Seattle and Boeing again. I would be very happy to say a lot more, inflicting reminiscences and prophecies on you. But I shall, I am sure to your grateful relief, postpone that pleasure to when I come back as promised!'

In 1984, evaluating J.R.D.'s contribution, Anthony Sampson in his book, *Empires of the Sky* wrote:

The smooth working of Air-India seemed almost opposite to the Indian tradition on the ground—not least at its home base, Bombay Airport, with its chaotic huts, shouting porters and primitive equipment, which was one of the least alluring gateways to the East. Tata, a domineering but far-sighted tycoon, could effectively insulate Air-India from the domestic obligations to make jobs and dispense favours; he could impose strict discipline and employ highly-trained engineers, of whom India has a surplus; and Air-India could make more intensive use of its jets than most major airlines. Tata continued as Chairman until 1977, the most long-lasting of all the pioneer aviators.[1]

PART III

CAPTAIN OF INDUSTRY

HUMATA · HUKHTA · HVARSHTA

CHAPTER I

Chairman

'As a schoolboy,' says J.R.D., 'when I was watching games of polo being played in Pune, I was hoping one day I too would play polo. On another day I wanted to be a good tennis player or a motor racing driver—all those ordinary urges of young people who want to do something with their personal lives. So really there is nothing that I did or showed signs of doing until I was suddenly put in a position where I was on test and testing myself. From 1926 to 1938 I considered myself really on test.'

Even so Sir Nowroji's death in July 1938 was a shock to J.R.D.: 'You know, we never think that people are going to die tomorrow. I thought Sir Nowroji would go on for at least five years more for he was not very old.'

J.R.D. is fond of saying that his co-Directors appointed him Chairman of the Board of Tata Sons, 'in a moment of mental aberration.' I've questioned him three times about the decision to appoint him Chairman and he's given me a different answer every time. The first time he replied, 'Maybe they found I was hard working.' On a second occasion he said, 'I was appointed Chairman probably because I was the only surviving Director of Tatas (other than Lady Tata) who was permanent under the firm's constitution, and I presume I had made some mark on them. I did not have a high opinion of myself then nor do I even now, except that I am now recognised.' On a third occasion he replied, 'Perhaps they preferred me to Sorabji Saklatvala (Jamsetji's nephew) who was old and not too bright.'

How did the other Directors, all senior to him, react to working under a young man? J.R.D. was to learn later that the only one sceptical of his abilities to be a suitable Chairman was Ardeshir Dalal, though Dalal had seconded the proposal to appoint him Chairman.

The first Indian to be Municipal Commissioner of Bombay, Dalal was a cold, austere figure, feared rather than loved. Dalal felt that J.R.D. had become Chairman purely by virtue of his being a member of the Tata family. 'He thought I would be a non-working young man, sort of a playboy liking his racing cars and aeroplanes but he found that I was working as hard as anybody else. About a year after my becoming Chairman he made some remarks that were repeated to me. He had admitted to a friend he was wrong in his assessment of me.'

In 1938, when he was appointed Chairman of the largest industrial conglomerate of industries of India, the tradition round the world was that seniority counted in the selection of top posts. The Kennedy era, when bright young men would hold high office, was still far away. So J.R.D. had no real precedent to follow. In any event the Tata empire was unique to India and J.R.D. needed to figure out how he would run an empire of steel, aviation, electric power, insurance, cement, oil and soaps and textiles. There was one thing he was clear about from the start—he was not going to be like his predecessor, Sir Nowroji, rushing from one Board meeting to another with hardly any time to think or to do creative work.

'This decision,' he says, 'was the watershed.' Until then the Chairman of Tata Sons was the Chairman of all Tata companies. 'I decided to give up the Chairmanship of certain companies. Having once presided at the meeting of the Hydro-Electric, I decided to give up that Chairmanship to Sir Homi Mody. Then Textiles. There were able men to handle these.'

Some Chairmen are nominal Chairmen, who play no active role in the day to day working of a company. Not J.R.D. Chairman of a number of Tata companies—though not all—he was a very active Chairman of the parent company, Tata Sons, and of Tata Steel, the companies he knew best. With his twelve years of experience in aviation he had a considerable grip on the affairs of Tata Air Lines, then still a division of Tata Sons.

If he was not going to be like Sir Nowroji, who was he going to be like? What were to be his guidelines as Chairman and his value system as a captain of industry?

'We mould ourselves on what other men expect of us,' says André Maurois. The personality of Jamsetji still exerted a strong spell on the House of Tatas. Inadequate as he may have felt by comparison, J.R.D. decided to try and emulate Jamsetji's philosophy.

It was Jamsetji who gave the House of Tatas its unique position in the nation. Jamsetji's conduct shows that in his later years he did not ask

'What enterprise is the most profitable?' but, 'What does the nation need?' If the answer was steel, hydro-electric energy or a University of Science, Jamsetji would make best efforts to fulfil that need.

J.R.D. did likewise. 'What does India need?' I have heard this question asked by him at meetings of the Sir Dorabji Tata Trust and I am sure his fellow Directors have heard the same question at Company Board meetings. Alfred Sloan said, 'What is good for General Motors is good for America.' J.R.D. thinks the other way round. 'What is good for India is good for Tatas.'

Jamsetji said one other thing that J.R.D. has not forgotten:

> We do not claim to be more unselfish, more generous or more philanthropic than other peopie. But we think we started on sound and straightforward business principles, considering the interests of the shareholders our own, and the health and welfare of the employees the sure foundation of our success.[1]

During the course of a discussion, J.R.D. explained why Jamsetji meant so much to him. Said he: 'Jamsetji was a man of great intelligence, a man of extraordinary vision. There are some very intelligent people but they have no sense at all of the future. Jamsetji had that sense. His vision of the future gave him a sense of what needed to be done for the country. And then he had integrity. Not only money-related integrity. Jamsetji had integrity of thought and mind. The final attribute was his great humanity—the way he thought about workers nobody in India or abroad thought at that time.'*

J.R.D. was also impressed by Andrew Carnegie, the Scotsman who came to America a poor man and rose to be the Steel King of the United States, then spent the last years of his life distributing the enormous wealth that he had amassed. J.R.D. quotes Carnegie:

> This, then, is held to be the duty of the man of wealth; to set an example of modest, unostentatious living, shunning display or extravagance...to consider all surplus revenues which

---

* For a more detailed appraisal of Jamsetji by J.R.D., see his Foreword to the second edition of *J.N. Tata: A Chronicle of his Life* by F.R. Harris, quoted in a slightly abridged form in Appendix F.

come to him simply as trust funds, which he is called upon to administer...the man of wealth thus becoming the mere trustee and agent for his poorer brethren.[2]

These words, thought J.R.D., could well have been spoken by Jamsetji. J.R.D. concludes his Foreword to Jamsetji's biography:

That he was a man of destiny is clear. It would seem, indeed, as if the hour of his birth, his life, his talents, his actions, the chain of events which he set in motion or influenced, and the services he rendered to his country and his people, were all predestined as part of the greater destiny of India.

Jamsetji headed Tata and Sons (as the firm was then called; it became Tata Sons—a private limited company—in 1917) for only seventeen years. So while the original vision was his, it was his successors who completed his mission for India. Of those who succeeded him, J.R.D. has had the longest innings as Chairman. What were the tools he used to fulfil Jamsetji's dream?

On the plus side there was Jamsetji Tata to inspire him and he made it a point to analyse Jamsetji's career and emulate him. On the minus side he felt keenly his lack of technical education, which he would have attained had he taken Mechanical Engineering at Cambridge, as he had intended. But fate and his father intervened to summon him to India. 'My father really in many ways had ruined my career by depriving me of a University education. I would have been an engineer and I've always felt the loss of that. I've always felt inferior to my own self because of what I could have been if I had an education.' However, when it is pointed out to him that but for his father inducting him into Tatas at that point of time (eight months before his death), he may not have been so well placed in the Tata hierarchy, he admits to the wisdom of his father's decision.

As a result of a lack of a university education J.R.D. had to rely on self-education, the thorough period of apprenticeship he was put through at TISCO and his latent genius for man management that came to the fore as he settled into his role as Chairman. Says he, 'I had no training in management but when I started in 1926, some books on management were being written. Not having had an academic training in engineering and technology, my only contribution to management had to be in handling men who had been so trained. Every man has his own way of doing things. To get the best out of them is to let them exploit their own instincts and only intervene when you think they are going wrong. Therefore, all my management contributions were on the human aspect through inducing, convincing and encouraging the human being. The exception was in the field of aviation, where I knew the technical side and perhaps half my love for aviation comes from the fact that it was the only field in which I have felt competent.

'One thing I regret is never having been in line-management except in the airlines. In other fields decisions I took had to be executed by someone else. As I had no technical training, I always liked to consult the experts. At times I felt like a soldier catapulted to be a General who has never been an officer. When I have to make a decision I feel I must first make sure that the superior knowledge of my advisers confirms the soundness of my decision; secondly, that they would execute my decision not reluctantly but be convinced about it; thirdly, I see myself in Tatas as the leader of a team, who has to weigh the impact of any decision on other Tata companies, on the unity of the group. I think this policy has paid off.'

For fifty years J.R.D. has worked on a consensus basis. Many able men are lonely men because they do not consult, they do not share their problems. In their insecurity some men are open, consult others and work out a consensus. Others in their insecurity turn inwards, become secretive and keep their cards close to their chest. They have little or no interaction with their peers. Lonely, they rob themselves of the richness of the experience of others. Charles De Gaulle often said he felt 'the chill of loneliness.' With his hauteur he had reconciled himself to, 'accept the loneliness which is the wretchedness of superior beings.' President Carter titled his book about the years in the White House: *Lonely*.

J.R.D. did not have a state to rule, but he did have an industrial empire and till 1970, under the Managing Agency System, he had the dominant voice in the management of all companies managed by Tata Sons or Tata Industries. And although Directors of all the Tata companies were appointed at his discretion, he genuinely sought out and paid heed to the views of his colleagues. It was in this manner that J.R.D. converted his lack of confidence at not having had a formal university education, into a strength. He never stood on false dignity but tried to arrive at decisions on the basis of the information and opinion he received. In other words, he was a great synthesizer.

In the later years of his life he has been criticized for being too much of a consensus man. This criticism does not worry him for he has worked out for himself the advantages of decision by consensus. He trusts people and holds that suspicion is the mark of little minds. 'I like people and trust them unless they show themselves unreliable or incompetent.'

*

One can seek the joy of achievement or one can seek the pursuit of power and control. J.R.D. chose achievement. His desire for consensus did in no way affect his drive for achievement. Ordway Tead in one of the earlier books on leadership, published in the 1930s, said:

> It may be useful for a leader deliberately to widen his range of active interest so that his will to power can find expression in more than one direction or outlet.[3]

J.R.D. was fortunate. He had two major outlets. The first was industry; the second was aviation. At an early age, when his ambition was at its height, his will to power widened. In aviation he had a world he could call his own.

'With Tatas you are dealing with very large companies—the organisation is there. You have to be careful that you don't do something by taking a line that might destroy or damage something important that has taken years to build. With Air-India I was creating something entirely new and therefore I did not owe anything to the past. Tata Steel was the creation of Jamsetji Tata. I couldn't throw my weight around there. There were so many important people, intellectually powerful people for whose judgement I had respect. With Air-India I was the creator. I was the founder so I could afford to make mistakes without undoing the good that was done by others in the past.

'Though I always took the opinions of others with Air-India too, I had the feeling "Look my judgement will prevail."' He felt earth-bound by the established traditions of companies like Tata Steel. With aviation, he felt, he could glide like an eagle over an immense valley, at times flapping his wings, at other times riding almost steady, tipping his wings gently here or there, buoyed up by the warm air beneath.

*

He took the same flair he had in dealing with people working for Tatas to his relationship with shareholders. As Chairman he had to preside at the Annual General Meetings of many companies each year. J.R.D. took the trouble to be well-prepared and briefed but there can be no preparation for unexpected questions. At a meeting of the Tata Oil Mills in the early seventies, a faithful but agitated shareholder said: 'Sir, the quality of Hamam soap has deteriorated sharply.'

'Who says that?' J.R.D. inquired.

'My wife, sir, she has been using this soap for several years.'

'If your wife is so discontented she must change the soap she uses.'

'Sir,' replied the shareholder, 'she won't do that, she will *never* do that.'

'Then,' said J.R.D., 'you must change your wife!'

At another meeting a shareholder was annoyed that even after a Tata company, New India, was nationalized another Tata company was still giving a lot of business to New India. Dramatically the shareholder declaimed 'I ask you, sir, "Why? Why? Why?"'

J.R.D. came back, 'And I ask you, sir, "Why not? Why not? Why not?"'

CHAPTER II

The Making Of A Family

When J.R.D. became a Director of Tatas on the death of his father in 1926 he was about half the age of the other Tata Directors. When he became Chairman of Tata Sons in 1938 he was still by far the youngest member of the Board. Senior men already present on it included Sir Homi Mody, Sir Ardeshir Dalal and Sorabji Saklatvala. To these distinguished members he added his own advisers like J.D. Choksi and Dr John Matthai. Naval Tata came on the Board at the instance of Lady Navajbai Tata in 1941. (Naval was the adopted son of Lady Navajbai Tata, wife of the late Sir Ratan Tata.) In addition to Directors there were senior executives like Sir Jehangir Ghandy and Kish Naoroji in Jamshedpur and P.A. Narielwala who looked after Mithapur. Knitting them into a family was J.R.D.'s job and with a couple of exceptions he managed to gather the diverse Tata luminaries into a family. He bestowed on them his friendship and, in some cases, his affection, and they responded.

Perhaps no colleague of his earlier period as Chairman came closer to him than Sir Homi Mody. In the years to come J.D. Choksi and Colonel Leslie Sawhney (who was married to J.R.D.'s sister Rodabeh) would also become very close to him but Sir Homi was his first friend among the Tatas, after J.R.D. became Chairman. Sir Homi was already a well-known figure in India's industrial life. Elected Chairman of the Bombay Millowners' Association five times, he was many years senior to J.R.D. and gave his young Chairman the care of an elder brother. Sir Homi was extremely conservative in his tastes and manners. He wore Savile Row suits and Sulka ties. He once wrote to Jeh accusing him of, 'approving nothing that was not modern and ugly.' 'Why?' Jeh wrote back, 'After all I even approve of you and although quite handsome in your own way, you can be hardly called modern.'[1]

Some idea of the great and varied talents of the top Tata men can be gleaned from the fact that they were "borrowed" by the Union Government every so often.

Within three years of Jeh becoming Chairman, Sir Homi was invited to join the Viceroy's Executive Council, to be in charge of supplies during the Second World War. 'It is a real wrench for me to see you go and I shall personally miss you very much,' wrote J.R.D. in his own hand on 18 August 1941 to Sir Homi noting that, without Sir Homi, Tatas, 'will be much more work and less fun than it is.'

Sir Homi returned to Tatas in 1943. Soon after, Sir Ardeshir Dalal was picked by the Viceroy to start the Planning Department. In 1948 Prime Minister Jawaharlal Nehru asked J.R.D. if Sir Homi Mody could be spared for the Governorship of Madras. J.R.D. protested half-jokingly that the government was repeatedly raiding Tatas for talent. He took it upon himself to say that Sir Homi was not likely to accept the offer which is exactly what happened when the offer was communicated to Sir Homi, then in the US. Later Sir Homi accepted the Governorship of Uttar Pradesh. After he shed the Governorship, Sir Homi returned to Bombay early one morning and turned up at Bombay House that very day, to the consternation of his Tata colleagues. He said he did not want to take any chances!

Although Sir Homi was one of his closest advisers and friends, J.R.D. preferred to have him in Tatas as adviser. He sized up the strength of Sir Homi as a wise friend but did not load him with the work of running companies. Sir Homi was not unaware of this fact. When he was given a farewell on the occasion of his departure to U.P. as Governor in 1949, he complained that,

> none of my colleagues thought I was capable of doing any work, with the result nothing was passed on to me...when I insisted very hard that I should be given some work...my colleagues decided that with proper safeguards the Taj Mahal Hotel be passed on to me. The safeguards were that the Chairman (Sir Homi) was to keep in his own hands all questions relating to alterations and interior decoration. Sabavala (A.P.) was to look after the hotel. Banerjee was to look after the kitchen, Darab Tata was to look after the Dutch Suite and Choksi (J.D.) was to look after the crooner! Outside these things I could damn well do what I liked. I could run

the lunches at Bombay House and I could appear in police courts to answer prosecutions.[2]

Sir Homi remained for many years Director of Public Relations and Chairman of the prestigious Sir Dorabji Tata Trust, which was effectively run by J.D. Choksi's brother, Professor Rustom Choksi. In a handsome tribute to Sir Homi, at his final farewell reception in May 1959, J.R.D. said:

> His generosity, his loyalty to his friends and to any cause that he espouses—and he espouses many of them, sometimes lost causes—his political courage, his truculent independence of views, his legendary wit which he uses as often at his own expense as at that of others, his refusal ever to be cowed down by anybody or by any event however calamitous it may be, all these form an important and characteristic part of Sir Homi's life. Above all, to me he is one of the staunchest of friends and one ever ready to share one's burden.[3]

As Sir Homi rose to reply he looked round the audience and said:

> The very large numbers in which you have gathered here today proves to me that my impending departure has aroused a good deal of enthusiasm.[4]

Speaking of J.R.D., Sir Homi said:

> He has a very keen intellect, a little too keen. His versatility is truly amazing. Whether it is a blast furnace or an ice-cream freezer, an aircraft engine or a cigarette lighter, he is equally at home with all of them. He affects to be his own doctor and a damned bad patient. All in all, I have great admiration and affection for our Chairman, and I think Tatas are singularly fortunate in having as their Chief a man of such wide vision and such a fine sense of right and wrong.[5]

Sir Ardeshir Dalal was the very opposite of Sir Homi. No humour escaped his lips and seldom did his perpetual cigar yield place to a smile.

He was, says Jeh, 'a very fine man.' Director-in-charge of Tata Steel he carried a good part of the then Tata empire's burden. Whenever he went to Jamshedpur he struck such terror into everybody, people would whisper, "Dalal is here," even though he might be far away in another part of the plant. Russi Mody recalls that when Sir Ardeshir went to see the Commissioner or some other British officer he would walk in front with his sola hat, baton in hand, not looking to the right or to the left, while a *chaprasi* followed behind wearing a belt with a shining brass plate and carrying the files. Dr Homi Bhabha who once had a quiet dinner with Sir Ardeshir at the Great Eastern Hotel in Calcutta later told Russi Mody that Sir Ardeshir, 'discussed nuclear physics with me as if he was my equal.'

Sir Ardeshir was tall and sparsely built, his eyes were heavy-lidded. Noted Michael Brown in the *Illustrated Weekly of India*:

> There was an indefinable atmosphere of preciseness about him. Even his cheroot seemed trained to scatter its ashes in the ashtray.

As we've seen earlier Sir Ardeshir's services were called for by Lord Wavell during the Second World War and it was he as a member of the Viceroy's Advisory Council who pioneered planning and development in India.

Sir Ardeshir introduced the profit-sharing bonus scheme for Tata Steel in 1934 at a time when labour relations were still turbulent. It took the Government of India thirty-one years more to introduce the Bonus Act, making it mandatory for companies to give an 8.33 per cent bonus.

Sir Ardeshir was very efficient but J.R.D. felt he took too much time doing needless things like opening all his letters carefully and placing them back in the envelope! He was a man of the highest business integrity and once when he felt that a colleague in Tatas, A.D. Shroff, was cutting corners he had a flaming row with him. Only J.R.D.'s intervention prevented either from resigning.

In complete contrast to Sir Ardeshir was Kish Naoroji, grandson of Dadabhai Naoroji. Kish did not believe in hard work. In England when the First World War broke out, Kish was promptly recruited into the Somerset Regiment and dispatched to France. His battalion was behind the front-line trenches. Six days a week Kish and his mates participated in exercises. Come Sabbath day they were woken up early and marched off for service to the nearest Protestant church, a couple of miles from the camp. As Kish and a few of his colleagues were believers in horizontal worship on Sunday mornings they came up with an idea to avoid Sunday morning worship. They marched up to the Commanding Officer and with a straight face, declared that it was against their conscience to go to the Protestant church on Sundays. 'Why?' barked the Commanding Officer. With eyes raised to heaven Kish declared, 'Because we are Catholics, sir.' The Commanding Officer opened a map and located a Catholic church nine miles away. 'Okay, ' said the Commanding Officer, 'you will now march to the Catholic church every Sunday morning.'

After the war, having saved the world for democracy, Kish came to India to save Tatas. For many years he was posted at Jamshedpur. Thereafter, he was sent to New York at the end of the Second World War. Later he was shifted to look after Tata interests in Delhi.

Kish was an amazing character. He was not bothered about files or procedures but given a job he would get it done. He was a superb public relations man. He had an air of casualness and every afternoon in New Delhi he was at the golf course. During the game, over drinks and at dinner, he got his work done. Feroza Narielwala, who knew him well, said: 'Kish made the impression of being nothing much but there was a lot of sense in his charm.' Posted to New York for about five years he received glowing tributes when he left for India. At a luncheon given by the Far East and American Council of Commerce and Industry, New York in January 1950, Mildred Hughes, Executive Vice-President, turned to J.R.D. and said, 'I know of no man more highly esteemed or loved in business circles….He did a magnificent job, not only for your company but your country.'

Even after several years in Tatas, Kish had still to be taught the elements of correspondence. Writing to him in New York in 1946, J.R.D. wrote: 'I must, therefore, once again request you not to mix up different subjects in single letters, as this causes both inconvenience and annoyance at this end. If you find it difficult to remember these instructions when dictating letters, standing instructions to your secretary should do the trick! I hope you won't drive me to returning your letters pertaining to different subjects before they are considered at this end.' J.R.D. never wrote off a person because of one or two irritating habits.

Kish and J.R.D. were good friends. 'How is your golf?' J.R.D. wrote to him in 1949. 'Do you find that your inability to see the ball past your waistline, coupled with an equal inability to bend from the waist, inter-feres with your game? If so, may I suggest a portable mirror on wheels as a remedy?' In reply Kish wrote, 'As to my golf, while the vision past the waistline is very considerably better than it was, the bending part is still liable to interfere with the game! I am asking the Corning Glass Works, who, I believe, made the two hundred inch mirror for the world's largest observatory at Palomar, to make a portable mirror on the lines laid down by you!'

While Kish could play the buffoon, Dr John Matthai was dignity personified. Sir Homi said of Dr Matthai:

> He has got many natural advantages: face, figure, manner, voice, all that and invested with an air of profundity in everything he said. Even if Dr. Matthai said "Good Morning" it sounded like a Papal benediction.[6]

In view of that Sir Homi always addressed Dr Matthai as "Brother John" and many were the times at the Tata lunch table when exchanges between the two regaled the others. Once they received the news that a friend had passed away. 'Brother John,' inquired Sir Homi, 'what do you think you and I will be doing when we are dead?' Dr Matthai removed the cigar from his mouth and replied in his deep voice, 'Roasting, I suppose.' The lunch table was also served with a sprinkling of smutty jokes. Yet the usual level of discourse was of a very high order. Sumant Moolgaokar more than once recalled that with men like Matthai and Mody at the Directors' lunch table such was the level of discussion that, 'you talked of national issues. You felt that Tatas belonged to the nation.'

Dr John Matthai was a Minister in the Interim Government under Jawaharlal Nehru. He joined as Minister for Railways and Transport in 1946, then became Minister for Commerce and Industry and was elevated to the rank of Minister for Finance. He resigned in 1950 due to differences with Pandit Nehru. The Prime Minister was keen to press on with the Planning Commission and although Dr Matthai was the one who drafted the Bombay Plan in 1943–44 he felt India was not ready and that the government should first husband its resources before implementing large schemes.

He returned to Tatas in 1950 as Director-in-charge of Tata Steel and worked for the company for the next seven years.

He was also for some years Chairman of the Sir Dorabji Tata Trust and left his mark on the institution by initiating two Trust institutes. In association with the United Nations he helped the Trust to start the first Demographic Research Institute in 1956, which is today the International Institute for Population Studies (IIPS) in Bombay. It was during his Chairmanship that the same Trust in collaboration with the Royal Commonwealth Society started the Tata Agricultural and Rural Training Centre (TACEB) for the blind.

He retired in 1957 to Kerala, his home state from which he had been away during his working life. J.R.D. wrote to him there: 'Throughout the years we have been together, you have been a wonderful source of strength and comfort to me. I always felt that I could turn to you when troubled or uncertain about any matter and get from you advice which would never compromise on fundamentals, particularly where the prestige of the firm and the country was involved. We all miss you here and I feel more alone, as a result of your going, than I can say. I wish, at least, that you had settled somewhere more within reach!'

From Kerala Dr Matthai replied: 'How good of you to write. I spend my time...doing a certain amount of desultory reading. I am discovering good books deserve to be read many times over and can be a continuing source of pleasure....I enjoy reading and also thinking. But the prospect of writing frankly bores me. Will you let me say that in surveying the past, there is no period that gives me such unmixed satisfaction as the time I spent with you in the firm. I wish in some way, some time I could repay the debt.'

As Director-in-charge of Tata Steel, Dr John Matthai had occasion to deal with young Russi Mody, now Chief of Tata Steel, who was then quite a restless, anti-establishment character. At the Directors' bungalow in Jamshedpur, J.R.D. had asked Dr Matthai 'John, what would you have done if you were Russi's age?' Dr Matthai replied in his deep voice, 'I would have been a revolutionary.'

It was Russi Mody who called on Dr Matthai when he was dying of cancer at the Tata Memorial Hospital. 'As I was leaving,' says Russi Mody, 'I said, "Get well soon." What else could I have said? But Dr. Matthai gripped my hand and said, "Russi, have I gone down so much in your estimation that you think I am such a fool?" I did not say much. I was almost on the verge of tears myself because I was very fond of him.' A short time after, Dr Matthai passed away.

J.R.D.'s longest serving colleague on the Board of Tata Sons was

Naval Tata, adopted son of Lady Navajbai Tata (widow of Jamsetji's son Ratan Tata.) In 1941 Naval Tata was appointed on the Board of Tata Sons and remained a Director till his death in 1989. He was involved at various times in the management of the aviation department, the Tata Oil Mills where he played a valuable role, the textile and the electric companies. Naval Tata was best known for his interest in employer and labour issues both in India and at the International Labour Organization (ILO). He held the record of having been elected to the Governing Board of the ILO thirteen times over a period of thirty-eight years. Wilfred Jenks, a Director-General of the ILO, at one point, spoke of Naval Tata's "distinctive stature in the ILO". Jenks noted that Naval Tata got the ILO interested in the issue of population and was, 'the first to urge us to embark on an active programme of management development.'

Along with Sir Homi Mody, Naval Tata was also closely connected with the Employers' Federation of India. In his last Presidential address of the Federation, Sir Homi said: 'If I am giving up my Presidentship, it is only because unlike some American Presidents I don't want to come in by the ballot and go out by the bullet.' As Sir Homi's successor, Naval Tata was often involved in negotiations with union leaders. On one occasion when the Labour Minister G.L. Nanda was presiding over a union meeting, a heated debate ensued on the issue of workers' participation in management. S.A. Dange, co-founder of the Communist Party and President of the All-India Trade Union Congress (AITUC), forcefully repeated his plea that workers should come on the Board of Directors. The management was opposed to such an idea. The meeting was getting out of control. At that point Naval Tata retorted to Dange, 'Okay! Okay! You can come on my Board but will you take me on the Executive Committee of your Union? As you want to know what is happening in my company, I also want to know what is happening in your union and it should be reciprocal.' The voluble Dange was lost for words.

J.R.D. says of his colleague, 'In our personal relations he was always affectionate but in some official matters we did differ.' For instance, J.R.D. wanted Tatas to keep out of politics but Naval stood for parliamentary elections from South Bombay as an independent candidate and lost, despite having given a good fight. Naval agreed with J.R.D. that the Swatantra Party, started by C. Rajagopalachari, needed support but felt it unwise for J.R.D. to have openly supported him, thereby incurring the displeasure of Nehru.

When I called on Naval Tata in 1981 he spoke warmly of J.R.D. and said it was the latter who had appointed Ratan Tata as Chairman of the prestigious Tata Industries. It is to the credit of both these men—J.R.D. and Naval Tata—that they worked together for forty-eight years to uphold the unity of the House they belonged to.

CHAPTER III

The War And The British

In 1938 Adolf Hitler marched into Austria. 'The man with the umbrella,' as J.R.D. remembers Neville Chamberlain, went to Munich. On returning to the UK, he waved a piece of paper, his agreement with the German dictator, and said that he had brought "peace for our time". From Heston airfield, he drove to 10, Downing Street. His Foreign Secretary, Lord Halifax, described the car journey,

> ...flowers were being thrown into the car, people were jumping on the running board, seizing his hand and patting him on the back.[1]

Despite the euphoria Halifax was realistic:

> No one who had the misfortune to preside over the foreign office, at that time, could ever for a moment of the twenty-four hours of each day, forget that he had little or nothing in his hands with which to support his diplomatic efforts.[2]

Lord Halifax says that if at the time of Munich, Britain wanted to go to war, it would have been without the support of the Commonwealth, for in 1938:

> South Africa had decided to remain neutral, a powerful opposition in Australia had declared against participation, and the attitude of Canada, to say the least of it, was uncertain.[3]

In March 1939, Hitler invaded Czechoslovakia. It was clear by then that Adolf Hitler was not merely interested in reassembling racial elements from German stock which had spilled over into neighbouring nations. According to Lord Halifax:

> Something much larger than this was being born and had taken shape in that evil mind.[4]

Britain issued guarantees of support to Poland and Romania. On 1 September 1939, Germany marched into Poland. On 3 September 1939, a downcast Neville Chamberlain rose in the House of Commons to declare war on Germany. South Africa, Australia and Canada promptly joined the war alongside Britain. On behalf of India, the Viceroy, Lord Linlithgow, declared war on Germany, without consulting the elected Central Legislative Assembly. The British Government proclaimed it was fighting for democracy and freedom.

'Whose freedom?' Jawaharlal Nehru promptly asked.

The Congress party, which was the leading party in the Central Assembly, had joined in the governance of the provinces in 1937. The Congress complained that the Viceroy had plunged India into the war without consulting the peoples' representatives. But there was scarcely any doubt in the circumstances that existed, that no consultation would have yielded satisfactory results.[5]

The Congress party, foreseeing the war, had decided in late August 1939 that it was prepared for whole-hearted cooperation with the British rulers, if an understanding could be arrived at with the British Government on the country's political future. Unfortunately, no such understanding seemed possible.

K.M. Munshi records that at the Working Committee meeting of the Congress party, 9–15 September 1939:

> Gandhiji at first, was for supporting the British unconditionally, but ultimately yielded to Jawaharlal Nehru.[6]

The Congress demanded from the British a clear declaration of their war aims, a declaration that India would get Dominion Status after the war, as well as an assurance that the Congress could participate in the war efforts at the Centre.

Almost two months of negotiations followed between the Congress and the British rulers. When no agreement was reached, the Congress ministries in the provinces submitted their resignations. Though opponents of British rule, both Mahatma Gandhi and Nehru were sad at the thought of England suffering the ravages of war. Gandhiji spoke of how tears came to his eyes as he pictured bombs being dropped on the lovely buildings at Westminster. Nehru wrote to the Viceroy:

> I am sorry for in spite of my hostility to British Imperialism and all Imperialisms, I have loved much that was England and I should have liked to keep the silken bonds between India and England. These bonds can only exist in freedom.[7]

As far as J.R.D. was concerned, he had earlier rather taken to Mussolini. 'I thought he was a buffoon but he at least did something. The Communists (in Italy), don't forget, were causing tremendous damage. When Mussolini came, things changed. He ran the trains on time.' But from the start J.R.D. had only contempt for Hitler. 'He was such an ugly man and talked a hateful language against the Jews. His Brown Shirts I found revolting.'

J.R.D.'s emotional thinking was very close to that of Jawaharlal Nehru. Like Nehru he supported the war against Hitler and at the same time he wanted to throw off the British yoke. In a preface to his book *Keynote*, J.R.D. says:

> I attended a Congress meeting at which fiery speeches by Jawaharlal Nehru and others led to some arrests.....That meeting was a dramatic one for me—for I was then torn between an urge to be personally involved in the freedom struggle and the realisation that I could not do so meaningfully without deserting the heavy management responsibility entrusted to me. I've never regretted my decision to stay out of politics which I rationalised to myself at the time by concluding that I could do more for the country in business and industry than in politics, for which all my instincts, in any case, made me unfit. I had no doubt that freedom was on the way and that when it came Jawaharlal Nehru would lead the Government. Who knows, I thought, I might one day have an opportunity to serve in more useful ways than going to jail today!

When asked whether this conflict in him arose before or after he became Chairman in 1938, he replied 'I think it was after.' J.R.D. does not remember the year of the meeting but *Fortune* magazine in its February 1944 issue stated:

At the celebrated meeting of the Congress in Bombay in 1942, J.R.D. Tata occupied a seat near the dais.

It was the Quit India session.

More than half-a-century earlier the same choice between politics and economic development confronted Jamsetji Tata. Jamsetji was present at the inaugural meeting of the Indian National Congress at the Tejpal Hall, Bombay, in 1885. He too had to choose. As Dr Zakir Hussain, a former President of India later said, 'While many others worked at loosening the chains of slavery and hastened the march towards the dawn of freedom, Jamsetji dreamed of and worked for life as it was to be fashioned after liberation. Most of the others worked for freedom from a bad life of servitude. Jamsetji worked for freedom for fashioning a better life of economic independence.'[8]

Through aviation Tatas played a significant role in the war effort. With the outbreak of the Second World War, civil aviation activity was virtually suspended and only passengers with government priority could fly. Tata Airlines operated internal services for the Government of India and the RAF Transport Command with DC3s and Beechcrafts (Expeditors) on loan from the RAF.

Asked what his personal attitude to the British was during the independence struggle, J.R.D. replies, 'Correct but cool.' He remembers spending a weekend during the war at the Viceregal Lodge with Lord and Lady Linlithgow. Lord Linlithgow was a massive figure, towering six-foot-six-inches in height. When he wore his favourite grey top hat with his long coat, it made him seem the tallest man in the world. His wife was also about six feet tall. One Saturday evening there was a dance at the Viceregal Lodge, and as a guest, J.R.D. gallantly offered to dance with Her Excellency. He escorted her to the dance floor. 'As I led her with the first dance steps I realised that she was so tall I could not even look over her shoulder to ensure I did not drive her into another couple. So I tried to peep under her raised arm as I led her on the dance floor.'

Over drinks one of the aides of the Viceroy got talking about His

Excellency's desire to upgrade the pedigree of Indian cattle; to this end he had got the government to import prize stud bulls from England. The aide told J.R.D. of a day when Lord and Lady Linlithgow went on an official visit to a well-known stud farm. Lord Linlithgow was being led around by the manager who explained various details as they toured the farm. The assistant manager, who followed with Lady Linlithgow, pointed out the farm's stud bull to her and said, 'Your Excellency, this is our champion bull. He has performed 365 times last year.'

'Is that so?' said Lady Linlithgow, 'You must go and tell His Excellency that.'

So the little man ran up to the Viceroy and almost breathless, spluttered, 'Your Excellency, Your Excellency, Her Excellency wants me to tell you that this bull that we've just passed has performed 365 times last year.'

'What! With the same cow?' asked Lord Linlithgow.

'No, sir,' stammered the assistant manager.

'Go and tell that to Her Excellency,' replied the Viceroy firmly.

When war broke out Lord Linlithgow ruled New Delhi and Jehangir (Jo) Ghandy (the first Indian to be appointed General Manager of the Tata Steel plant) ruled Jamshedpur. Later to be known as "the uncrowned king of Jamshedpur" Sir Jehangir was busy raising money for the war effort. He was a short, dynamic, strong-willed man with a twinkle in his eye. When the battle for Britain began in the British skies, September 1940, he set about raising money for one Spitfire aircraft which cost £5,000* for the RAF and another for the IAF. He was delighted to receive a cable from Lord Beaverbrook, 'Heartfelt thanks for a gift that brings to all of us in Britain encouragement and inspiration for the stern task that lies ahead.' Spurred on by this cable, Jehangir Ghandy decided as late as 6 December 1940 to raise funds for another Spitfire for the British by Christmas. This time it was a "heave and a ho" for the citizens of Jamshedpur but they managed to rally around their chief and just on the eve of Christmas another Spitfire was presented to beleaguered Britain.

Tatas by now were forging bullet-proof plates and rivets to be mounted on war vehicles nicknamed "Tatanagar". Tatas were proud about the first report from the 8th Army in the Western Desert that when a 75 mm shell had burst against a "Tatanagar", the metal plates had buckled but had not been holed and all its occupants had escaped uninjured.

---

* Around £105,000 in 1991.

Heading the desert war in the early years was General Sir Archibald Wavell. In 1942 Wavell was shifted to India and made Commander-in-Chief of India.

> The memory of his great victories over the Italians, that had lightened the gloom of the winter of 1940/41, had not yet faded, and many Englishmen still regarded him as their best general.[9]

In August 1943 Wavell was appointed Viceroy of India.

Wavell was determined to break the deadlock between Britain and India. In pursuance of his decision, Wavell, as Viceroy-Designate, wrote from England to his private secretary in India, on 20 August 1943, that he would like to summon, 'with the greatest possible secrecy' ten men to meet him. The first nine were political heavy-weights of various hues including Gandhi, M.A. Jinnah and B.R. Ambedkar. Last of all he put 'Representative of big business (? Birla or head of Tatas).'

*

Wavell was a man of the highest integrity, a sensitive man who would read and write poetry in the midst of war. His book *Other Men's Flowers*, a collection of much-loved poems with interesting footnotes by Wavell, was to become one of J.R.D.'s favourite books. Nevertheless his personality was not one that reached out to others.

J.R.D.'s encounter with the Desert General was somewhat like Archbishop Theophilus' encounter with a Desert Father called Abba (Abbot) Pambo. Abba Pambo was visited in the North African desert by Archbishop Theophilus. But Abba Pambo did not speak to him. When the Archbishop finally said to Abba Pambo, 'Father, say something to the Archbishop so he may be edified.' Abba Pambo replied:

> If he is not edified by my silence, he will not be edified by my speech.[10]

One of Wavell's eyes had been replaced with a glass eye. As J.R.D. entered the Viceroy's spacious room, they shook hands and Wavell waved J.R.D. to the chair on his right where J.R.D. settled down. For sometime not a word was spoken.

Ranged before Wavell on his large desk, like troops on parade, were four rows of paperweights. In the ensuing silence, Wavell lifted the paperweights and moved them from one place to another. J.R.D. had the bright idea of stretching out his hand, lifting a paper weight and surprising him by saying 'Checkmate! General.' At that bright idea a wide grin appeared on J.R.D.'s face. Just then Wavell lifted his single eye to find the character in front of him grinning for no apparent reason. J.R.D. does not remember the conversation on that occasion. But J.R.D. thinks the comments that Wavell published in his diary may have been coloured by this first encounter. The comments in question are:

24th January 1944:

> I did not get on very well with young Tata, with whom I had a short chat after dinner, a pity as I think he is able and influential about Indian business, but he seemed to be conceited and unhelpful. I expect I did not take him in the right way.[11]

21st April 1944:

> Young Tata came to lunch when I met him down at Bombay. I found him rather a supercilious and tiresome young man, but got on with him rather better today. Sir Ardeshir Dalal had spoken to him about the offer I made him yesterday*. I think Tata was pleased the offer had been made but was doubtful whether the firm could spare him.[12]

J.R.D. was forty when he first met Wavell. Perhaps he was somewhat abrasive at the time and therefore was unable to find a personal equation with the Viceroy. Several of Tatas' top men, Homi Mody, Ardeshir Dalal, Sorabji Saklatvala, Jehangir Ghandy had been knighted by now, and feelers were sent to J.R.D. on how he would respond to being knighted

---

* Sir Ardeshir Dalal, Director-in-charge of Tata Steel, was offered the post of Executive Councillor to the Viceroy.

217

for the considerable support of Tatas to the war. But J.R.D. conveyed that he was not interested in the knighthood and the matter was dropped. On his impressions of the British of the Raj period J.R.D. says: 'For those people who came here as ICS and became government leaders and administrators in India, on the whole their integrity, their honesty I admired. But I always had angry opposition to their continuing as rulers and was always hoping that we could break away or they would get away...their intellect made them realise that historically they are a great people. The top people were tops. The people in the middle jobs, the middle class, they were easily prone to racism. At the lower ranks they were not a very admirable people except in times of crisis. The manner in which they responded to the war was superb. I realised some of the good things they did for India...but I was terribly anxious to see them out. In some ways they went off rather too easily and left the burdens...As people to deal with, I feel more attuned intellectually or emotionally to the British than to the Americans.'

As the war was advancing, J.R.D. was looking beyond to the years of peace when he knew that the British would have to leave India and the country would be able to exercise her own discretion in planning her future. He felt that leading men of industry should meet and plan the future. He invited G.D. Birla, Sir Shri Ram, Kasturbhai Lalbhai and Sir Purshotamdas Thakurdas to join him and three of his colleagues from Tatas. From the Tatas' think-tank he selected Sir Ardeshir Dalal, A.D. Shroff and Dr John Matthai. Up to this time economic planning was the preserve of communist Russia. These men who represented the interests of capitalism were to be the first to produce an Economic Plan for India.

CHAPTER IV

The Bombay Plan

The father of planning in India was neither Jawaharlal Nehru nor J.R.D. Tata nor G.D. Birla—it was Sir M. Visvesvarayya, the enlightened Diwan of Mysore, who planned the industrialization of Mysore State, and published his book, *Planned Economy for India*, in 1934. Four years later, inspired by a visit to the Soviet Union, Jawaharlal Nehru established the National Planning Committee with Professor K.T. Shah, the economist, as Member-Secretary of the committee. The National Planning Committee produced some reports (which incidentally, were published with the financial assistance of Tatas.)[1]

In January 1944 a comprehensive document entitled, "A Plan of Economic Development for India" was published, under the signatures of five eminent industrialists and three technocrats. The industrialists were: J.R.D. Tata; G.D. Birla; Kasturbhai Lalbhai, the textile millowner from Ahmedabad; Sir Purshotamdas Thakurdas from Bombay and Sir Shri Ram from Delhi. The technocrats were all from Tatas—Sir Ardeshir Dalal; A.D. Shroff; Dr John Matthai. The plan came to be known as the Bombay Plan or the Tata–Birla Plan.

Soon after publication, Lord Wavell wrote to the Secretary of State, Leopold Amery on 25 January 1944:

A considerable stir has been created by the Rs.10,000 crore Economic Plan for India.[2]

The plan was reprinted a couple of times the same year and in May 1944, Penguins in wartime Britain also published it. Wavell again writing to Amery on 16 May 1944 in a letter marked "Private and Secret" said: 'I see the Bombay Plan has come out in the Penguin series. Sir Gregory (Member, Viceroy's Executive Council for Finance) who takes criticism very much to heart, thought we should at once produce a rival pamphlet and broadcast it through the India Office. I doubt if this would be a success; but we are getting Holburn of the *Times* to write a special article about the progress made by the Government of India with planning and development and I do not think we can go beyond something of this kind.'

The London *Economist* commented on the plan from time to time. On 17 June 1944 it said that the plan was,

> a commendable piece of enterprise....But it will take much harder thought, much more laborious effort and, it is to be feared, a much longer span of time to reach the goal.

Apart from the discomfiture it caused the British rulers in India—that Indians were ahead of them in thinking for the country—the plan created a storm in India. Leftists attacked it for being reactionary; old-fashioned businessmen thought it too radical; and Gandhians thought it too un-Gandhian. The plan called for massive investments in power, mining, roads, railways. Almost fifty per cent of the plan outlay was slated for industrial development, twenty-five per cent for housing, and only ten per cent for agriculture.

Shriman Narayan Agarwal came out the same year with *A Gandhian Plan*, while the radical humanist M.N. Roy came out with *The Peoples' Plan*.

The Bombay Plan proposed a fifteen-year period of development, divided into plans of five-year spans. The first plan was to start modestly with an outlay of Rs 14,000 million* the second plan with Rs 29,000 million** and the third plan with Rs 57,000 million.ᵞ

The planners hoped that the per capita income of the Indian would double at the end of the period even keeping in mind the estimated rise of population. It was in many ways a remarkable document. Here was a group of capitalists who propounded that profits should be kept within limits through the fixing of prices and restriction of dividends. They advocated a steeply graduated income-tax with heavy taxation on un-earned income and even recommended death duties.

---

* Around 650 trillion rupees in 1991.
** Around 1,350 billion rupees in 1991.
ᵞ Around 2,650 billion rupees in 1991.

Written during the war, the planners were looking ahead to peace and took it for granted that a national government would soon take over from the British.

The authors of the plan suggested controls on production, distribution, consumption, investment, foreign trade, foreign exchange, wages and working conditions in order to achieve goals of development.

When asked how a capitalist could think of an Economic Plan, J.R.D. replies: 'I can't take credit for more than being the first businessman to see the need for a Plan. It went on to planning but it started with a feeling that Indian businessmen must prepare themselves for what was to happen after the war...when India became free as I was sure it would and we must do something to develop the country. It was (to start with) only a committee of businessmen....(I thought) there must be a role for us—we must accelerate development, then in the course of deliberations came the idea of planning in the modern sense...I knew Independence was bound to come: I knew the country's economy would have to be tackled—that economic prosperity needed to reach not only the few but the many. Businessmen and not only the Government should play a role.'

He went on to mention an event that was to prove a watershed in his life. 'Till I went to the Congress meeting, I thought this freedom movement could not work but then I saw the real impact of Gandhi backed by Nehru. I didn't know (Sardar Vallabhbhai) Patel then. I knew that the English would be economically destroyed by the war. I always took an interest in history—contemporary history, ancient Greece and more of Rome—that excited me. Later on, the Napoleonic era. Taking the historical view I knew England would never hold on to the Empire after the war.'

The credit for converting or shifting the committee from a deliberating body to an articulate one, according to J.R.D., goes to G.D. Birla. 'G.D. Birla was a man of high intelligence and knowledge. When we were floundering to find a structure in the first few meetings, it was he who suggested: "It is difficult to forecast what India should do after being free....So let's do it this way—first estimate to get the people the kind of standard of living that they want. What is needed? So many calories of food requiring so many million tons of grain, so many metres of cloth, housing—how many cubic feet of housing, so many schools, etc." That concept of quantifying made it easy and it was on that basis that Dr. Matthai wrote the Plan.'

Thanks to the Economics and Statistics Department built up by Tatas since 1940 and the statistics available with the government, comprehensive figures were available on food consumption, cloth consumption, the average life-span of an Indian in 1930 (twenty-six for an Indian compared to sixty for an American then*) and so on. The committee could fix targets on the basis of these existing statistics. They could quantify the items and on this basis work out the kind of investment needed. The plan even indicated where the finance could come from including about an estimated one thousand crore rupees worth of stock securities India had accumulated with Great Britain, savings, borrowings, created money and "hoarded wealth". The last was then estimated at Rs 10,000 million out of which, it was assumed, Rs. 3,000 million would be available.

'From this Plan it became clear that the more people you had, the more you had to have of those things—more of cloth, more of food, more of housing, and so forth,' says J.R.D. Strangely enough, the Bombay Plan laid the minimum emphasis on population control.

In order to raise the average life-span of Indians the plan pin-pointed the importance of sanitation and water supply, a dispensary in every village and a small hospital in every town. The planners had looked at the Soviet model and learned from its mistakes, especially its neglect of the consumer goods industry by its concentration on heavy industry. 'Planning without tears,' said the Bombay Plan, 'is almost an impossibility. But we can learn some lessons from the Russian experience and avoid the errors to which planners in their enthusiasm are liable.' Part II of the plan appeared in December 1944 dealing primarily with "Distribution—Role of the States".

When Dr Matthai, who wrote both parts of the Bombay Plan, asked H.V.R. Iengar, ICS, who was later to be the Governor of the Reserve Bank of India, what he thought of the document, Iengar replied:

> I was tremendously impressed with the broad sweep of the document. The manner in which various parts had been worked out in detail and have been integrated into a whole.

Thirty years later H.V.R. Iengar, by now retired, delivering the Dr John Matthai Memorial Lecture at the Kerala University said:

---

* An average Indian's life-span in 1991 is about sixty years.

I have been reading the Bombay Plan in the last few days and have been amazed by the fact that conceptually it is very modern.

Iengar further commented:

In the approach of the (Government's) Fifth Five-Year Plan, the Planning Commission pointed out that the fault in the four plans was that they concentrated just on the total growth without bothering about the distribution of the fruits of economic growth, that this had resulted in a growing disparity between the few rich and the many poor in the country and that this should be corrected on the basis of what may be called a consumption-oriented plan...it is extraordinary to see that this was the very basis adopted by the authors of the Bombay Plan.[3]

According to A.H. Hanson,

In some respects the methods envisaged (by the Bombay Plan) anticipated those of the three Five-Year Plans.[4]

One immediate result of the publication of the plan was that Lord Wavell requested J.R.D. for the services of Sir Ardeshir Dalal to be a member of the Viceroy's Executive Council and start the Department of Planning. Sir Ardeshir accepted the offer and hence could not be a signatory to Part II of the Plan entitled "Distribution—Role of the States".

Geoffrey Tyson, editor of *Capital*, says in his work *Nehru: The Years of Power* that the Bombay Plan's

chief merit was that it focused public attention on India's needs, and by reason of the distinguished auspices under which it was put forward, it implied the commitment of the Indian business community to the principle of planned economic growth. There is no evidence that Nehru ever regarded the Bombay Plan very seriously, but he doubtless took due notice of the fact that the leaders of Indian trade and industry had pledged their support to the national economic plan. Thus, when the details of the First Five-Year Plan were announced, it was to a public whose mind was well conditioned to planning. The only thing out of the ordinary was the seemingly astronomical dimensions of the sums of money involved.[5]

J.R.D. admitted later that the Bombay Plan did not pay adequate attention to agriculture. As we've seen, the Bombay Plan's recommendation was fifty per cent allocation for industry and ten per cent only for agriculture. The First Five-Year Plan corrected this imbalance and allotted 17.4 per cent to agriculture and 8.4 per cent to industry.

In 1986, at the launching of J.R.D.'s book *Keynote*, R. Venkataraman, then Vice-President of India, recalled the Bombay Plan as one of J.R.D.'s contributions to India. J.R.D. in his reply said, 'My only contribution to it was to arrange for the Bombay Plan to be written. It was done mainly by Dr. John Matthai, but after considerable discussion. My regret is that far beyond the period we visualised for the per capita wealth of the Indians to be doubled, we have not reached that stage and, even if we continued as we did, then in another fifty years we would be barely half as poor because we were growing only about 1 per cent a year per capita.'

The record of capitalism in India is little better than that of capitalism in other countries but the Bombay planners can at least claim that they were the first—and perhaps the only capitalists in the world—to draft an Economic Plan with a strong social concern for their country. Besides, they did it before anyone else barring the Soviet Union.

CHAPTER V

Mission To The West

As the war in Europe was winding to a close, and post-war prospects opened before India, there was a zest for life and a vision of the industry of the future in free India seized many Indians. Tatas too were full of enthusiasm, especially as they had done their bit in the war and now hoped to lead the country's industrial advance.

By the end of 1944, as the allied troops pressed on in Europe, the Germans gave stiff resistance. In the Pacific the initiative was wrested from the Japanese as Americans advanced island by island. The war machine in India was considerably strengthened with the arrival of the Americans. At this time, looking to the future and following on from the magnificent contribution of Indian industry in the war years, on 10 October 1944, the Government of India announced:

> The Government of India have invited a group of Indian
> industrialists and businessmen to visit England and America,
> as soon as war exigencies permit, with the object of studying
> the present industrial organisation of these countries, the
> technical advances made by them during the last few years,
> and their post-war industrial plans. The Mission will be
> unofficial in character and its members, all Indians of inde-
> pendent views and position, will be free to arrange their
> programme and discuss any matters unfettered by terms of
> reference or any form of control by Government. They will
> be accompanied by their own technical advisers and will bear
> their own expenses throughout the trip.

> The Government will arrange facilities for them to visit
> industrial establishments and to contact leaders of industry

225

and prominent businessmen in Britain and the United States. It is believed that the Mission's study on the spot of latest developments in the industrial sphere, and the knowledge and ideas which they will bring back with them will be of great value in the further industrialisation of the country after the war.

The mission was to be composed of the foremost industrialists of India including J.R.D. Tata, G.D. Birla, Sir Sultan Chinoy, Kasturbhai Lalbhai, Krishnaraj Thackersey and Sir Padampat Singhania.

As often happens on such occasions, a lot of preliminaries had to be gone through. The task of coordinating arrangements fell on G.D. Birla because J.R.D. Tata was recovering from pneumonia at Ooty. 'I am not too much worried about the climate,' wrote G.D. Birla to J.R.D. Tata. 'I have been to Europe in winter and I have never disliked it. I do not know if it would be possible but I want to take advantage of staying at some sanatorium when we go to America.'[1] He also wrote to Sir Azizúl Haq, Member of the Viceroy's Council for Commerce in Delhi, who was coordinating the tour with Britain (where wartime rationing was stringent.) In his letter of 18 February 1945, he wrote: 'Myself, Mr. Kasturbhai, Mr. Krishnaraj and Sir Padampat we four, are strict vegetarians. As one cannot live on vegetables alone, or on bread, for protein food we need plenty of milk, sugar and butter. I think we shall each need about a quart of milk, about three to four ounces of butter and similar quantity of sugar per day. Besides this, we shall need vegetables and bread for which, of course, I understand there is no difficulty. The chief thing is milk, butter and sugar. I hope it will be arranged before we leave the country. If you will kindly write to London and make these arrangements and let me know, we shall feel assured that there will be no difficulty about food.'

When the time came for the mission to leave, Birla was on board but three of the four "strict vegetarians" dropped out! Whether they could not go for personal reasons or were daunted by reports of the food shortage in wartime Britain or the flying bomb attacks on London, is not known.

Thorough and methodical in his preparations Birla had written to H.M. Patel, ICS, Secretary of Industries and Civil Supplies about another matter on 19 February 1945: 'I take it that we shall be given the most comfortable type of plane for the journey.' He laid down further conditions. 'Except for special reasons we shall generally start work at 10.00

a.m. It is desirable that we should not rush things but work at regular time with free days on Saturdays and Sundays...where it is a question of formal dinners and lunches, lunches would be preferred. Late nights would not be welcome to some at any rate.'

Birla telegraphed J.R.D. on 19 March 1945: 'Appears you are still unwell. Unless your health recovers hundred per cent strongly advise should not take risk leaving country. Health more important than anything else. Wire if you have improved.'

Again on 19 March, G.D. Birla wrote to J.R.D: 'While the visit of the delegation is very important, the health is the first consideration and, therefore, I will strongly advise you not to stir out of Ooty until your health is All.'

The drama preceding the visit continued for a while with Krishna (Betty) Hutheesingh, sister of Jawaharlal Nehru, adding her little bit. On 15 March 1945, she wrote asking whether J.R.D. could consider her husband Raja for the post of Secretary to the delegation. To add weight to her request, she quoted the authority of her brother, Jawaharlal (then in jail), saying that he had suggested Raja should apply for "some such post". She stated confidently that a fellow member of the proposed mission, Kasturbhai Lalbhai, had agreed that her husband was certainly fitted for the job.

On 21 March, J.R.D. observed to A.D. Shroff that this agreement by Kasturbhai was, 'a case of passing the dirty work on to someone else!' J.R.D. continued, 'Raja is quite unsuitable for this particular job and I am sending Betty a polite reply.'

Meanwhile events were developing on the front. On 7 May 1945, General Jodl surrendered to General Eisenhower and the following day General Von Keitel, who was in charge of the eastern front of the Soviet Union surrendered to Marshal Zhukov, Commander of the Soviet Army. The mission was originally slated to leave in wartime but just as it was about to leave, the war with Germany ended.

On 7 May, on the eve of their departure, the members of the mission got the shock of their lives. The *Bombay Chronicle* of that day carried headlines:

GANDHIJI'S BOMBSHELL FOR INDUSTRIALISTS

Ask them to wait till leaders are free. Freedom will come only after big business forego crumbs from Indo-British loot.

The statement was issued by Gandhiji from Mahabaleshwar. On 6 May, enthusiastic people were celebrating the first anniversary of Gandhiji's release from the Aga Khan Palace, amongst cries of "Mahatma Gandhi *ki jai*". Thinking perhaps of his Congress colleagues languishing in jail, Gandhi was probably in no mood to support anything that had any links with the British. Gandhiji said, among other things: 'All the big interests proclaim with one voice that India wants nothing less than her own elected National Government to shape her own destiny free of all control, British or other. This independence will not come for the asking. It will come only when the interests, big or small, are prepared to forego the crumbs that fall to them from partnership with the British in the loot which British rule takes from India. Verbal protests will count for nothing so long as the partnership continues unchecked.'

He advised the members of the mission: 'The so-called unofficial deputation which will go to England and America dare not proceed, whether for inspection or for entering into a shameful deal, so long as the moving spirits of the Congress Working Committee are being detained without any trial for the sole crime of sincerely striving for India's independence without shedding a drop of blood save their own.'

Both Tata and Birla were shaken, perhaps Birla more so. The public were aghast because Gandhiji's blue-eyed boy, G.D. Birla, was the second most important person in the industrial delegation. Birla telegrammed Gandhiji on 7 May. 'I am very much pained and I refuse to believe that you could have given a public expression of distrust in the bonafides of myself, Tata and Kasturbhai whom you have so well known...we are intelligent enough to know our limitations and we know that we have no authority to enter even into a good deal to say nothing of shameful stop your statement sure to be construed as strong denunciation of our motives whereas you usually refrain from expressing any opinion when you do not know full facts....'

J.R.D. issued a rejoinder on 8 May to Gandhiji which he released to the press* and which he forwarded to Gandhiji with a covering note, in which he said 'I cannot tell you how unhappy I was by the views you expressed about our trip and by the strong language you used. What made it worse was that I or some other member of the group was not given an

---

* For the complete text of J.R.D. Tata's rejoinder to Mahatma Gandhi's statement to the press on 8 May 1945, see Appendix E.

opportunity of removing misapprehensions which you evidently entertained about the purpose of our trip. In the circumstances, I was driven to
issue my statement in the Press in order to make my position clear.'

Gandhiji, said J.R.D., had done the mission a grave injustice. There
was no question of any "shameful deeds" and every member was travelling at his own expense. J.R.D. wished Gandhiji had called for clarification before making the statement. The purpose of the trip, reiterated
J.R.D., was to gain such knowledge and experience abroad which would
'enable us to play a better part in India's economic development.' J.R.D.
concluded his statement by saying that India could not afford to stand still
while other nations, great and small, were forging ahead.

Gandhiji, having rocked the industrialists with the public statement,
softened the blow by sending individual messages to J.R.D. and
G.D. Birla. In his handwritten postcard to J.R.D. he wrote on 20 May
1945: 'Bhai Jehangirji (in Gujarati; he thereafter switched to English for
J.R.D.'s benefit) I have your angry note, if you can even write anything
angry. If you have all gone, not to commit yourselves to anything, my note
protects you. My answer is to the hypothetical question. If the hypothesis
is wrong, then naturally the answer is wrong and, therefore, it is protective
of you all. There was no question of my referring to anyone of you and I
was dealing with an assumption. I hope I am clear.'

Even Aristotle with his logic could not have found a better way out of
the situation.

*

The mission left within a week of the surrender of Germany. The final
group included J.R.D. Tata, G.D. Birla, Nalini Ranjan Sarker,
A.D. Shroff, Sir Sultan Chinoy, Laik Ali and Ajaib Singh.

N.R. Sarker was one of the three Viceroy's Executive Councillors who
had resigned in protest against the government's refusal to release
Gandhiji during his fast in 1943. A.D. Shroff of Tatas had been India's
unofficial delegate at the World Monetary Conference at Bretton Woods
in July 1944. Sir Sultan Chinoy took a leading part in bringing broadcasting to India in the early 1930s.

They were accompanied by nine technical advisers, with Dr P.S.
Lokanathan as Secretary to the mission. Dr P.S. Lokanathan was the
Editor of the *Eastern Economist*. Among the technical advisers were

Sir Jehangir Ghandy of Tata Steel; Sumant Moolgaokar, an innovative young engineer from the Associated Cement Companies (ACC) and B.W. Figgins, who ran the Flying School for Tatas at Pune.

When they landed in Britain, mid-May 1945, J.R.D. was asked, 'What would you like to see?' His British hosts were taken aback when he said that he would like to fly over German cities devastated by war to study the damage; he wanted to do this in order to see for himself the destruction India had been spared. Planes were obtained, and the delegation flew so low over German cities in rather bumpy weather that most of them ended up air-sick. J.R.D. saw what he wanted to and the grim reality of the destruction of Germany was an unforgettable experience for him and his party.

The mission visited the Federation of British Industries, and various Chambers of Commerce and the State Department in the US. 'Our visit appeared to create much interest in both countries. This was expected in Britain,' the delegation's report said, 'but we were glad to notice in the States also a considerable amount of interest in Indian affairs and in the future possibility of trade with India.'

In Britain they found that the British leaders were reconciled to the inevitability of India's political and economic independence in the years to come. In the US, knowledge of India 'was not equal to their evident interest in our country's affairs.' The mission felt the need for the establishment of a permanent organization for the dissemination of appropriate and true facts about India.

The delegation had a strenuous programme of visiting factories and Moolgaokar remembers how the Americans even opened up their high-security factories for the Indian delegation.

A German newspaper carried the headlines:

USA GIVING THE GLAD EYE TO TATA. LIVELY INTEREST IN INDIAN HEAVY INDUSTRY

America's efforts to find a market in India for her export and for the investment of her capital are primarily directed towards obtaining a foothold in Indian heavy industry and, by giving support to the Indian wish for autonomy, to neutralise the political advantages enjoyed by English enterprise. This is particularly evident in the interest shown in the USA in the greatest national Indian industrial concern, the Tata Works, which have been built up with strong support from the USA....

Tatas were not unknown in America. Even prior to the visit of the delegation, *Fortune* magazine had written in January 1944:

> In any economy—even in the US—the House of Tata would be noteworthy....To the Indian, the natively created House of Tata represents a source of national pride, a signpost on the rocky road toward an industrial future. The story of Tata both mirrors and colours the history of the growth of Indian industry. Virtually every Tata enterprise today is engaged directly or indirectly in India's war effort....Tata will remain what it is today—one of India's greatest national assets, significant in itself, but still more significant as the promise of a far-off industrial future.

J.R.D was more inclined to buy machinery from America than from Britain because the delivery time was shorter, as its factories were (unlike Britain's) both intact and its products more modern.

The delegation learnt the importance of keeping machinery up-to-date by comparing the records of British and American industry. In the immediate pre-war years Britain, emerging from the Depression, had not renewed her machinery. Consequently Britain's war production was inefficient compared to America's but, 'in the field of research...it seemed to us that Britain was as advanced as America.'

During the visit, there was subtle propaganda from the British in the USA that Indian sterling balances of about $ 1,000 million were the result of high war profits made by Indian business. The delegation refuted this charge by stating that they were the result of sacrifices made by individuals in the war and urged that the sterling pool be made convertible into dollars to finance India's industrial growth. Unless Britain agreed to convert part of the reserve into dollars, India could not import American machinery.

The delegation's report concluded, 'We have returned from our trip enriched with first hand knowledge...and with better informed appreciation of the significance, scope, needs and complexities of modern industry.' The delegation members recognized the need, 'for massive scientific research and education,' in India, both of which were in slender evidence at that time. Fortunately, only a few months before taking off for the West, J.R.D. had helped the Sir Dorabji Tata Trust to approve Dr Homi Bhabha's proposal for the Tata Institute of Fundamental Research (TIFR) which a decade later became, 'the cradle of India's atomic programme.'[2]

*

Some months after the mission's return, at a time of intense political activity, Herbert L. Mathews of the *New York Times* interviewed J.R.D. Tata, G.D. Birla and Walchand Hirachand. J.R.D. Tata is quoted as saying:

> In future India's link will be with the USA rather than Britain. We have all the resources except oil, and we have inexhaustible manpower, which has proved that with training it is as good as any. I hope there will be a coalition, not a one-party government, and that it will place the development of the country in the foreground. Enormous returns would follow investments of revenues in roads, sanitation, education and agricultural improvement in villages. The villager has never had a chance; but by raising the standard of living, there can be created vast purchasing power in our internal market alone. The British need not go, but they have antagonized us by their die-hard tactics and done so much to retard Indian independence that I am afraid they will find it difficult to carry on.[3]

His concept of development is as relevant today as it was nearly fifty years ago.

CHAPTER VI

The Two Top Captains

For almost half-a-century two men held the commanding heights of Indian industry: J.R.D. Tata and G.D. Birla. Even in their eighties both were without peer. Both men were charming conversationalists. Highly intelligent, they respected each other for their accomplishments, but their ways were different and since the mid-1940s, when they collaborated on the Bombay Plan, they seldom came together.

Tatas was the earlier established industrial house. Birlas were primarily a trading concern, which entered industry mainly after the Second World War. Birlas had only twenty companies when the war ended in 1945. By 1965, they had grown to 150 companies with interests in areas as diverse as aluminium, rayon, and car manufacture. By 1985 Birlas had 200 companies—twice Tatas' number. In assets and turnover the two houses raced neck to neck in the 1980s.

The two captains of industry emerged from totally different backgrounds. G.D. Birla was born in a village in Rajasthan where hardly six men could read and write and where camels were the only transport. J.R.D. Tata was born in Paris, a city renowned for its sophistication and glamour.

Birla was educated in a village school, mainly in arithmetic. His education ceased at the age of eleven; at twelve, he was inducted into the family business, which included trading in opium, wheat, silver and oilseeds. J.R.D. received his education in France, India, Japan and England and winds blowing from many windows stimulated his young mind. J.R.D. entered the family business at twenty-one. Both men felt the lack of an educational base and, with determination, equipped themselves with well-furnished minds. The reading of both was vast and varied.

At sixteen, Birla came under the spell of Gandhiji and the Mahatma became a close confidant. Unusually, Birla could both do business with the British and, at the same time, be an active participant in the freedom struggle. J.R.D. kept business and politics strictly apart, though his sympathies for the Congress were known to the British.

In the Archives of the India Office Library, London, lies an old stencilled document marked "Strictly Secret", entitled "BIG BUSINESS AND CONGRESS CIVIL DISOBEDIENCE, 1942". This Intelligence Bureau report is a study of five companies including Tatas and Birlas.

While they could find little on Tatas, except one unconfirmed donation for the relief of families of workers in jail, the report says of Birla that 'He has in the past helped Congress greatly with considerable, if irregularly paid donations.' Despite this, the Viceroy, Lord Wavell liked Birla, more than J.R.D. and even suggested to Leopold Amery, the Secretary of State, that Birla be preferred to Tata for lunch with the Queen:

Wavell to Amery   *12 June 1944*
                                    Viceroy's Camp, Simla

I think Queen Mary would find G.D. Birla better company than J.R.D. Tata if she wishes to invite one of them to lunch. Tata is a pleasant enough fellow to meet, but I have not found him communicative, and as a casual acquaintance he is very much the same as any other wealthy young man who has a conventional education and turns himself out well. Birla on the other hand is a less conventional type. He has plenty to say and whatever one may think of Marwari businessmen and their ways, he is well worth talking to. I think Queen Mary will have a very dull lunch with Tata and quite an interesting one with Birla.

As we've seen earlier on in the book, in 1946 the year after the mission to the West, came the first difference of opinion between Birla and Tata. The Minister of Civil Aviation was liberally issuing licences for starting new airlines. J.R.D., foreseeing the danger, was against the policy. Birla wanted to get into aviation. That, for J.R.D., was bad enough. What was worse, J.R.D. got written reports that Birlas' Bharat Airways had tried to entice away five Tata-trained pilots.

On 23 August 1946, J.R.D. wrote to G.D. Birla: 'You will remember that on more than one occasion in the past you expressed in clear terms the view that it should be the policy of leading business firms in India not to add to the many difficulties already facing all of us as a result of social and political trends, by entering into avoidable competition with each other or doing anything which would unnecessarily add to the others' difficulties.'

He pointed out that Birlas' entry into aviation, where the field was limited, was not consistent with this policy and would strengthen the case for nationalization in the future.

Further on in the letter, J.R.D. said: 'Apart from the general aspects of the question, I am even more sorry to find that Bharat Airways have been making a determined attempt at enticing some of our staff, and particularly our senior pilots, to join their organisation. I am sure you will agree that we have ground for considering this a most unfriendly act. In view of our friendly personal relations in the past, I do not believe that this step has been taken with your approval.'

The case of air transport, replied Birla on 23 September 1946 'stood on a different footing' (from the general principle). Stating that air transport held no attraction for him in terms of money making, he nevertheless felt it was closely allied with the defence of the country and said, 'if my firm has decided to enter this new field, it is purely on patriotic grounds...I don't take the view, as you do, that for many years to come there will be room only for few persons. India is such a big country and the demand so large that there is ample room for all the newcomers.' Birla promised an inquiry into the charge of enticement and assured, 'If any truth is found, suitable action will be taken.' On 28 September he wrote, denying the charge of enticement.

J.R.D replied on 22 October 1946: 'I can only agree to differ (on the question of the potential for air transport). There is simply not enough business for the number of companies which you visualize. As regards my complaint that Bharat Airways had tried to entice our men, I can assure you there is no doubt in the matter at all. For your information, here are the names of the pilots who were definitely approached: Captains Jatar, Dhuru, Naralkar, H.P. Mistry and Shirodkar.'

J.R.D. concluded: 'Finally, may I say that I was most disappointed and perturbed at the editorial which appeared recently in your paper, the *Hindustan Times* in which the older airlines—which obviously included Tata Air Lines—were most unfairly stigmatised as anti-national. I submit in the friendliest spirit, that this action is not in keeping with the views which you expressed to me in the past, and to which I referred in the first paragraph of my letter of the 23rd August. I am sure that if Tatas owned a newspaper in which such an unprovoked attack had been made on one of your companies, you would have felt as aggrieved as I do in the matter.'

Birla replied on 8 November 1946: 'I would be very sorry indeed if you believed that editorials in the *Hindustan Times* or in the *Eastern Economist* are written under my instruction or advice....Please believe me, however, when I say that I would be the last person to associate myself with any attack on the House of Tatas for which I have got great regard, although I disagree with them on many points.'

Birla's Bharat Airways did fly. There was not enough business and J.R.D.'s worst fears of nationalization came true in 1953.

After independence, G.D. Birla assiduously cultivated several Congress party leaders, bureaucrats and MPs. J.R.D. made courtesy calls but was not as attentive as the other; their ways were different. Speaking to *India Today*, J.R.D. said of G.D. Birla:

I did not approve of his ideas. We never got together in anything except that I recognised him as a man of vision.[1]

Ved Mehta, in his book, *Mahatma Gandhi and His Apostles,* compares the heads of the two companies—(a comparison that has often been made) one, a modern, Westernized business executive, and the other,

a moneylender and merchant of the old school, an Indian tycoon, secretive about his wealth and power.[2]

Birla told Mehta:

The difference is that I am religious and Tata is not. He is therefore frustrated, and I am content. I do not know about his business methods, but my business runs on auto-pilot, I am only called in to make the big, momentous decision. I can't speak for Tata, but I am a simple man, with simple habits....I have always gone it alone.

J.R.D.'s relationship with Birla after the airline episode was cool. In

1958, Birla sent him a letter with press clippings which endorsed Birla's views on smoking, coffee, alcohol, apparently to convert J.R.D. to his beliefs. J.R.D. replied on 24 January 1958: 'I have given up nicotine, and hardly ever drink coffee or alcohol. According to others, I should give up meat. What I would like best to give up is work! I note with some relief that neither Professor Hochrein nor Dr. Schleicher have condemned sex too!'

According to his biographer Ram Niwas Jaju, Birla confided that after the death of his second wife, when he was thirty-two, he never looked at another woman.[3] That is certainly another area in which the two men diverged. Even in his eighties, when a pretty face comes within his ken, the eyes of J.R.D. light up.

When G.D. Birla died in 1983 in London J.R.D. said in his condolence tribute: 'Although I had known Mr. Birla for some fifty years, we met only at relatively rare intervals and to my regret our mutual relationship was therefore not a very close one.' Referring to the Bombay Plan, J.R.D. spoke of Birla's outstanding intellectual accomplishments, his luminous intelligence and his far-seeing vision of India and the world. 'Although he was the product of a highly orthodox social and family environment in which he was born nearly ninety years ago, he had a remarkable combination of both a Cartesian Western mind and a traditional Hindu mind steeped in Hindu religion and philosophy. He must also have had a marvellously flexible mind for he was able to approach problems and issues with both parts of his mind and come up with practical solutions that compromised neither.

'It was also a matter of regret to me that in the decades that followed, until his lamented death, our paths did not cross more often or more purposefully. We nevertheless retained a high esteem for each other and years later I had reason to be grateful to him for his instant and generous response to the appeals I made to him for support to the Family Planning Foundation of which I was the Founder Chairman...another indication of his clear and wide ranging mind.'

CHAPTER VII

Cultivating Trust

Tata Steel (TISCO) opened in 1911, establishing an eight-hour working day when steel mills in other parts of the world operated for ten to twelve hours. In 1920, Tata Steel introduced leave-with-pay, another first.[1] In the 1920s news of the victory of the Communist revolution in Russia blew over India. Literature in Indian languages, printed at foreign instance, flooded into the homes of industrial workers. The trade union movement gathered momentum.

TISCO was not behind others in its benefits, but because it was in the front rank of Indian industry, it became the target of attack. It was hit by a flash strike in 1920 that stunned the management. Another came in 1922. A third strike took place in 1924. Mahatma Gandhi came with Dr Rajendra Prasad and C.F. Andrews to settle the strike. A dismissed General Secretary of the union was re-employed. R.D. Tata recognized the trade union with C.F. Andrews as President.

The next strike could be squarely blamed on the management. Said J.R.D., in a speech in Jamshedpur in 1963: 'My association with Jamshedpur began when I came here with my father in the second decade of the century. And I still remember the tremendous excitement it was for me to come here and to be allowed to go into the Works. In those days, for a few years more, the dominant figure at Jamshedpur was the General Manager, T.W. Tutwiler. He was known, respected, feared and loved as the uncrowned king of Jamshedpur, and he behaved like a king...

'Tutwiler was also known for his lurid language, which, on one famous occasion, had unexpected and unfortunate consequences, for it was largely the cause of the first major strike at Jamshedpur. At that time, there was a Parsee gentleman who was a Foreman in Coke Ovens and used to come to work in a tie and collar. That upset Mr. Tutwiler who thought that a steel worker should not wear a tie and collar. One day, something had gone wrong and Mr. Tutwiler, in an angry mood, caught hold of his tie and said: "If there is a god-damned son-of-a-so-and-so in this town who can wear a tie and collar, it's me." Mr. Maneck Homi—now to name him—promptly resigned and formed a labour union. And this led ultimately to quite an interesting strike and sequel.

'Steel prices were incredibly low then, and I well remember that joists were imported in India for Rs. 110 a ton: and then, we still had the pre-war plant (World War I); the new plant had not been completed. Life was carefree and living was cheap in those days. Rupees 1,000, which was a little more than I got, was a princely salary. But the masses were miserable; poverty and ignorance were rampant in the country and there was an underlying sense of frustration and bitterness. In 1930, there was a devaluation of sterling, and the rupee followed automatically. This led, for a period of ten years, to an exodus of gold from the country. That gold came mainly from the peasants' ornaments. It was distressed gold, a symptom of the very depressed and miserable economic condition of those days.

'Life was simple in those days, there was very little stress and strain. The British regime was supreme; the coming disobedience movement, the struggle for Independence and ultimately the loss of the Empire, was something so far away that nobody thought about it. In 1932, we saw the beginning of really active civil disobedience which cast its shadow ahead. Then we come to 1938—the year I had to become Chairman.

'For me it was a fateful year. I was only thirty-three years old, and, with Sir Nowroji Saklatvala, the then Chairman, hale and hearty, I hoped I would be in the background for quite a few years more. But he died that year, and my colleagues elected me Chairman. The happy and carefree life that I had led up to then suddenly came to an end. I found myself shouldering responsibilities which I would have been delighted to see in other hands.'

Among his responsibilities as Chairman of Tata Steel, was to occasionally deal with Professor Abdul Bari, the fiery, intemperate President of the Tata Steel Workers' Union. Talking about their relationship, J.R.D. said: 'When I became Chairman of Tatas, I had not, until then, been actively involved in labour problems though I had, from the start, sensed the vital importance of good relations in industry. Professor Bari was a man of violent emotions, and liable to explode into equally violent anger. As I listened to his angry and quite unjustified onslaught on our

Management, I was somewhat upset, to put it mildly. Yet, when the next morning I first met him face to face, I found to my surprise that I could not help admiring and even liking this extraordinary man, uncontrolled, but totally honest. As I greeted him with a friendly smile and shook hands with him, I saw a change take place in his face which I interpreted rightly, as I found out later, as astonishment at finding not only that I felt no resentment for his violent attack on us the previous night, but was genuinely friendly towards him. We laughed together and when I gently asked him whether we really deserved his lambasting of the previous evening, he replied, "Tata Saheb, I am sorry, but when I get on to a platform I get excited and can't control myself!" Mutual esteem blossomed between us at that moment, that lasted for the rest of our relationship, unfortunately cut short, when he died in tragic circumstances in 1947.'[2]

The tragic circumstance was that Abdul Bari, also President of the Bihar Pradesh Committee, was travelling by car. At a check-point the police were inspecting the trunks of all the cars for contraband and weapons (this being the time that the war of nerves between the Indians and the British had reached breaking point). When Bari's car was being inspected, he entered into an argument with a policeman and grabbed his rifle. In the scuffle the rifle got triggered and poor Bari was killed.

J.R.D. concludes: 'I have told you of this little episode merely to illustrate my conviction that there is really no difficulty in establishing good relations with people in general and between Management and workers and their Union in particular if one cultivates a liking for, and trust in, those one deals with.'

It was under Bari's successor, Michael John, that labour relations settled down. In the 1950s the Communists made a determined bid to wrest control out of the hands of Michael John. The co-founder of the Communist Party, S.A. Dange, came and installed himself in Jamshedpur. In response J.D. Choksi, Vice-Chairman, also went from Bombay to Jamshedpur to support the management. Dange failed. Tatas stood firmly with Michael John's union.

On labour relations J.R.D. says, 'I came long ago to the conclusion that the three most important requirements for getting along with people were, first, communication. The importance of frank and sometimes continuous discussion between people or groups is clearly seen in the example of the splendid relationship established and maintained, year after year, between the Steel Company and the Union at Jamshedpur. Its success and effectiveness is due largely to the continuous dialogue between the Company, its workers and their Union, which ensures that all matters and decisions of consequence are fully discussed and understood by both sides. The second is the need for total honesty and sincerity in dealing with people. Human history is replete with examples of the disastrous consequences of mistrust and suspicion amongst people and nations. The third is to trust and, if possible, to like the people with whom one deals and to inspire a similar response in them.'

He did not believe in establishing a personal equation with the union leader alone; his vision held that communication should take place across the board. In a significant note to his colleagues in Tata Steel, he spoke of the need of establishing a Personnel Department in Tata Steel at a time when very few industries had such departments by compulsion of law. In a letter in 1943, he wrote; 'One of the inherent drawbacks of industrial history with its large and concentrated labour forces, is the difficulty in maintaining personal touch between management and employees. As a result many petty grievances, negligible individually, but substantial in the aggregate, are allowed to build up....' In his note he was searching for ways in which he could associate labour with the task of management, although he did not believe that the management of factories should be handed over to labour.

The Personnel Department of Tatas was established in 1947.

In the mid-1950s, Tatas took another significant step in associating their workers with the running of their departments. Two historic agreements were signed in June and August 1956, which set out the basic rights and obligations of management and employees and provided for closer association of the employees with departmental councils where management and employees could discuss their problems regularly. The Chairmanship of the discussion would alternate between representatives of management and employees. Each council would have an equal number of representatives from management and employees.

J.R.D. ascribes this development to discussions he had with Minoo Masani, his 'chef-de-cabinet' at that time. Today, the President of the Tata Workers' Union, V.G. Gopal, proudly says that although other large industrial houses and public sector undertakings have attempted similar councils, nowhere else has it worked as well as in Tata Steel. The joint councils decide on the entire spectrum of labour and industrial problems as they arise except where collective bargaining is concerned. Wages of steel plants are, in any case, settled on a national scale for all the steel plants, as the government is the major producer of steel.

*

In December 1978, two Cabinet Ministers of the then Janata Government proposed the nationalization of Tata Steel and a couple of other enterprises. The two were George Fernandes and Biju Patnaik, Ministers for Industry and Steel respectively.

The Tata Workers' Union registered their protest by cable to Prime Minister Morarji Desai before the management could lodge its protest. The workers from Tata Steel plants, mines and collieries resolved in January 1979 that nationalization would be detrimental to all employees of the company. The government had to think twice before touching Tata Steel and in any case most of the senior ministers were not for nationalization. Later the same year the company celebrated fifty years of industrial harmony.

Speaking to a Jamshedpur audience in 1985, J.R.D. said:

> I claim no personal credit for the outstandingly good and cooperative relations which have prevailed within the Company since those early days. If credit is due to any person in the Company for what has been achieved by these very means, it would be to your present Chairman, Mr. Russi Mody, whose human qualities and extraordinary ability to arouse love and friendship in others have been a source of inspiration to all, including myself. Co-operation is never a one way traffic. From the Union we have in Mr. Gopal, a worthy successor to the traditions of Abdul Bari and Michael John, and we all admire the manner in which he is carrying the torch.

> On the Management side Russi Mody's role in shaping industrial relations has been the most outstanding. When he was in the collieries he established a unique rapport with the workers. Whenever he saw that there was a need or an injustice which the company could set right, he promptly acceded to the demand of the workers and when the line had to be held, he would hold it.[3]

Napoleonic in his size, his drive and his energy, Russi Mody as Managing Director would sit every week in shops of the steel plant where any worker who had a complaint could meet him. He would listen to them and receive their written petitions, to most of which he could not accede because the company had its own rules and regulations. But whether his reply was "Yes" or "No", he sent a personally signed letter to every petition. In a normal year he would send out 10,000 letters. He has the flexibility to sit one week with the workers on the shop floor and share their cup of tea, and the following week holiday on the Riviera, sipping French wine. Golfer and gourmet, Mody can complete nine holes of golf and end up with twelve eggs for breakfast. A pianist and a pilot he was educated at Harrow and Oxford. He says that as Harrovians Winston Churchill and Jawaharlal Nehru are no more the responsibility on him is now considerable. Indeed it is! For Tata Steel and its eighteen associate companies account for almost a third of Tatas' turnover. Mody thinks on a national scale and is one of the most forthright spokesmen of Indian industry.

Mody executed the ideas J.R.D. had for labour and their bond comes from this factor of a common intent. Mody says that in his early years in Tata Steel there was a wide gap between him and J.R.D. but inch by inch the gap narrowed to the friendship they enjoy today.

Though Russi Mody's role in shaping the industrial relations of Tata Steel has been outstanding, the following incident is typical of J.R.D. In Bombay House in October 1986, there was a luncheon in honour of Frank Thompson, President of the Ford Foundation. Present at the table were some Tata Directors, including Russi Mody. Speaking of Tata Steel, J.R.D. said to Thompson of Russi Mody: 'Do you know *his* company has enjoyed industrial peace for over 50 years.' At this point a senior Director J.J. Bhabha interjected, 'Mr. Tata has failed to add that out of these 50 years, Mr. Tata was Chairman for 46!'

CHAPTER VIII

The Birth Of A Giant

Involved with steel, J.R.D. felt that Tatas were competent to establish in India an engineering complex the like of which the country did not have. Tata Steel provided the springboard for this new complex.

Within one-and-a-half months of its birth in September 1945, the Chairman J.R.D. Tata explained to the Board, 'The Company was promoted to manufacture not only boilers and locomotives but also heavy engineering equipment such as road rollers, tractors, earthmoving equipment, diesel engines, etc. The original intention was that the manufacture of such heavy engineering should be undertaken only after the Company had successfully manufactured boilers and locomotives. But, in view of the immediate post-war demand for the former type of equipment, the Agents felt that it might become necessary to commence the manufacture of such equipment at an earlier stage....'[1]

For any new project they undertake Tatas look for an anchorman. Just prior to the launching of Tata Engineering and Locomotive Company (TELCO), J.R.D. had taken on his mission to the West, the forty-year-old Executive Director of the Associated Cement Companies (ACC), Sumant Moolgaokar.

Moolgaokar had qualified at the Imperial College of Science and Technology in London in Mechanical Engineering. On returning to India in 1929, Moolgaokar had difficulty in finding a job. Finally, the C.P. Cement Works in Dwarka, Gujarat, offered him one without salary! Accommodation was generously provided. Only after six months was he given a salary of Rs 250, a princely sum in those days.

Even from far away Dwarka the brilliance of this engineer shone through to the Directors of ACC in Bombay and, a decade later, when the Second World War prevented the import of cement machinery from abroad, Moolgaokar designed and fabricated India's first cement plant and erected it for Tatas at Chaibasa in Bihar.

Tatas were among the pioneers of the cement industry. Back in 1912 Indian Cement was formed by Tatas; Shahabad Cement followed in 1921. In 1936 Tatas joined three other companies in starting the Associated Cement Companies.

Moolgaokar had become the Executive Director of ACC when J.R.D. selected him for the industrial mission to the UK and USA. Shy and self-effacing, Moolgaokar would slip away in the evenings while the others partied. 'I was more cut out to visit factories than to attend cocktail parties,' he would say.

One day J.R.D. turned to Moolgaokar with the words, 'How long are you going to make the glue that sticks the bricks together?' He hoped to interest the young engineer in the new company coming up in Jamshedpur. On his return from the West, J.R.D. requested the Chairman of ACC, Sir Homi Mody, to release Moolgaokar. Sir Homi declined saying, 'We need Moolgaokar.' J.R.D. did not take "No" for an answer. He approached Sir Homi a second time. Sir Homi was Chairman of ACC, by virtue of his representing Tatas on its Board. Noting that the Chairman of Tata Sons was determined, Sir Homi relented. In 1949 Moolgaokar joined TELCO as Director-in-charge.

TELCO's project started with boilers, went on to underframes of wagons and then to locomotives. Britain was the main exporter of locomotives to India and was not keen to part with its technology—and its customer. So Tatas scouted around in Germany. Soon enough, Krauss Maffei, headquartered near Munich, agreed to assist with technology for boilers and locomotives.

TELCO had a shaky start. For the arduous task of boiler-making, the crew was entirely made up of tough Pathans from the North-West. The first boilers had hardly come out when in 1947 Partition rocked India and with it the fury of Hindu–Muslim riots. The Pathans fled from Jamshedpur to Pakistan overnight. The boiler plant was paralysed.

In the time it took to train a fresh crew, Moolgaokar used the workers to keep the place spick and span. Sir Jehangir Ghandy visited the plant and remarked that he had never seen such a shining workshop. He suggested to Moolgaokar that if TELCO could not make a profit manufacturing boilers they could, 'at least show visitors around and charge them for seeing the showpiece.' It took six months to train the next crew. 'The first boiler to emerge was,' said Moolgaokar, 'the most expensive in history. The company had yet to settle down and we produced more General Managers than boilers.' English General Managers succeeded one another till a German finally came and held the floor.

Locomotives steamed out from the Tata plant next and in a little over a decade more than 1,000 Tata locomotives joined Indian Railways. Soon Tatas realized that they were in a most vulnerable industry. Their sole customer was the Indian Government and the Railways could beat down their prices to any level they wanted. They were vulnerable to demand fluctuations too. Just at that moment, destiny opened a door.

Daimler-Benz, which till the 1950s, had not extended itself beyond Germany, asked Tatas if they were interested in manufacturing trucks. J.R.D., his legal advisor J.D. Choksi, and Moolgaokar went to Geneva for negotiations. The Daimler-Benz lawyer had drafted terms which were disputed by J.D. Choksi, whom the Germans called "Yoksi". It appeared as if the negotiations would break down. Just then the Chairman of Daimler-Benz broke the ice. He looked at J.R.D. and said '*You* draft the agreement and we'll discuss it.'

In that moment of trust a partnership was born in May 1954. The Tata draft was accepted. 'We derailed the locos and got on the track with trucks. After that we never looked back,' said Moolgaokar.

The next hurdle was to tackle the Indian Government for permits and licences. The mid-1950s were a difficult period for industry and the basic thinking of Prime Minister Nehru was that the state should get an increased control in heavy engineering. With some trepidation J.R.D. and Moolgaokar went to see T.T. Krishnamachari, the Industries Minister. Much to their surprise T.T.K. quickly got the point of Tatas entering the field of road transport and without reference to anybody said 'Go ahead.' 'But that's not enough. We need the permits and the licences,' said J.R.D. 'You will get it,' said T.T.K. and instructed one of his secretaries to make everything available to Tatas speedily.

'T.T.K. could be arrogant, ill-tempered and undemocratic but he was an extraordinary decision-maker,' says J.R.D. Before they knew it the project was cleared. The initial collaboration between Benz and Tatas was to be for fifteen years. The original name of the company was the Tata Locomotive and Engineering Company. It was changed to the Tata Engineering and Locomotive Company on 24 September 1960, when the primary objective of the company shifted from locomotives to trucks and general engineering.

Within a few years of signing the successful collaboration with Benz, J.R.D. wrote to his friend George Woods (later President of the World Bank) in 1959: 'Our TELCO's locomotive division is never likely to make any money in view of the purchasing policy of the Railways. We have to keep up with it both as a matter of service to the country and in recognition of the fact that the industry was established with Government's support. We hope to switch over to diesels in the years to come and have put forward a project for the manufacture of both the chassis and the engines but, as I have said, all we can hope to do financially is to break even and spread our general overheads. We have everything to gain by strengthening the automobile division. We shall of course continue to develop truck manufacture to the maximum extent possible, but we feel the time has come to move into the small car field where there is likely ultimately to be an unlimited market. You may, however, be relieved to know that there is no certainty about this scheme as there will be no doubt be strong opposition from our competitors and possibly from other sources.'[2]

The Benz engineers were sticklers for perfection, as were J.R.D. and Moolgaokar, so it turned out to be a happy collaboration. Once every week all the parts rejected by the German technicians in Jamshedpur were displayed and a post-mortem held. This "major attention to minor details" was a painful exercise for the Indian engineers but it was through such discipline that TELCO's standards were jacked up.

Moolgaokar said, 'There is a belief in our country that our culture and our Indian character cannot allow our people to attain consistently high standards, that shoddiness and carelessness are our God-given, unalterable way of life. But if, with faith in them, you ask our men for their best, they rise to your belief and in their worth and create a momentum towards improvement. Often I have seen men who are considered ordinary, rise to extraordinary heights. Do not accept second-rate work; expect the best, ask for it, pursue it relentlessly and you will get it.'

In the years that followed, Tatas decided to set up a TELCO factory and a research division in Pune. Moolgaokar was not interested in assembling trucks or only manufacturing them. He believed that a company should have "in-built strength" and he gave TELCO the capacity to design and manufacture its own sophisticated machinery, dyes and press-tools to create the machines to make the parts of a truck—a facility that very few automobile plants in the world have, as they depend on other tool manufacturers. When he was working on this aspect of TELCO, some Directors felt he was spending too much money for the future when the present was still to be secured. J.R.D. supported Moolgaokar.

TELCO farmed out ancillary parts to hundreds of suppliers, sending out TELCO experts to show them how to make them to satisfy TELCO's existing standards. The cycle of wealth widened and, with it, engineering skills were upgraded all over India.

In the late 1970s when Tata trucks were in short supply and sold at a premium of up to Rs 40,000 per truck, Tatas refused to raise truck prices. 'Profits should come from productivity and not by raising prices in a favourable market,' said Moolgaokar, 'Our greatest asset is customer affection.'

Soon, virtually all of the running of TELCO fell on the shoulders of Moolgaokar. J.R.D. remained more involved with Tata Steel. 'I realised early,' said J.R.D., 'that Sumant was a lone wolf. If I let him run it his way he would deliver the goods. And he did.' Left to himself, J.R.D. would have pressed ahead and perhaps expanded the range of TELCO to the extent he envisaged when the company was born. Moolgaokar had his own vision which was not to go wider, but deeper, mastering the field of engineering to lay the foundation of an engineering complex whose products the world would seek. By giving men like Moolgaokar the freedom to grow J.R.D. helped to create not only a superb corporate head but one who could build in J.R.D's words "an edifice for India".

In a more general context J.R.D. told me 'If I have any merit, it is getting on with individuals according to their ways and characteristics. In fifty years I have dealt with a hundred top directors and I got on with all of them. At times it involves suppressing yourself. It is painful but necessary....To be a leader you have to lead human beings with affection.'

Moolgaokar's style reflected the man. While J.R.D. can be tolerant of human frailty, Moolgaokar was not. To Moolgaokar the job came first. They were amazingly similar in some ways, in other's very different. Both had a passion for precision and both loved speed. At a young age, J.R.D. loved fast cars, Moolgaokar fast motor cycles. Both had workshops in their homes. J.R.D. loved carpentry. Before air-conditioning was common, J.R.D. found it uncomfortably stuffy concentrating on woodwork as a ceiling fan would throw up the dust. And so he turned to metalwork. Moolgaokar was J.R.D.'s adviser and supplied the castings J.R.D. needed for his workshop. When each of them went abroad each would get as a present for the other something for the workshop.

Together the two men complemented each other and made a formidable team, one which elevated TELCO to the commanding position it occupies today as one of India's two largest private sector companies (1990 turnover: Rs 25,000 million) TISCO being the other (1990 turnover: Rs 21,000 million.)

Three times did TELCO attempt to produce a good car. First a Mercedes-Benz in the fifties, then a DKW in the sixties and, in the 1980s, a Honda. In 1960 TELCO loaned four Mercedes-Benz cars to K.B. Lal—Secretary for Commerce and Industry, to use for a year before making a decision. One car went to the Defence Minister, V.K. Krishna Menon. The cars were much appreciated, no decision was taken, and were returned. After waiting, Daimler-Benz went to Singapore and Singapore earned the foreign exchange—India lost it and TELCO lost the benefit of producing a good car for the country.

In the 1980s Tatas tried again, hoping for permission from the government to make Honda cars in India. They had the in-built strength to indigenize and begin manufacture quickly. Tatas were given hope in 1985 that the Union Government would accept their proposal and a Memorandum of Understanding was signed with the Japanese company. The permission never came through. Though disappointed, J.R.D. is not bitter. He said, 'I honestly think TELCO has the capability of indigenising very fast. We do our own designing, our own engineering, so, I suppose, we could have indigenised the Honda car pretty soon. But I suppose the government having initially gone too fast, is now hesitating before it permits so much foreign exchange outgo.'

TELCO's light commercial vehicle, 407, designed by its own team, appeared in the market one year after the sleek Japanese models of Nissan, Toyota, Mazda and Mitsubishi. Within twelve months, TELCO captured a larger share of the Indian market than any of the Japanese models. After the 407 came the Tatamobile and for some years TELCO could not produce enough vehicles to meet the demand. The estate car and the sports car are to follow and an indigenously designed and produced Indian car is expected.

While thinking for the company, Moolgaokar did not forget larger concerns. For instance, TELCO has made a singular contribution to environment. When rocky and barren land was purchased at Pune for a second TELCO plant, Moolgaokar first worked out a plan for blasting the rocks for the planting of thousands of trees. But trees need water, so an artificial lake was constructed for storage of water. 'I was criticised for spending Rs. 15 lakhs,' said Moolgaokar. J.R.D. supported Moolgaokar. For decades the TELCO nursery has supplied thousands of fruit trees to neighbouring villages bringing greenery and light to the barren villages around. Today at the TELCO lake at Jamshedpur, birds come to nest in winter from as far away as Tibet. They return home and the following year they come back to collect the young ones and each wren recognizes its own mother.

'We did not have to create a lake and plant trees to produce a truck. But we did,' says J.R.D. proudly. 'What I am most proud about is not the making of steel or trucks but our social concern.'

J.R.D. observes: 'Planning a new factory in Pune (in the mid-1960s) needed courage. Money was then difficult to raise. He built the factory, he built the men, he built the technology—created a research department. He began to build a new edifice for the future. The Moolgaokar era was creative. A magnificent factory and a magnificent force that elicited pride.'

Moolgaokar is J.R.D.'s only colleague whom he honoured by comparing him to Jamsetji Tata, in having a vision and setting about the task of fulfilling it.

In December 1988 Moolgaokar stepped down in favour of R.N. Tata, the son of J.R.D.'s distant cousin, Naval Tata. A few months later Moolgaokar died. He, who had shunned the limelight all his life, was featured on the front pages of all the newspapers. The *Times of India* headed its editorial on him "Nation Builder". The *Financial Express* wrote:

> There are two ways of making money in India. One is to manipulate politicians and bureaucrats to get licences, permits and then search for scarce products and reap the economic rents (windfalls) that go with them. The second is to concentrate on increasing skills, technical prowess, quality, and hence total productivity. The majority of Indian businessmen have since Independence followed the first route, because it is so much easier. Mr. Sumant Moolgaokar followed the second route, and took his company to the threshold of world class. Under his guidance, TELCO developed a well-deserved reputation for investing in men as much as in machines, in nurturing and developing creativity instead of

simply buying technology from abroad, in making products appropriate for India and yet good enough for international markets....[3]

The secret of Moolgaokar was revealed by his close colleague (once Deputy Chairman of TELCO) N.A. Palkhivala, who said at a condolence meeting on 10 July 1989 that Moolgaokar 'was not building a factory, he was not building trucks, he was building a nation.'

Moolgaokar often said: 'I want Tatas to belong to the nation,' meaning they should think for and make their decisions in the context of the nation's needs.

Towards the conclusion of a memorial address on Sumant Moolgaokar, J.R.D. said, 'To me Tatas and India are in many ways one and the same, and one of their purposes and the objective is to build India.'

CHAPTER IX

An Empire Expands

The decade following J.R.D.'s appointment as Chairman in 1938, was the most creative for the Group. On 1 January 1939 Tata Chemicals was launched. In the early 1940s he had a thought to bring together leading men of industry to plan for India's industrial growth after the war. It resulted in the Bombay Plan 1943–44. In 1945 TELCO was born and in 1948 Air-India International spread its wings abroad.

Tata Chemicals was born beside the sea in the dry and rugged area of Okhamandal, Saurashtra, on the western tip of India beside the Arabian Sea. At Okhamandal only cactii and *babul* trees brave the dry weather. Water is scarce and herds of camels roam wild, while other camels pull carts. A tough breed of people, called Waghirs, inhabit the area. Tall, they usually wear white turbans and sport fierce moustaches. In the past, quite a few of them used to make their living robbing pilgrims journeying to and from the holy city of Dwarka.

The Gaekwad of Baroda, Sayajirao, invited Tatas to take over a small salt works in this somewhat inhospitable region, perhaps foreseeing the potential of the little enterprise then being run, not too successfully, by an engineer called Kapilram Vakil. As Tatas had pioneered steel and electric power, the Maharaja thought that they could probably do the same with an inorganic chemical industry especially given the fact that this far India had failed to produce soda ash. On its production depended India's self-reliance in the manufacture of glass, ceramics, textiles and a host of other industries.

Tatas had the opportunity to start something new for India. The manufacture of soda ash was then a closely-guarded secret—only six companies in the world knew the method. The formation of soda ash required eighteen boilers each operating at a different temperature, and the companies which held the monopoly deliberately had false gauges on show to confuse visitors.

Tatas knew there would be hurdles to the successful start up of the company but they had no idea how many there would be. 'Of all the Companies with which I have been concerned, none has had to overcome so many difficulties compounded with bad luck, as has been the lot of Tata Chemicals,' says J.R.D. He decided to go ahead anyway and Tata Chemicals started in January 1939.

The salt works at Mithapur around which the complex was to grow, had hardly any roads. A railway line to Okha port passed nearby but there was neither a railway station nor a stop. The train-drivers were, therefore, inveigled to halt the train by a company man waving a tea kettle on the single track. In the semi-desert a cup of tea was always welcome and as the driver sipped the piping hot beverage, passengers tumbled out or struggled up the steps of the compartment.

The local population was none too obliging either. At a speech to the Duke of Edinburgh Conference at Oxford in 1956, J.R.D. recalled: 'When we built (the Mithapur) plant, land had to be acquired largely from the local Waghirs. Though this land was very unproductive and very adequate compensation was paid, when it was acquired compulsorily by Government, it naturally created some disturbances. I remember the case of an old gentleman who felt that the acquisition of his property was nothing less than an insult to himself and his ancestors which could only be washed away in blood.

'Now, apparently in order to reward the peaceful ones, the old British Government of India used to grant arm licences to some of them whom they thought loyal and trustworthy. This gentleman was one of them and toted a formidable muzzle-loading rifle. As the General Manager of the factory was the representative of the institution that had acquired his land, he considered that his blood was the blood with which the insult was to be atoned, so he began to stalk him, until the Manager who normally wasn't easily troubled, went to the police and said this was becoming a nuisance. The police asked: "What can we do until he shoots you? Come again when that happens!" The General Manager solved the problem by employing the old gentleman as his personal bodyguard! After that there was no more trouble, at least for the Manager. Instead a few other characters in the town found themselves in danger and I seem to remember that ultimately the old gentleman had to be moved to a somewhat less lethal occupation.'

As in Jamshedpur, Tatas had again to create a township, this time in a barren area. As the town was being built, machinery, including a boiler, and a generator were ordered for the factory and the township. To complicate matters the Second World War soon broke out and the first consignment of goods bound for the factory was torpedoed. Tatas ordered a fresh consignment from a neutral country, Sweden, in 1941. As the shipment left Sweden on its journey via the Soviet Union, Germany invaded the Soviet Union.

At this time Hitler's Panzer divisions had cut through the Ukraine and were pressing on towards Moscow. The Russians fell back before the Germans who were soon at the gates of Stalingrad. Thinking that the second consignment was also lost, J.R.D. quickly ordered a new one from America.

Sometime later, company officials heard that their Swedish consignment was passing through the railway yards of Moscow *en route* to the Caspian Sea. After that there was no word again for a while until, one day, Tatas got a cable saying that the consignment had arrived in the Persian Gulf through Iran.

One of J.R.D.'s activities at that time was making speeches for the war effort. 'I remember saying at one of these speeches that a Government, a country (the USSR) that is supposed to be flat on its back and already defeated but in which the railways worked so well for ordinary commercial traffic to go through from North to South, could not be in that shape.' Many then believed that Hitler would win, but the dogged onward progress that the Tata Chemicals' shipment made gave J.R.D. the deep conviction that Hitler, then at the height of his success, would one day be defeated. That conviction shaped his thinking and planning for the remaining period of the war and for the post-war period.

When both the Swedish and the American shipment arrived it was found that the know-how available in India was totally inadequate. The quality of the soda ash was not right nor was the quantity produced optimum. Finally, after the war, Dr Ho, an expert from America, was called out. He told Tatas that they were in the wrong place and in the wrong industry. J.R.D. rather defiantly said that Tatas specialized in that: gloom enveloped the senior management at Dr Ho's prognosis.

In those days J.R.D. used to fly his own small plane to Mithapur. One of these visits was to prove decisive. At the plant he met an enthusiastic young chemical engineer. Words cascaded down his lips as he mentioned his dream of doubling production. 'I did not understand very much of what he said for I was not trained as a chemist or as an engineer.' But, J.R.D. says, he perceived that the young man might be just the person to get the plant moving.

He found, on inquiry, that the engineer called Darbari Seth had excellent academic qualifications and had worked with Dow Chemicals in the United States. Seth had helped to design the latest soda ash plant in Holland. But at Mithapur he was relatively low in the hierarchy and his plans were brushed aside by his seniors. J.R.D. asked the top men to put Seth in charge of the soda ash plant.

'Did you put him over the head of the General Manager?'

'No. I didn't but I forced him (the GM) to give Darbari a clear straight hand to do everything he wanted. He nearly demolished the place but things changed.'

Seth had claimed that he could raise the production from 200 to 400 tonnes without the German assistance the management was contemplating. Despite the Cassandras, of which there were legion, he proved he could. Working with a team of young engineers through the day and late into the night he doubled the plant capacity to 400 tonnes and within a few months even touched 500 tonnes on one occasion. Mithapur—the City of Salt—was secured for the future. In the years to come another challenge—this time to do with drought—was also faced by Seth and his men![1]

In 1970, J.R.D. made Seth the Managing Director, and in 1982 in keeping with his policy of stepping down in favour of those who made a significant contribution to the growth of the company they were recognized for, he handed over the Chairmanship of Tata Chemicals to Darbari Seth.

Seth continued to lead the onrush of Tata Chemicals. The company was the promoter of Tata Fertilisers, which was later merged with Tata Chemicals. And, in 1990, Seth scored his biggest triumph—he succeeded in getting a contract for Tatas for a Rs 3,000 crore petro-chemical project at Haldia in West Bengal—the largest of its kind awarded to private industry till then.

Tata Chemicals was just one company. There were to be several others. Some were started afresh; others were taken over and expanded. As in life, so in business there is a time to start new things and a time to consolidate earlier gains. The 1950s (as we shall see in the next chapter) were a difficult and decisive time for India. The first firm steps were taken in State control over enterprises, the airlines were nationalized in 1953, and insurance in 1956, both measures plucking bright jewels from the Tata crown. For the first time permits and licences came into operation. Despite such constraints in the 1950s, Tatas increased the production of their steel plant from one to two million tonnes and TELCO launched out into collaboration with Daimler-Benz.

In 1954, Tatas took over the engineering side of Volkart Bros., a highly respected Swiss company which had begun trading in India exactly a hundred years earlier. The engineering side was small, mainly air-conditioners and some agencies, but very soon Tatas expanded it into a large-sized engineering complex that could air-condition a frigate and manufacture machine tools, plastic mouldings and dyes.

In 1962 Tatas entered the tea business jointly with James Finlay and set up a tea blending and packaging company. Fourteen years later, when the British management of James Finlay pulled out of India, Tatas took over their interests—fifty-three tea estates—in eastern and southern India, making Tata Tea one of the largest tea companies in the world.

In the 1960s, the government made concerted efforts to control big industrial houses, without taking into consideration their contribution in the progress of the nation. The Constitution of India had clearly laid down (Article 39C) that there should be no concentration of economic power "to the common detriment". The government did not ask whether or not the holding and the expansion of Tatas was "to the common detriment" or "to the common good". Also, the year 1966, when Indira Gandhi was Prime Minister, witnessed a devaluation of the rupee. Tatas sensed that they would have to branch out into fields where the government would be unable to create obstacles and also figure out ways to survive in the depressed economic climate.

The two major assets of Tatas were, first, their holdings in reputed companies and, second, the skills they had developed over the decades. In 1962, Tatas launched their first consultancy service—in this project the Tata Electric Companies worked along side Ebasco in America. In 1968 it came to be known as Tata Consulting Engineers and provided its services to a broad spectrum of industries in the fields of power, chemicals, fertilizers and engineering.

J.R.D. was quick to recognize the importance of computers, perhaps because of his association with Dr Homi Bhabha, who built the first computer in India at the Tata Institute of Fundamental Research (TIFR). The first computers were enormous in size and occupied a whole room.

The first four Directors of Tata Sons: (from left) Jamsetji Tata, R.D. Tata, Dorab Tata and Ratan Tata (standing)

J.R.D. on his appointment as Chairman of Tatas in 1938

With Jawaharlal Nehru

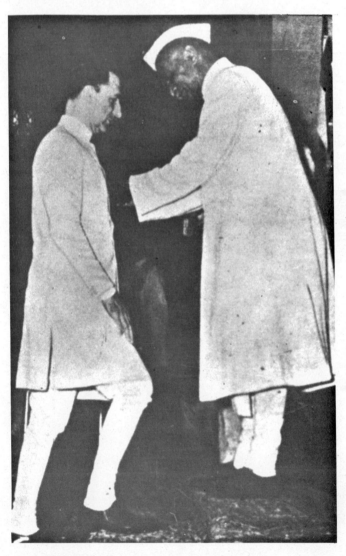

Receiving the Padma Vibhushan from President
Rajendra Prasad in 1955

Wearing the Knight Commander's Cross with the
Badge and Star of the German Order of Merit

Accepting the honorary degree of Doctor of Law from
Bombay University

The Henry Bessemer Gold Medal awarded to J.R.D.
by the Institute of Metals, London, in recognition of
his contribution to the iron and steel industry

Receiving the Tony Jannus award in recognition of his
contribution to the development of scheduled
commercial aviation

Cardinal Gracias confers the honour of Papal Knight
on J.R.D.

At the Founder's Day celebrations
in Jamshedpur

Thelly Tata garlanding V.G. Gadgil, one of the first
four to join J.R.D. in Tata Air Lines

At an Air-India reception: (from left) Bobby Kooka,
Thelly Tata and J.R.D.

J.R.D. at Jamshedpur

J.R.D. with Indira Gandhi and Rajiv Gandhi

In the Tata Boardroom after he stepped down as
Chairman of Tata Steel. On his right are S.A. Sabavala,
Russi Mody and Nani Palkhivala

J.R.D. with his Bombay colleagues in 1982: (from left)
Darbari Seth, Ratan Tata, Ajit Kerkar, S.A. Sabavala,
J.J. Bhabha, S. Moolgaokar, Nani Palkhivala and Naval
Tata

A letter from Mahatma Gandhi

A letter from Indira Gandhi

My dear Jehangir,

Within the next few days you will get an invitation which probably you will not be able to read as it will be in Hindi and Urdu. So I hasten to tell you that this is about Indira's marriage which is fixed for March 26th in Allahabad. It is perhaps too much to expect a very busy captain of industry in these strenuous times to attend to such functions. But if Jeanne & you and Thelma could come we would be happy.

Yours Sincerely

Jawaharlal Nehru

A letter from Jawaharlal Nehru

Seated in the chair of Jamsetji Tata, with a portrait of
the founder of Tatas next to him

In the Boardroom. Behind him is a bust of Jamsetji Tata

By the late 1960s, a decade before the rest of India became aware of computers, Tata Sons opened a software division, the Tata Consultancy Services. It began with a dozen experts and grew to 2,000 high calibre professionals constituting a think-tank unrivalled in this country, giving its services to several highly developed countries worldwide. Later, Tatas got permission for a link-up with the Burroughs Corporation of America, then the second largest in the computer field, to start Tata Burroughs, now Tata Unisys.

A company started in the 1980s, whose performance impresses J.R.D., is Titan Watches. A sizeable part of the nine million watches sold annually in India were smuggled into the country. Watches with renowned names from Japan and Switzerland were popular and there was an initial scepticism whether the Tata project would succeed—especially as HMT (a government firm) was already producing good quality watches.

In association with the Tamil Nadu Government and in collaboration with France's Ebauches and Citizen Watch Company, Japan, Tatas launched Titan Watches. The quality and the marketing techniques of Titan Watches was excellent and gradually Titan edged out smuggled watches.

The one business the government encouraged was exports. In the 1960s Tatas began their own export trading house, Tata Exports, presenting India as an industrially reliable country. The firm has not only exported Indian products but given its superb after-sales service; country after country in Africa has turned to Tata Exports for development projects.

With the liberalization policy instituted by Rajiv Gandhi in 1985, there was a considerable expansion of Tatas. Between 1985 and 1989 fourteen new companies came to be born. When J.R.D. took over as Chairman, the entire group consisted of fourteen companies though most of them were formidable companies even at the start. In spite of all the restrictions placed by the government on the expansion of big firms, when he laid down office fifty-two years later, there were ninety-five companies. Their turnover had risen from seventeen crore rupees to an estimated Rs 10,000 crore.

The *Economic Times* of 27 December 1990 listing the top companies of India mentions three Tata companies in the top ten, assets-wise. They rank Tata Steel as No. 1, TELCO as No. 4 and Tata Chemicals as No. 10.

CHAPTER X

The Decade Of The 1950s

International recognition came to J.R.D. at an early age. A sketch of J.R.D. was included in the book, *The 100 Most Important People in the World Today* published by Pocket Books in 1952. The *Sunday Times*, London, featured him in its Portrait Gallery in 1951. The article gives a glimpse of what he was like in his mid-forties:

As an employer, he in humane and liberal. Some say almost socialistic. Unspoilt by success, easily approachable, sympathetic in manner, he is admired and respected by labour leaders, who have often appealed to him to intervene in disputes between management and unions. Such personal popularity as he enjoys could hardly be paralleled in an organisation employing, as his does, over 1,20,000 workers.

Highly strung, but tenacious of purpose, he works quickly and intensively. His passion for accuracy and interest in detail are often disturbing to his colleagues and to his health. He does not disguise his contempt for superficial knowledge and glib opinions. A slide-rule is his constant companion. He will never put his signature to a document of any importance without personally revising the draft.

Outside the office, J.R.D. Tata's chief recreation is golf, swimming, skiing and books. Amid his wide reading he has a deep love of poetry, both English and French. A life-long and devoted nationalist he is a trustee of the Mahatma Gandhi National Memorial Fund and a member of public bodies

258

concerned with industrial planning and research. If it be true, as has been said that "a great society is one in which men of business think greatly of their function" J.R.D. Tata is surely helping to make India a great society.[1]

In the years that followed independence the government did try to use J.R.D.'s talent, but this phase was not to last as we shall see.

*

The decade of the 1950s was to prove decisive for India. The country was settling down after the upheaval of Partition. Initially, the government in New Delhi was in a cooperative mood. J.R.D.'s proposal for Air-India International was cleared speedily in March 1948. On 22 June 1950 Harekrishna Mahtab, Union Minister for Industry, requested J.R.D.'s help to tackle problems in the government's industrial establishments. He wrote: 'The other day when we met I mentioned very briefly some points to you and sought your advice. The first point is what, according to you, should be the form of management of the industries which have already been started by Government. Statutory corporations perhaps have to be set up. In that case, I would like to know a few names from you who will constitute the corporations. Will you be able to spare some of your trained personnel to work on the corporation or on the Board of management in addition to their work in your industries?

'The next point is how to set up some basic industries such as steel. As you know, both on account of finance and personnel, it will not be possible to set up the industries in the coming few years. Will it not be possible to start these industries on the basis of cooperation between Government and private capital? Will you come forward and join the steel industries which are in my view at the present moment? Without waiting for the Government policy in these matters I shall be much obliged if you kindly let me know on what basis you can join so that I may turn it in my mind and find out if it fits in with the Government policy or not. I hope you will not please mind going a little out of the way and suggest to me if any other industrialist can be approached in this connection. I would most earnestly request you to kindly consider these problems as if you are in the Government and you have to solve these things. I take the opportunity of approaching you in this spirit because you have allowed me to do so.'

In reply to Mahtab, J.R.D. sent a detailed note on the structure and form of management needed for government's industrial establishments. Then, in 1951, the Defence Minister Baldev Singh offered J.R.D. the Chairmanship of Hindustan Aircraft Limited (HAL), Bangalore—on whose Board J.R.D. had been for some years. Regretting his inability to take on the Chairmanship, J.R.D. spelt out his reasons on 16 August: 'First, I do not think it is a sound principle for the head of what is, to all intents and purposes, a defence industry to be an active businessman or industrialist.

'Second, as head of Air-India, a Company which has dealings with HAL, it is not desirable that the two Companies should have a common Chairman.

'Third, I think it is important that the Chairman of the Company should, as far as possible, be a resident of Bangalore or within easy reach of the management. For the Chairman to be in Bombay is hardly better than for him to be in Delhi.

'Fourth, I am frankly not satisfied that there aren't perfectly satisfactory alternatives to my appointment. I have previously recommended that Sir R. Mudaliar should be appointed Chairman of the Company...'

In 1952, Girijashankar Bajpai, former Secretary-General in the External Affairs ministry and Governor of Bombay, probed J.R.D. about joining the Indian delegation to the United Nations. J.R.D. declined the invitation to the UN for the second time.

By the mid-1950s, everything had changed, and J.R.D. would no longer be viewed with as much favour by the government. In December 1954, the parliament accepted the socialistic pattern of society as the objective of its social and economic policy. In pursuance of this decision an Industrial Policy Resolution of 1956 was presented to the nation. According to the resolution, industry was to be divided into three categories:

- Those which are the exclusive responsibility of the State like air transport, railways, arms and ammunitions and such like.
- Those to be progressively State owned.
- Those left to private industry to develop.

Meanwhile, in January 1956, by an ordinance, life insurance was nationalized overnight and the New India Assurance Company, which Tatas had nurtured and developed since 1919, was taken over. Later, in 1973, General Insurance was nationalized. More than three decades later, in 1990, J.R.D. was invited to a luncheon at the nationalized New India Assurance Company. He spoke about how upset he was with the nationalization of New India. He called it "a fraud". He recalled that when General Insurance was nationalized in 1973, he went before the Committee of Parliament with J.D. Choksi and N.A. Palkhivala, his legal advisers. 'I pointed out to them that they were taking over New India at less than the cash value of its assets and thereby in effect depreciating the currency.' In other words, he said, they were reducing the value of ten rupees to nine rupees.

The Parliament Committee replied, 'It is for the good of the people.'

'Then why do you pay anything if it's for the good of the people?' he replied.

By the mid-1950s, J.R.D. was both hurt and disillusioned. Three companies that Tatas had nurtured—the domestic airline, the international airline and New India were nationalized within four years of each other. Was it worth taking the trouble and showing enterprise when the fruits of it were snatched away with "confiscatory compensation?"

In 1953, when nationalization of the airlines was taking place, Dinshaw Daji, Tatas' old solicitor, brought to J.R.D.'s notice Gandhiji's words on the subject:

I look upon an increase in the power of the State with the greatest fear, because, although while apparently doing good by minimizing exploitation, it does the greatest harm to mankind by destroying individuality which lies at the root of all progress.

The State represents violence in a concentrated and organised form. The individual has a soul, but as the State is a soulless machine, it can never be weaned from violence to which it owes its very existence.

It is my firm conviction that if the State suppressed capitalism by violence, it will be caught in the coils of violence itself and fail to develop non-violence at any time.

261

What I would personally prefer, would be not a centralization of power in the hands of the State but an extension of the sense of trusteeship, as in my opinion, the violence of private ownership is less injurious than the violence of the State. However, if it is unavoidable I would support a minimum of State ownership.

What I disapprove of is an organisation based on force which a State is. Voluntary organisation there must be.[2]

J.R.D. wrote back to Daji: 'As usual Gandhiji saw things with a deep intuition. In the almost universal cry for nationalisation in this country, no one seems to think of the potential danger of concentrating enormous economic power into the hands of a small political-cum-bureaucratic minority.'[3]

J.R.D.'s fears were well founded, for soon enough, the concept of the government taking control of the "commanding heights of economy", as Pandit Nehru put it, made politicians and bureaucrats think they were a superior breed who could dictate to private enterprise. This superiority they extended even to heads of so-called Autonomous Government Corporations.

Speaking to Tata Steel shareholders in August 1956, J.R.D. observed: 'Faced with an unprecedented array of political, economic and social problems, it (the Government) has borne with ability, devotion and success an appalling burden of responsibility. We also realise that its task in the coming years will remain a formidable one; that the success of India's plans of economic development is essential to the future prosperity and happiness of our people. Where we differ is in the respective roles of the State and of the individual in the economic activity of the nation. We sincerely feel that some of Government's policies are based on economic theories of capitalism and socialism which no longer fit the facts, if ever they did.'

Even so he was not against the government starting the public sector. What he was against was the monopoly of the public sector which he foresaw would result in inefficiency and state capitalism. When the first steel plant was being put up and the government was thinking of a half-a-million tonne capacity, J.R.D. told the government that they were thinking too small.

The problems with the "political-cum-bureaucratic minority", whose passion was not the creation of wealth, but its control, continued as the years went by. The Finance Minister, Morarji Desai, delayed Tata Steel's application for a new capital issue, insisting that Tatas should charge a premium on the new issue, which Tatas wanted to bring out at par. The hundred rupee share of Tata Steel quoted at Rs 200 had come down to Rs 170. The dividend had come down from 8.25 per cent to 6.80 per cent. In J.R.D.'s view, the premium the Finance Minister wanted on the capital issue was not justified. The Minister would not budge. In a letter to Desai on 23 July 1959 J.R.D. noted: 'I hope you will not mind my expressing some surprise that the Government should take such detailed interest and intervene in a matter which involved no important question of policy or principle but only commercial and financial judgement on a simple business issue. Would it be wrong of me to regret the passing of the days when businessmen with a good record of efficiency and integrity could be trusted to make such decisions themselves?'

There was more trouble to come. As we've seen, in his time with Air-India, J.R.D. had several run-ins with the apparatus of the State. A bureaucrat, M.M. Philips, then Secretary of the Transport and Communications Ministry, thought himself important enough to totally by-pass J.R.D., the Chairman of Air-India, and issue a gazetted notification appointing as a member of the Board of Air-India, Air Commodore (later Air Chief Marshal) P.C. Lal.

J.R.D. wrote on 25 January 1958: 'I am all the more surprised at the Government's decision as I find that, while the General Manager of I.A.C. (Air Commodore P.C. Lal) has been appointed to the Board of A.-I. I., the General Manager of A.-I. I. has not been simultaneously appointed a member of I.A.C. I trust that immediate action will be taken to remedy the anomalous and discriminatory situation. I need hardly add that my objection to the appointment of Air Commodore Lal in the circumstances in which it has been made has nothing to do with him personally. As a man and as one for whom I have the greatest regard, I welcome him unreservedly as a member of the A.-I. I. Board.'

Philips' reply did not meet J.R.D.'s objections and in a letter dated 9 February 1958, J.R.D. went over him to the Minister for Civil Aviation, Professor Humayun Kabir. J.R.D. wrote: 'While I certainly do not wish to be troublesome in this matter, I feel there is an important question of principle involved and I am quite frankly very reluctant to accept the continuation of the present situation.'

Professor Kabir replied on 26 February stating it was the government's exclusive responsibility to appoint members of the Board under the relevant Act. To this J.R.D. shot back: 'I entirely agree but I have myself in my letter to you of the 19th referred to Government's right to make such appointments to the Board as they please.

'I, however, stand by my view that Government should recognise the accepted convention under which the Chairman of a Board is extended the courtesy of being consulted before Government make a new appointment to a Board.' J.R.D. went on to state: 'If the Government wants non-officials, such as myself, to undertake the task and responsibility of being the Chairman of a Government Corporation, they must, in fairness, treat them as any head of an organisation expects to the treated. I do not see what Government can lose by sounding a Corporation's Chairman before making an appointment when the final decision is wholly and indisputably theirs.'

A bureaucrat is not to be treated lightly.

Within a month, Philips sharpened his bureaucratic scissors and tried to clip the Chairman's powers. He wrote to J.R.D. on 25 March, that every case of promotion in Air-India involving a salary of Rs 1,500 or more, 'should require the prior approval of the Government,' (namely of the Secretary of the Ministry himself!)

J.R.D. replied that this would 'amount to a declaration of no-confidence in the Board and Executive Management of the Corporations. As Chairman of Air-India International, I consider this proposal totally unacceptable.' He pointed out to the bureaucrat what he had told the Minister in an earlier letter: 'I shall always be willing to serve but not to waste my time. I feel that with the growth of the State Sector in industry, it is time that Government made up their minds as to whether they want State Companies or Corporations operated like Government departments or as commercial concerns. If the former, there is no room or need for men like myself. The Ministries concerned should take them over and run them departmentally. If Government want them run as commercial concerns, they should pick the best men they can get and let them get on with the job, subject only to general policy control.'

Because he fought these battles early on and did not succumb to the control of the bureaucrats, he was given a reasonably free hand in later years in Air-India—but even so he had to stand up every time the bureaucrats indulged in their passion for control. In spite of these provocations he felt keenly the injustice of keeping the bureaucrats on a shoe-string salary. He was particularly worried when a bureaucrat asked an old college friend of his in Tatas whether he could arrange to have his car regularly tanked up because he could not afford it otherwise. Decades later, J.R.D. was still fighting the bureaucrats' cause. He went to Indira Gandhi, the Prime Minister, and presented the case of the bureaucrats quite forcefully in the early 1980s. He pointed out to her that bureaucrats were actually getting less and less each year and this was a ridiculous situation. She did heed him and took some corrective action though it was not adequate in terms of the cost of living.

*

Looking back over that time, J.R.D. says he hoped Air-India would be the first of the joint sector enterprises between government and the private sector. Unfortunately, the great creative energies of a person like J.R.D. that could have gone into the creation of the likes of Air-India, were often frittered away in fighting the political-cum-bureaucratic machine, which like an octopus was getting hold of India. Even so there was some cause for cheer.

In this period T.T.Krishnamachari cleared speedily both the expansion of Tata Steel from one to two tonnes and for TELCO to go in for truck manufacture.

Meanwhile, at home J.R.D. was finding pleasure, and perhaps some solace, in his workshop.

The workshop, an area of 250 square feet, has rows of shelves, with instruments arranged on them as neatly as in a dental clinic. Once, when J.R.D. was in a wheelchair, with a hairline leg fracture, and wanted a particular spanner he directed a friend thus: 'Second shelf, third item.' He knew where it was, much to his friend's amazement.

One of his American friends had a son, Rufus Burton, called ("B. Boy"). B. Boy wanted a nose wheel to fit to his toy Super-Constellation and the order was sent to J.R.D. J.R.D.'s reply was prompt:

J.R.D.T. Home Workshop Incorporated:

14th December 1957   The Cairn,
                                                     Altamount Road,
                                                     Bombay 26

We have today despatched to you by airmail the nose wheel ordered by you from the undersigned for your Air-India Super-Constellation Model Airplane. If the nose wheel does not fit properly, we suggest one of two alternative solutions; either you throw away both the airplane and the nose wheel or you send the nose wheel back to us with the airplane for adjusting and fitting here. We are glad to inform you that we can guarantee delivery in approximately twenty-four months.

As the undersigned had the good fortune of taking Mr. Burton Senior for a ride in a game of golf, as a result of which he disbursed more than the value of the nose wheel, we have decided to make no charge in this instance.

Faithfully yours
J.R.D.T. Home Workshop Inc.

(J.R.D. Tata)
President

The fourteen-year-old boy wrote back, 'Dear Uncle Jeh, the wheel looks nice on the aeroplane. Thank you for putting springs in the wheel. I think you are very clever to make it.'

CHAPTER XI

From Empire To Commonwealth

During the 1960s, India was ruled by three Prime Ministers and she participated in two wars. It was a decade when a giant was cut down to human proportions and a diminutive-sized man was projected as a man of great stature. The Chinese invasion of 1962 reduced Jawaharlal Nehru's stature, while the Indo-Pakistan war of 1965 magnified Lal Bahadur Shastri's.

The edifice of Nehru's foreign policy was built on peaceful co-existence. A pillar of that edifice was friendship with China. A crack appeared in that edifice the day China marched into Tibet. The crack became a chasm when the Dalai Lama stepped across the border into India in March 1959. From that moment onwards the Indian people refused to accept the validity of Nehru's policy on China. One event led to another, but no one expected China to strike across India's north-eastern frontier. V.K. Krishna Menon, India's pro-Communist Defence Minister, had neglected to properly equip or even clothe the Indian Army for such an emergency. Unsurprisingly, China cut through India's defences up to Bomdila and then withdrew as suddenly as she had arrived. The war left India shaken and Nehru shattered. Nehru vehemently resisted efforts to oust Krishna Menon, but finally bowed to the public outcry and appointed Y.B. Chavan Defence Minister and General J.N. Chaudhuri the Army's Chief of Staff.

When China invaded, J.R.D. promptly offered the resources of the Tata organization to the government. The offer wasn't immediately taken up but the new Defence Minister requested J.R.D. for a confidential report on India's aviation needs in the long term. The report of the Tata Committee covered "Aircraft and Ancillary Electronic Equipment Required by the Defence Services", and was compiled by J.R.D. along with the Defence Secretary, the Foreign Secretary, the three Army Chiefs of Staff, and Dr Homi Bhabha, Secretary of the Department of Atomic Energy. It made detailed recommendations on the equipment and expansion of the Indian Air Force for the next ten years including fighter bomber squadrons, transport, guided missiles, ground radars and electronic equipment. J.R.D.'s wide experience with Air-India meant that the brunt of the work fell on him. As Chairman of the airline he was in a position to make a comparison of the methods of organization, management and control of stock keeping between Air-India and the Indian Air Force. As recently as 1986, Air Chief Marshal La Fontaine, head of the Indian Air Force, said: 'This report of Mr. Tata's was compulsory reading for all Chiefs of the Indian Air Force.'

On 4 October 1966, J.R.D. received a telegram from the President of India to say that:

IN RECOGNITION OF YOUR MERITORIOUS SERVICES IN THE FIELD OF AVIATION FOR MANY YEARS I HAVE GREAT PLEASURE IN APPOINTING YOU TO THE HONORARY RANK OF AIR COMMODORE IN THE AIR FORCE STOP PLEASE ACCEPT MY HEARTY CONGRATULATIONS

S. RADHAKRISHNAN

Air Chief Marshal Arjan Singh, D.F.C., wrote to say, 'I am sure that your elevation to the rank of Air Commodore, (the rank which Sir Winston Churchill held) will be appreciated by all enthusiasts of aviation.'[1] J.R.D. replied: 'While I do not suffer from excessive modesty, I must confess I was a little embarrassed by this news as I do admit to a considerable difference between my stature and that of Winston Churchill who had the same rank in the R.A.F.!' Observing that he had not done enough for the Air Force to deserve the honour, except perhaps working on the report on Air Force requirements, he went on to say 'Considering that practically none of our recommendations have been implemented by Government, the value of even that service is doubtful!'[2]

Three years after the Indo-China war came the war with Pakistan. This time J.R.D. offered the help of the country's industrialists to the war effort. Subsequent to his meeting with Prime Minister Lal Bahadur Shastri on 7 October 1965, J.R.D. was advised by him to meet the Army Chief of Staff and others. J.R.D. wrote to the Prime Minister, 'A group of senior engineers from Jamshedpur are coming to Delhi this week for more detailed discussions, from which I hope will emerge concrete action which will materially help the war effort. You will remember that in the course of my discussion with you, I urged that you should take full advantage of the unity of purpose and the urge to serve the national effort which the recent conflict with Pakistan and the continuing emergency has created amongst all sections of the people of our country and particularly in industry. I also urged that with this object in view, Government should consider lightening the burden of controls on industry and postponing or abandoning some proposed measures which would serve no essential purposes in the context of the present emergency.'

In reply, Lal Bahadur Shastri wrote on 3 November 1965: 'We shall certainly give careful thought to the suggestion you have made about lightening the burden of controls in some sectors of industry.'

Shastri did not live long enough to undo the havoc wrought by the controls for within ten weeks of writing to J.R.D., after signing the accord with Pakistan at Tashkent, he died. It was to take the Union Government twenty-five years to implement J.R.D.'s recommendation, learning from experience rather than from advice.

In January 1966, Indira Gandhi became Prime Minister with the support of powerful Congress party bosses who were against her main adversary, Morarji Desai. Morarji Desai was known to be inflexible and fixed in his views. The party bosses—and others—thought that Indira Gandhi would be a pliable person. At that time Indira's aunt, Vijayalakshmi Pandit, told J.R.D. that Indira "has a mind of her own". The Congress bosses were to learn this to their cost.

In the election campaign of 1967 the crowds came to cheer the good-looking daughter of Jawaharlal Nehru and when the results came in she had scraped through to remain Prime Minister but with the support of the leftist parties. A number of states were lost to the Congress. Indira Gandhi decided, then, to cultivate popular support through populist measures such as the abolishing of the privy purses of princes. The measures worked and she was able to successfully emasculate her old supporters, the Congress bosses, and to beef up her own party. Some time later she nationalized the nation's banks, though she had no real belief in socialism.

Following on from the socialist stance she had chosen to adopt, she accelerated the campaign against big business that was set in motion under her father's rule. Far from "lightening the burden of controls" that Shastri had spoken of, the constraints grew. Between 1964 and 1969 four committees or commissions were constituted to examine one aspect or the other of the economic power in hands of large private houses, although J.R.D. repeatedly pointed out that the real danger was concentration of power in the hands of the State—the ministers and the bureaucrats. The committees were:

- The Mahalanobis Committee, 1964.
- The Monopolies Inquiry Commission, 1965.
- The R.K. Hazari Committee, 1966.
- The Industrial Licensing Policy Committee or the Dutt Committee, 1969.

As a result of the committees' deliberations the Monopolies and Restrictive Trade Policies (MRTP) Act of 1969 was born. It brought under its purview all undertakings whose assets singly or interconnectedly were about Rs 200 million (raised to Rs 1,000 million in 1985 under Rajiv Gandhi's policy of liberalization. The assets limit and all the pre-entry restrictions were removed in September 1991 by the P.V. Narasimha Rao government). Dominant undertakings were defined as those which had more than one-third of the market share lowered to one-fourth in 1984. Both these kinds of undertakings would henceforth have to take the permission of the Central Government before substantial expansion, establishing new undertakings, merger or amalgamation, and acquisition or takeover.

In a memorandum submitted to the Government of India on 17 May 1972, J.R.D. Tata said: 'It is strange that while so much has been talked so often about "concentration of economic power", no attempt has been made to define what precisely is meant by it, either in terms of its actual characteristics or of the uses to which it has been put.'

At the Annual General Meeting of Tata Steel, held on 24 August 1972, he was to say: 'I believe there has been a real confusion of thought in regard to the true nature and extent of the economic power about which so much fear and suspicion, genuine or politically convenient, has been expressed....Deprived of the right to decide what and how much to produce, what prices to charge, how much to borrow, what shares to issue and at what price, what wages and bonus to pay, what executives to employ and what salaries to pay them and in some cases, what dividends to distribute, directors and top management from the Chairman down have hardly any economic power in our country. Taking my own case, I doubt that there is anywhere in the world outside India any industrial executive in charge of a major enterprise with less real power than I have....

'While Government's restrictive measures have been directed mainly at curbing or destroying managerial freedom, initiative and policy-making powers, thus nullifying the principal advantages of private enterprise management, the real and justifiable target of criticism was, or should have been, the illegitimate accumulation of wealth, particularly in the form of tax-evaded funds in India and abroad and its use for anti-social purposes, including bribery and corruption. I submit that the right policy for a socialist Government, such as ours, armed with the immense powers that it has, should be, on the one hand, to take every possible step to curb the acquisition and anti-social use of illegitimate funds and to punish those guilty and, on the other, to release for nation-building purposes all the resources, energies and dedication of those in the private sector whose past record has shown them to be worthy of the nation's trust.'

Unfortunately, the Central Government appeared not to have the least intention to sort out the sheep from the goats. The lowest common denominator would be taken as the mean and an ideology, that was to be disowned in the land of its birth, would continue to hold sway. The axe fell heavily on 2 April 1970. 'The largely mythical bogey of concentration of economic power in private hands which protagonists of the left had so assiduously propagated for years,' as J.R.D. called it, also resulted in the termination of the Managing Agency System.

The system had begun when British financial interests controlled, from a distance, companies based in India. These financiers appointed agents to manage for them. In the 1850s when the first large scale industry—cotton textiles—was starting up in India, capital was scarce. Money was concentrated in the hands of traders and some wealthy merchants and there was no investing class and no industrial leadership. However, when the opportunity to promote industry on a wide scale arose, some of the merchants became promoters of industries and invested capital. Others followed suit and on their guarantee banks sanctioned loans. This was how most of the major players in Indian industry got their start.

Tata Sons, and later Tata Industries, were Managing Agents of all Tata companies. If a company did well they got a share in the profits. If it did badly, Tatas never took a percentage of turnover as some others did. Tatas, in fact, nursed their unprofitable companies (like Tata Chemicals) for years till they became viable, using the funds of their profitable companies to prop up an unprofitable one. A Managing Agent of repute went out of his way to help in times of crisis, as Sir Dorab Tata did, when he pledged his personal fortune and his wife's jewellery for a bank loan to pay the wages of Tata Steel employees.

'The Managing Agency system is liable to be abused for selfish purposes and was, in fact, so abused in some instances,' said J.R.D. 'But I do not know of any form of management, whether in the public or private sector, which is not open to dishonest practices or corruption if those at the helm are so inclined.'[3]

Until 1970 J.R.D., under the Managing Agency System, ruled an Empire carrying on his shoulders its burden and its glory. From 2 April 1970 the Empire became a Commonwealth. J.R.D. found himself in the role of the Head of a Commonwealth where he had to rule with persuasion and influence. His writ no longer ran over all Tata companies, for the Board of Directors of each company was independent of the parent company (though many or some on the Board are from Tatas and Tatas still manage them on behalf of the shareholders of each company.)

Not only for J.R.D., but for his officers trained by him to think for the whole of Tatas—the new development was a traumatic experience. The resident Director of Tata Industries in Delhi, who had one visiting card 'Tata Industries', got eighteen printed in the names of the different companies he represented.

In sum, when the Managing Agency System was abolished, an era ended. As the years passed, powerful Managing Directors stamped their own identities on the companies they ran. While J.R.D. is still alive there is a sense of unity continuing to knit the various companies together and, what one might call the Tata ethos, still holds sway—a Group identity that is greater than the sum of its many parts. This is so because when J.R.D. ruled his Empire he did so not as an autocrat but as a democrat who never imposed himself on his chief executives. His successor, Ratan Tata, inherits the task of holding the companies together, and time alone will tell to what extent he will be successful.

CHAPTER XII

The Indira Gandhi Era

From Anand Bhavan, Allahabad, on 4 March 1942, Jawaharlal Nehru wrote a handwritten letter: 'My dear Jehangir, within the next few days you will get an invitation which probably you will not be able to read as it will be in Hindi and Urdu. So I hasten to tell you that this is about Indira's marriage which is fixed for March 26th in Allahabad. It is perhaps too much to expect a very busy captain of industry in these strenuous times to attend to such functions. But of course if you and Thelly could come we would be happy.'

In his reply of 11 March 1942, thanking him for the invitation, J.R.D. said: 'It was nice of you to take the trouble of writing personally. Thelly and I would naturally love to take advantage of this opportunity of meeting you again and of congratulating, and giving in person, our best wishes to Indira. The difficulty, however, lies in the fact that it would take up three days in all. I am afraid I am not much of a "captain of industry", but I am certainly being kept as busy as if I were one!

'I am particularly heavily booked during that week. I hope, therefore, that you will forgive me if we don't manage to come and will not think that we don't want to. Apart from that I fear that you will have your hands so full with all the people who will be coming from all parts of India that we would be a nuisance to you. Unless, of course, you allowed us to stay at a hotel. In fact, if later on I found it was possible for me to get away from Bombay and, at the last moment, we flew over, it would be only if you promised to let us stay at a hotel: If we are unable to see you this time, I sincerely hope we shall see you in Bombay soon after.'

Twenty-four years later, the young girl J.R.D. knew became Prime Minister.

In the interim period they met occasionally when Indira Gandhi accompanied her father to the Juhu shack of J.R.D. or whenever J.R.D.

was invited to Nehru's home as a family guest.

Whenever possible J.R.D. tried to give his ideas to Nehru and Indira Gandhi but with little success. Nehru's technique for gently discouraging him was to look out from his window at his garden at Teen Murti Marg or to invite him to see his Chinese panda. Indira had another technique, equally gentle and impolite. When J.R.D. tried to interest her in a discussion on matters of economic policies, after a few minutes she would start doodling or opening mail envelopes. J.R.D. remembers on one occasion when Sharokh Sabavala was present he whispered to him, apparently loudly enough for her to hear him, 'She's not interested. We are boring her.' Mrs Gandhi looked up and said, 'No, No. I'm not bored. I'm listening, Jeh.'

Except for a brief period during the Emergency, Indira Gandhi ensured and did maintain a friendly equation with J.R.D. She wrote to J.R.D. once to say how thrilled she was with the perfume he had sent her. 'I don't normally use perfume and I'm so cut off from the "chic" world,' she said, adding, 'It was good to see you. Please do not hesitate to write or to come and see me when you want to convey any views—favourable or critical.'[1]

Indira Gandhi did see but did not heed J.R.D. on economic matters but she did listen to him on matters concerning Air-India and often included him as a link between herself and the leaders of France and some other countries. When any French dignitary came to Delhi J.R.D. was sure to be invited, whether it was President Giscard d'Estaing, Prime Minister Jacques Chirac, or the Minister for Culture, André Malraux, whom J.R.D. admired.

When Jacques Chirac was on his way back from China in 1976, he stopped in Delhi. At a reception hosted by Mrs Gandhi for him, which J.R.D. attended, Chirac asked him to look him up on his next visit to Paris. When J.R.D. took no action thinking Chirac had merely intended to express courtesy towards him he received a reminder from the Embassy and assuming that the subject the French Prime Minister wanted to discuss was probably the Government of India's ban on Concorde flights across India, he called on him in Paris. He found, however, that the latter's only concern seemed to be India's and China's mutual enmity which he deplored, considering the common historical importance, ancient civilization and large population of the two countries. He had apparently mentioned the matter to Mrs Gandhi but seemed to think that J.R.D. should speak to her in the then fairly common belief that he had an influential relationship with her. Jacques Chirac mentioned to J.R.D. that he had expressed the same view to Chou En-Lai, on a recent trip to China. The latter had replied that while he agreed with Chirac, the Indians had behaved very badly towards the Chinese, blaming them for the then brief military confrontation between the two countries in the Himalayas. When on his way back from France J.R.D. repeated this conversation to Indira Gandhi, she replied that she agreed with Chirac but the Chinese had behaved very badly with the Indians!

In the course of their meeting Chirac told J.R.D. an interesting and amusing story about China's presence in East Africa, where it had built an important railway for Mozambique. On being asked by Chirac why Communist China had decided on a presence in East Africa, and why, after having taken the trouble to do so China had totally pulled out from Africa, Chou En-Lai replied that China had gone to Africa because Mao Zedong felt that Communism should not be left exclusively to Soviet Russia to propagate. China pulled out for three reasons: first, they found that while they created no revolutionary impact, the expenditure on maintaining their presence in Africa, and particularly on the construction of the railway, for which they had sent over 20,000 Chinese workers, was excessive; second, they found that the African authorities did not seem to understand what was said to them by the Chinese and third, the Chinese themselves found it difficult to understand what the local authorities tried to tell them!

J.R.D.'s interest in his relationship with Indira Gandhi had mainly to do with economic development while Indira's had mainly to do with political power, in the pursuit of which she split the Congress, abolished the privy purses of the princes and nationalized the banks, with all of which J.R.D. disagreed.

In 1967, Darbari Seth, the head of Tata Chemicals, had prepared a major multi-fertilizer project which aroused so much interest in Indira Gandhi that she flew to Mithapur for a presentation of the project which appeared to thrill her for reasons of its conceptual grandeur and the vital role it would play in India's agricultural development. Some people around her, however, strongly objected to supporting Tatas whose application was rejected.

There was much confused and politicized thinking on economic matters in government circles when J.R.D. was invited to an international round table conference on economic growth and social justice in February

1964 by Dr V.K.R.V. Rao. J.R.D. replied: 'I sometimes feel that we have so hopelessly confused ourselves with words, slogans and clichés that we have lost sight of simple truths, the simple objectives and the simple means of attaining them....To the extent that socialism or any part of it demonstrably promotes such welfare and brings about economic growth, we should accept it, suitably adapted to our particular needs. Similarly, to the extent that capitalistic enterprise serves the people's welfare as it has done in all the most prosperous countries in the world, it should be accepted and encouraged.'

Perhaps next to Nehru, the longest association J.R.D. had with a political figure was with Indira Gandhi:

> We were good friends socially but as with Jawaharlal, I never was able to break through the fence which they both built around themselves. While I regretted the limitation in my relationship with them which deprived me of the chance to communicate freely with them on matters of national interest about which I felt I had something worthwhile to say, such relationship as was accepted by them did permit maintaining, unspoilt, a friendship which persisted till the end...Indira was a very complex person whose character and deeds will fully emerge only when sufficient time has passed after her death for uninhibited reminiscences and biographies to be published about her.[2]

As with Jawaharlal so with Indira; while there was warmth and friendship in their association with J.R.D., what was lacking and was much missed was confidence and trust. They liked J.R.D. as a person but he nonetheless represented for them the world of private industry of which they had the deepest suspicion.

*

A dramatic political situation emerged in the country when Indira Gandhi's election to parliament as Prime Minister in 1971 was challenged in the Allahabad High Court and the decision in that court went against her. It was a time of great suspense as to whether she would resign as Prime Minister. Instead, she appealed to the Supreme Court and asked

N.A. Palkhivala, Legal Advisor and a senior Director of Tatas, to take up her brief.

On 13 June 1975, when Indira Gandhi was under considerable pressure to resign, J.R.D. wrote to her: 'I may disagree at times with the policies of the Government but it has never affected my personal regard and admiration and my affection for you.'

On 26 June Indira Gandhi declared a state of Emergency in the country and Palkhivala was so outraged that he wanted to return the brief that very day. It was an agonizing day for the House of Tatas because a single act by one of their senior-most Directors could have had severe repercussions on the House. When senior Directors of Tatas privately met to confer on Palkhivala's move and urged him not to do so J.R.D. did not express his view. Privately, J.R.D. had told Palkhivala that he disapproved of his returning the brief. However, he did not feel it morally right to prevent a valued colleague from following the powerful dictates of his conscience. (A senior Director of Tatas says that having Palkhivala as a colleague for over thirty years has helped in stimulating J.R.D.'s own thinking on many national issues though, interestingly, their conclusions can differ. For example, J.R.D. prefers the Presidential System, N.A. Palkhivala says no system will make any difference, it is character that counts.)

Sometime later, when J.R.D., again accompanied by Sharokh Sabavala, met Indira Gandhi—at his request—the atmosphere was glacial. Indira Gandhi alleged that her information was that some Directors in Tatas were against her and her policies. Indira accused Tatas of being a "breeding ground" of hostility to her government—a charge that surprised and hurt J.R.D. very much.

The *Hindustan Times* on 2 January 1977 carried an interview with J.R.D. While making clear that he had earlier 'not hesitated to disagree,' with such of the economic policies of the government which he felt were counter-productive and that he would not hesitate to criticize the government again should the need arise, J.R.D. lauded the 'refreshingly pragmatic and result-oriented approach,' of the Emergency period which had led to 'conditions of discipline, productivity, industrial peace, price stability and widespread involvement necessary to achieve rapid economic growth.'

Referring to the political scene, J.R.D. said that he could well understand that the declaration of the Emergency, 'had had a traumatic impact on some intellectual circles wedded to Western democratic concepts going back to the Magna Carta and nurtured, over the years, by many historical pronouncements on human rights and freedom.' But he felt that in the present Indian situation, 'it was not the interests or the views of a small educated elite that should prevail but the interests of 600 million people for whom no freedom or right matters more today than freedom from want and the right to work and earn a decent living.' He expressed the hope that over the years and decades to come we could evolve our own democratic solutions to our political, social and economic problems.

When the above paragraphs were shown to J.R.D. he said: 'Something is missing, I think.' He said it should have been prefaced by words like 'Contrary to his beliefs in freedom and democracy which he has held through all his life....' Next, he read the words ascribed to him 'conditions of discipline, productivity and price stability and widespread involvement.' He stopped and added 'but it could not be a "widespread involvement" because there was none.'

I noted, 'I am not worried about the incidental wording because what the press reports could often be different from the actual words used. The question is: Would a person of your standard, with your views, support a regime that put people in jail without trial and imposed stringent restrictions on the press? It seems out of character that someone like you should make this statement. Were you carried away by the economic gains of the Emergency (reversing inflation) or by your personal feelings for Indira? Or were you concerned about the security of Tatas because you felt she was getting antagonistic to Tatas?'

Promptly he replied 'No, no. None of my opinions have anything to do with that (namely, effect on Tatas.) I am absolutely sure of that.' Nor did he think that his personal feelings for Indira could have influenced his judgement at that time.

'The economic gains you have mentioned,' I observed, 'the rest you have not touched upon.'

'Wait a minute. It should have been published that I expressed these ideas because despite my long held view of democracy and freedom the conditions that prevailed in the country at that time, the lack of peace and discipline, made me feel that they would never allow the country to progress unless they were brought to heel. But how to express that? If you could bring that out. I was so incensed, so worried about the conditions that did prevail at that time, the total lack of discipline, the reasons that Mrs. Gandhi gave.'

'She gave law and order as the reasons,' I said, 'but it was sparked off by her not stepping down after the Allahabad High Court's judgement.'

'Quite right—that's not brought out. It doesn't need to be—because it has nothing to do with my opinion.'

He continued: 'However, I would have liked to make the point, "despite his long held views".'

When I tried to sum up his own reasons which he had expressed, namely his great concern at the disintegrating conditions prior to the Emergency, and the need for some kind of discipline, he replied, 'Be careful. Do not try to cover up.' Then, with some force, he added: 'I think I was wrong.' He paused and went on to say: 'But it was my feeling at that time. You could put it that I regret it today but in those days the apparent disintegration, that could have led to anything, appalled me.'

Soon after, in March 1977, Indira lost the elections and her seat in parliament. In a handwritten note to Indira Gandhi, J.R.D. said, 'You have been much in my thoughts these last few days as the drama of the elections unfolded and reached its incredible climax. I can imagine the physical and emotional strain to which you have been subjected and my heart and Thelly's go out to you in your ordeal and distress.

'However much I may have felt at variance with some of the actions and policies of Government during the past twenty months, my personal affection and regard for you and recognition of your immense services to the country never dimmed. I am sure that you will not lose heart and will face this traumatic experience with the same fortitude and indomitable spirit you have so amply demonstrated all these years.'

A year later, when Morarji Desai removed J.R.D. from the Chairmanship of Air-India, Indira Gandhi wrote a handwritten note appreciating J.R.D.'s contribution to the airline*.

When the Janata regime came to power with Morarji Desai as Prime Minister, Tatas in turn were treated as friends of Indira Gandhi and J.R.D. was in his own words, "thrown out" of the Atomic Energy Commission (AEC), of which he had been a member since its inception, and a year later, from the Chairmanship of Air-India.

When J.R.D. met Indira Gandhi on 28 April 1980, after she returned to power, he said to her that it was a bit hard for him to find that Tatas, who had served the country with total honesty and patriotism and always

---

* Quoted in the chapter "Dismissed" in Part II.

extended full cooperation to government, were treated as *persona non-grata* by both successive governments.

J.R.D. tried to clear the air with her at this interview. He recalled his first contact with Motilal Nehru and his friendship with her father and mentioned how on Nehru's death, J.R.D.'s affection and devotion were transferred to her and were still with her, though unrequited. He told her that he felt sad at the coldness towards him and his firm. It was not fair to hold against the latter, the attitude and statements of one Director of Tatas. She replied: 'This is not the case today.' J.R.D. remarked, 'I am happy to know that.'

J.R.D. thanked her for putting him back on the Board of Air-India.

In her final term in office (1980–1984) the realization seems to have dawned on Mrs Gandhi that the leftist advisers of her earlier years had led her up the garden path and liberalization—which J.R.D. had strongly supported—was set in motion by her and accelerated by Rajiv Gandhi who came to office after her.

In a memorial volume, edited by Indira Gandhi's Press Secretary Sharada Prasad, J.R.D. wrote:

> There are really two very distinct Indira Gandhis to remember. First, of course, the great politician-statesman whose total dedication to the task and responsibilities of governing so large a country as ours, so diverse and so plagued by such intractable problems, evoked universal admiration even from those who disagreed with her.

> My contacts and relations with Indira Gandhi were rarely in her official capacity. Because I was critical of most of her Government's economic policies which I knew would not be changed, I usually avoided bringing up the subject when we met and sought her help only or intervention only in matters of public interest not directly concerned with the Government's policies. The quickness of her response was sometimes dramatic.

> It was the other "non-official" Indira Gandhi I knew well and felt close to, the gracious, cultured and compassionate human being for whom I felt the expression "civilised", in the best

sense of the term, was so particularly apt. On the non-political stage, her personality was a many-faceted and fascinating one which responded to the simple human interests and aspects of life, to beauty wherever she found it, to nature, to children, to music or to art. She repeatedly demonstrated to me her amazing capacity for remaining serene and apparently unconcerned by the tremendous pressures thrust upon her as Prime Minister. Busy as she was from morning till night she found time for acts of kindliness and compassion which added to the pressures on her already overburdened programmes.

The energy of her ever-alert and wide-ranging mind was as boundless as her physical energy, with which she astounded those she dealt with, often to the point of their exhaustion. I was a personal witness to this on the occasion of her visit to the site of the Pokharan explosion (1974). The members of the Atomic Energy Commission, including myself who accompanied her, were hard put to keeping up with her as she scrambled over rough terrain. I spent most of an exhausting day with her on that occasion, at the end of which she was as fresh and energetic as when she started.

Indira Gandhi may not have been a beautiful woman in the classical sense, but her charming smile, her poise and the intelligence that shone from her eyes were such that the memory I retain of her is indeed that of a beautiful human being whom one would have wanted to do more to help and who will be mourned not only for her total dedication to the country for which she ultimately sacrificed her life but also for her human qualities and her friendship.

As with Jawaharlal Nehru so with Indira Gandhi; J.R.D.'s differences of opinion on politics seem to have in no way impinged on his personal appreciation of the finer qualities of the two Prime Ministers, nor, except during the brief period of the Emergency, was his relationship with either of them affected.

CHAPTER XIII

A Dream Fades And Reappears

The history of liberty is the history of limitations of
Governmental power....

-Woodrow Wilson

If one enters J.R.D.'s office unannounced, he is invariably working on some paper with a pencil or a ballpoint pen in his hand. One evening when I walked into his office, he was not so occupied but was looking pensively at the wall in front of him. Before I could pull up a chair, he said, 'You know, Russi, my life has been a struggle—never once has any Prime Minister asked me what I thought of the economic policy of the country.' He paused, and added with some feeling, 'In no other country would that have happened.' The thought was obviously playing on his mind and he seemed upset. He said, 'The only occasion when I was asked for my views was by P.N. Haksar (Principal Secretary to Indira Gandhi.) Tatas prepared a careful paper and it was leaked to a Communist paper.'

J.R.D. had great dreams of what India could do and the part Tatas would have in developing her economy. In the post-war world when some countries went the Communist way and others kept to the capitalist road, J.R.D. felt:

India went one better. Under the leadership of Jawaharlal Nehru she adopted a mixed economy in which public enterprise would join with private enterprise in bringing about rapid industrialisation and economic progress. Despite misgivings in some quarters.... I, in my innocence, believed India would thus progress even faster than other countries which relied on only one source of entrepreneurship and investment.

I had the vision of a partnership between the public and the private sectors, each concentrating on those fields for which it was best suited, each complementing and supporting the other and both actively helped by Government in converting India rapidly into a great industrial power and, in the process, creating millions of jobs every year for its people. The vision dimmed as the concept of such a grand alliance for progress itself dimmed and was progressively discarded.[1]

J.R.D. was speaking at a Rotary Club on "19th & 20th Century Socialism" in 1970. There he pointed out that nineteenth century socialism believed in State ownership of all means of production while twentieth century socialism, which is symbolized by the Scandinavian states, stood for a welfare state. J.R.D. believes in the latter, although Nehru 'to his undying credit refused to be stampeded into extreme socialistic reforms,' says J.R.D.. Even so, within a few years of independence, in

a negation of Jawaharlal Nehru's earlier pronouncements, major sectors of private enterprise, including air transport, banking, insurance, coal and copper, were nationalised, all of them on confiscatory terms. The Constitution was amended again and again to subserve the economic policies of Government wherever they were incompatible with the Constitution.[2]

The citizen's fundamental rights were abrogated, culminating in the 25th Amendment which abolished the very concept of compensation. J.R.D. continued: 'Whole areas of economic activity in which the private sector had always operated were denied to it or severely restricted.

'A formidable panoply of laws, regulations and controls over the operations of private sector companies was created and constantly expanded until owners' rights and most of management's freedom of action were extinguished.

'India became the most heavily taxed nation in the world and taxation on individuals was made confiscatory.

'A virtually total Government monopoly of investible and lendable funds was created by nationalising banking and insurance.

'The privately owned funds so seized were used to acquire an ever-growing share in ownership of joint stock companies.

'A mandatory convertibility clause was attached to all term loans from financial institutions for the avowed purpose of bringing private sector enterprises under state ownership.

'This failure to recognise, in time, the real shape of things to come and to resist it,' says J.R.D., 'is at least, in part, due to the insidiously soporific effect of the gradualness with which momentous changes have been introduced in our economic life. I have no doubt that if they had been sought to be brought about all at once, when those who won India's freedom and the founding fathers of our Constitution were alive, the reaction would have been intense, swift and decisive.'[3] His letters of the 1950s and thereafter show his sadness at being unable to affect the situation.

Controls created shortages; shortages generated black money. For example, the private sector was not allowed to set up cement plants; nor would the government set them up. To cap it all, the government controlled the selling prices of cement, thereby ensuring that the manufacturers would not have the resources to reinvest in new machinery or expansion of their own. A roaring black market continued for years and years. Newspapers reported the case of a resourceful Chief Minister who found it expedient to openly collect funds for a Trust he set up against the issue of cement licences by his government. From the mid-1950s up to the early 1980s, India continued to be in the grip of dogma and ideology. When vast quantities of cement had to be imported, wisdom finally dawned on the government and permits to set up fresh plants were granted.

'Unrealistic and wishful plans, excessive bureaucratic control and an expropriate taxation have been the three principal factors responsible for our present economic plight,' said J.R.D. in 1966. Over the years men like him paid ninety-seven per cent of their earnings in income tax and, on top of that, they had to pay wealth tax which raised their taxation level to more than a hundred per cent.

Others, instead of fruitlessly protesting and holding to standards they knew to be right, found the easy way out: corrupting government officials and ministers, keeping some Members of Parliament on their payroll (as well as underwriting their elections) and trading in black money. The panoply of controls and permits and licences helped generate these funds. In the process of saving their own skins they stimulated this cancer of corruption. J.R.D. kept his companies clear of these shortcuts to prosperity. He told me in 1979, 'Had we adopted some of the methods that other industrialists have adopted, including having Members of Parliament on their payroll, we would have been twice as big as we are today.' He paused and added: 'But we would not have wanted it any other way.'

J.R.D.'s contribution during the dark decades was to hold the flag high so that when brighter days came there would be a standard others could rally to.[4]

He pointed out the dangers of their actions to those who blissfully believed that one could walk the democratic path politically and at the same time adopt a totalitarian economic system. He enquired:

> Once all economic power is centred in Government, will it not be found equally necessary to curtail other freedoms in the interests of the State? History shows that the danger lies in the very existence of unlimited power which may one day fall into less worthy hands.[5]

He quoted Woodrow Wilson in support of his perception:

> The history of liberty is the history of limitations of Governmental power, not increase of it. When we resist...concentration of power, we resist the powers of death, because concentration of power is what always precedes the destruction of human liberties.[6]

While the government was concentrating more and more power in its hands (even today the Union Government's permission is needed to set up a unit to manufacture razor blades or tennis shoes), a propaganda campaign was launched to prove otherwise—that big business concentrated economic power in its hands. The main beneficiaries of this campaign were politicians and bureaucrats who gathered more and more powers and controls in their hands.

Finally, J.R.D. had had enough. In 1968, at a meeting of the Planning Commission to which he was invited and where the Deputy Chairman of the Planning Commission, Dr D.R.Gadgil, was presiding, J.R.D. said: 'As the head of the largest industrial group in the private sector, I must be possessed of a tremendous concentration of economic power. As I wake up every morning, I carefully consider to what purpose I shall apply my great powers that day. Shall I crush competitors, exploit consumers, fire recalcitrant workers, topple a Government or two? I wish Dr. Gadgil or some other eminent protagonist of this theory would enlighten me as to the nature of this great power concentrated in my hands. I have myself totally failed to identify let alone exercise it.'

Tatas, he said, were in a difficult position. If they did not embark upon a new major industrial venture, they were accused of inactivity and lack of dynamism. 'If it seeks to diversify into a promising medium-sized venture, it is accused of attempting to crush, or block the growth of small entrepreneurs. If it wishes to embark on a major capital-intensive project, it is accused of monopolising capital resources and adding to its concentration of economic power!'

Every politician talks of socialism, no one of the welfare of the people, said J.R.D.

It was fruitless to wait for the government, he declared. Industry had to demonstrate its intentions on its own. The following year he decided that if government did not understand that socialism meant public welfare, industry could lead the way. It could look beyond the welfare of its own employees to the people.

He said:

> Every company has a special continuing responsibility towards the people of the area in which it is located. The company should spare its engineers, doctors, managers to advise the people of the villages and supervise new developments undertaken by cooperative effort between them and the company.[7]

He particularly singled out assistance in family planning. J.R.D. followed up his statement by amending the Memorandum of Association of some Tata companies which could authorize them to undertake this task.

Both in Jamshedpur and Mithapur, Tatas launched programmes to assist neighbouring villages. He asked industry to "adopt" villages:

> Is there a village in our country which does not need some improvement to its scant amenities—a school, a dispensary, a road, a well, a pump and, above all, opportunities for gainful employment?[8]

286

Twenty years later, Tata Steel alone has adopted 246 villages. The company trains people in earning a livelihood, helps to market their products, improves the crop yield of the farmers and has encouraged cooperatives of Adivasi women.

Jamsetji Tata cared for his workers and shareholders. Sir Dorab Tata invited Sydney and Beatrice Webb to organize social services for Jamshedpur. J.R.D. has gone beyond that to care for thousands who have no involvement in producing wealth for his companies.

Eight years after J.R.D. made his call for this social responsibility for the rural areas, the Janata Government came to power and gave tax concessions to industry to do exactly what Tatas were doing without tax concessions. Some companies were smart enough to get the tax concessions and siphon off the funds thus saved for other purposes. The government got wise to this ploy, tax concessions were withdrawn and many companies withdrew from rural work. 'If you are going to do rural uplift only because you are getting a tax rebate, you might as well not do it,' says Tata Steel's Vice-Chairman, S.A. Sabavala. While others backed out when tax concessions were withdrawn, Tata companies continued their work in the villages. Where some have tended to give the nation a diet of words, J.R.D. has chosen time and again to demonstrate that industry had a conscience.

It was only in the evening of his life that he found that some of his early exhortations to the government were being followed in the closing years of Indira's rule, and more so in the initial years of Rajiv Gandhi's prime ministership. Big companies were allowed to expand and diversify.

At eighty-two, speaking at the launch of his book *Keynote* before a distinguished audience in the capital, he said: 'My one sorrow and regret is that the Government had, from Jawaharlal Nehru onwards and at least up to a couple of years ago, not allowed many of us imbued with enthusiasm and hope to do enough. Today (1986) things have changed and now the last sorrow of mine is that I have reached an age where I am not likely to be able to participate purposefully in the better things that are happening, the better opportunities and the quicker progress that I visualised.

'But still, as I look back, at least age has brought some wisdom and a philosophical sense and, therefore, I am, as I have been for the last many years, a short-term pessimist but always a long-term optimist. Today at least I can say—not that I have contributed anything very significant to it—that we all can be optimists in the short-term as well as in the long-term. And I only wish that I will be spared long enough to see that we are on the march.'

In mid-1991 when, with its back to the wall, India announced it was moving into an era of economic liberalization, I observed to J.R.D., 'It must be a source of gratification to you that you have lived to see many of the policies you advocated being accepted by the government (on MRTP and other restrictions).' To my surprise he was lukewarm. 'Yes, but let us wait and see whether the bureaucrats allow its implementation.'

PART IV

THE PATRIARCH

CHAPTER I

The Professional

J.R.D.'s predecessor as Chairman, Sir Nowroji Saklatvala, had depended for the best professional advice from experts *outside* the House of Tatas. The firm of Wadia, Ghandy & Company, Solicitors, provided legal advice. The financial wizard, F.E. Dinshaw, came each day at noon to advise the Chairman. This system of outside-the-house consultations was followed at that time by companies in England and America too, and it still often is. J.R.D. was convinced that such capabilities should be available in-house. He convinced Sir Nowroji of this need. Only a month before Sir Nowroji's death, in July 1938, J.D.Choksi, a brilliant solicitor and partner at Wadia, Ghandy & Company, joined Tatas as legal advisor. In later years J.D.C., or 'Jinx' as he was called, grew to be the right hand man of J.R.D. and rose to be the Vice-Chairman of Tata Steel and Air-India. Twenty years later, J.D.C. brought into Tatas the brightest legal luminary he could spot, Nani Palkhivala. At the age of forty-one, Palkhivala was made a Director of Tata Sons and given the dispensation to continue his private practice. He has served Tatas for over three decades.

Spotting talent is the first step. Attracting it is the next. Holding on to talent is even more difficult. J.R.D. was to prove successful in all three. After securing the legal side with J.D.C, J.R.D. looked for a clever financier. He found one in A.D. Shroff, a partner of the sharebrokers Batlivala & Karani. Shroff had an early spell at the London office of the First National City Bank of New York and had also been a lecturer at Bombay's Sydenham College of Commerce.

Another lacuna J.R.D. detected was the lack of statistical information on which new economic initiatives could be based. He, therefore, started a Statistics Department. Y.S. Pandit, Statistical Superintendent in the Bombay Government Labour Office, left his job and came over to Tatas in 1940. For the next twenty-eight years, Pandit was to serve Tatas well.

Says Pandit, 'J.R.D. had a hunger for statistics. In those days—the 1930s—they were not readily available. Industrial houses did not bother about statistics. The only source that showed an interest in compiling them was the Government. In Bombay House we used to compile statistics on different subjects and stencil them on a rather unpresentable page or two. We soon found they were much appreciated not only in India but also abroad.'

A couple of years later, Dr John Matthai took charge of the expanded Department of Economics and Statistics. In the early days the information the department gathered was available only to Tata Directors. However, in time, *The Statistical Outline of India* put out by Tatas became a source of vital information to Indian business, industry and to government officials, including planners.

Within two years of J.R.D.'s Chairmanship, he had completed the appointments of the professionals he needed. Having obtained the services of professionals of a high order, J.R.D. now had to use their talents in the best way possible. He wanted them involved not just in running their departments or the companies but to think for the whole of Tatas and its growth. The Board of Directors of Tata Sons was a very select body at that time, comprising just six members. Foreseeing expansion in the wake of the war, J.R.D. established, in 1945, Tata Industries owned entirely by Tata Sons. Tata Industries then became the Managing Agents of the Tata companies, in place of Tata Sons, and on its Board sat the Directors of all the major Tata companies and the heads of some major departments.

The Tata Industries Board was at the Directors' level. But what about the Chief Executives and the numerous Heads of Departments? They too needed to think for the whole of Tatas. In pursuance of this objective J.R.D. started, in the early 1950s, the Inter-Departmental Conference. It met every month and as many as twenty-five key Tata executives would discuss and attempt to resolve problems like the wage structure or how to cope with government policies. Most of the issues had very little to do directly with their individual departments but the exercise gave them all a wider sense of involvement, a larger perspective and a concern for the whole of Tatas. A consensus man, J.R.D. was also trying to introduce the concept of integration and of corporate working amongst the officers of his companies.

J.R.D.'s initiatives were not always successful, often because they were ahead of their time. Looking ahead to the end of the Second World

War, J.R.D. was exploring ways in which to get the best young men into Tatas. Finally, he conceived of a body with the high sounding title of "The Superior Staff Recruiting Committee". In all solemnity Sir Homi Mody, Sir Ardeshir Dalal, Dr John Matthai and J.R.D. interviewed some brilliant young job-seekers. 'I was then much younger and innocent,' says J.R.D. 'Those were the days when recruitment was done on the basis of "My son is brilliant." They never said in those days, "My daughter is brilliant."'

The first three candidates chosen for the Superior Staff category were Messrs Kasbekar, Kakatkar and Bhavnagri. As luck would have it, all three of them disappeared in two years. 'I never knew why these three were in such a hurry to leave,' says J.R.D. P.D. Kasbekar joined the Indian Administrative Service (IAS) and rose to be Chief Secretary of Maharashtra; B.G. Kakatkar became Secretary-General of the Indian Cotton Mills Federation and Jehangir Bhavnagri joined UNESCO's Creative Film Division. In the lunch room, thereafter, says J.R.D., 'the team of Kasbekar and Kakatkar became the subject of a joke. My idea of creating a service, long-term and high grade, was treated a little as something to be made fun of at that time.'

But, undeterred, J.R.D. returned to the attack. Admitting that, 'while they (the first set of candidates) were not selected with proper preparations and study and that they did not last very long, I still was satisfied that this was the right thing to do and that we must do it again but do it properly.' He realized that some of his colleagues 'who may have been eminent in every other activity did not know anything about how to recruit people straight from colleges and universities and that we should get professional advice.' This was easier said than done. In the 1940s neither management courses nor management colleges were yet developed so there were no real parameters against which to judge prospective management candidates.

J.R.D.'s role model in management was the man who trained him—John Peterson. How was it, he asked himself, 'that a young Briton straight from college, could come to a foreign country and administer various departments with such distinction?' J.R.D. was convinced that 'Tatas would become a very big show,' and they would need a prestigious management school of their own—that would be Tatas' main source of top executives.

It was thus that the Tata Administrative Service (TAS) was formed. Dr F.A. Mehta, an economist trained at the London School of Economics

(LSE), joined Tatas in 1956 as the first entrant to the TAS. After selection Dr Mehta expected to be welcomed by the Chairman over a cup of tea. No such invitation came. Many months later Mehta, still in his twenties, wrote a signed article in the *Economic and Political Weekly* attacking the Union Minister for Commerce and Industries, T.T. Krishnamachari. Soon after the piece appeared, there was a call from the Tata representative in Delhi saying that the powerful Cabinet Minister, perhaps the most important man in the Cabinet for Tatas, was fuming that Tatas had goaded their young employee to attack him. J.R.D. çalled Dr Mehta in for a dressing down. Taken aback, the young man said he was sorry that Tatas had been affected adversely and that he would write a personal letter of apology to T.T. Krishnamachari, thereby absolving Tatas.

'I don't want these dramatics,' said J.R.D. 'Get a grip on your emotions when you write.' Dr Mehta says he became more circumspect thereafter. After about twenty years J.R.D. called Dr Mehta again: 'When you knock the (Government's) policy, you still keep its shape. See how Palkhivala and I go all out for the convertibility clause?'*

'Sir,' said Dr Mehta, 'It was 20 years ago that you said that I should get a grip on my emotions and write with dignity.'

'With dignity, yes,' replied J.R.D.,'but not with cunningness and subtlety.'

In 1959, when J.R.D. went to Japan to address a major business forum there, he asked Dr Mehta as an economist to give him a background note on Japan. Dr Mehta submitted a report of about thirty pages and promptly got a note to say that 'Your notes are excellent but your name should not be Mehta but "One-too-long". Please summarise for me in four pages your booklet.'

'He taught me,' says Dr Mehta, 'the art of summarising.' When J.R.D. came back from Japan he wrote a congratulatory letter to Dr Mehta and said how much some of his points were appreciated in Japan.

Dr Mehta says that, 'In business or industry the moment the number two man gets too much attention in the press or public and he gets glory, the number one tries to pull him down. JRD's greatest credit is he never assumed an adversarial role to anybody coming up in Tatas.

---

* A clause in the loan agreements of government financial institutions under which they reserve the right to convert a portion of the loan amount into equity capital of the borrowing company.

He would disagree on policy matters. He would even criticise the personality of someone but he would never assume an adversarial position.' Then Dr Mehta pointed out that J.R.D. had never sacked anybody. I asked J.R.D. if that was true of him throughout his over sixty years in industry. He said that he had moved people to jobs that better suited them but only on one or two occasions did he have to get rid of people and, 'that too always in consultation with my colleagues.' Dr Mehta says that 'Tatas are an unstructured organisation capable of doing great things and throwing up outstanding managers, while Levers by contrast are a well-structured organisation.'

Thirty years after joining Tatas, Dr Mehta is Chairman of the Forbes, Forbes & Campbell group, heading seventeen companies. Dr Mehta says that J.R.D. was the first in India to realize that, 'staff functions should have a knowledge-base,' which is why he established the Department of Economics and Statistics in 1940 and the Department of Public Relations (probably another first in Indian business) headed by M.R. Masani in 1943.

J.R.D. conceived of the Tata Administrative Service (like the ICS that John Peterson belonged to), as a mobile force. Different companies would invite them and they would move from one assignment to another. Security of service would be guaranteed and if they were not happy with one company they could ask for a transfer to a company of their choice. There would be no loss in emoluments. The whole idea of mobility was that people would begin to think for the whole of Tatas and not just their company.

This elite service today represents the first choice of MBA and IIT graduates wanting to go into management. For fifteen months entrants are given training in various Tata companies and then absorbed into the group. In the early years the scheme worked well where mobility was concerned. However, the excellence of TAS graduates made Managing Directors increasingly unwilling to release their blue-eyed boys to other companies and slowly the element of mobility got reduced.

Many TAS officers have spearheaded major enterprises. For instance, Xerxes Desai, who was Managing Director of Tata Press, was put in charge of Titan Watches.

Camellia Panjabi, Vice-President of Indian Hotels, has been one of the driving forces in the expansion of that company. She has noticed a qualitative difference in TAS graduates over the years. 'In the early days,' she recalls, 'we used to share confidences about our Managing Directors, measure the slowness of their reactions. We were always worried about how we could change things completely and totally. We never talked in terms of career paths. When I meet the younger ones today the first question seems to be "How do you see my career path?" I have to answer that very often. Maybe it's something to do with the different style of education in management and business. But things have changed in fifteen years and I think that people have been taught to believe that they have to make an analysis of the organisational structure and find out how fast they can weave their way in and out. I've tried to learn from them and see whether I can apply it to myself but I find it very difficult. I am always still thinking of how I can change things which exist—maybe with more wisdom and more maturity—but this career path problem does not seem to come to my mind and I guess it's a reflection of a different generation.'

Even so, it is J.R.D.'s hope that the TAS will be a force to keep the firm and all its companies together in the coming years. Speaking to TAS officials in 1988 he said, 'Loyalty is not only to a company but also to an idea, a loyalty to a life purpose, loyalty to the country, a loyalty to the opportunities that one may have to serve.'

At the time that J.R.D. was recruiting fresh talent for Tatas in the mid-1950s, he was also wrestling with the idea of how to give the right inputs to those who were already in important positions in Tatas. The first Indian Institute of Management (IIM) was to open at Ahmedabad in November 1961, in collaboration with Harvard University. Even before that, with the help of Dr John Matthai and Professor R.D. Choksi, Tatas decided on the idea of a Staff College in the 1950s. The Turf Club at Pune was hired for a month and Tata Directors and distinguished outsiders were asked to lecture to Tata officers and share their experience in higher business and management. Most of the top Tata Directors spent three to four days interacting with young and middle-aged Tata officials. 'It was an invaluable experience—interacting with men like Dr. John Matthai and having the pleasure of questioning them about various aspects of management,' says one of them. More often than not a sense of camaraderie grew between the participants and they returned to their jobs refreshed and inspired. Later, a gracious home in Pune with wide grounds was acquired by Tatas, and it was made the base for a permanent Tata Management Training Centre (TMTC) in 1964.

J.R.D.'s vision was fulfilled. The TMTC has been the focus of management training for different Tata company programmes and has trained personnel for other companies and the public sector as well. In 1987, at Rajiv Gandhi's instance, it was chosen as one of the three centres for advanced training for senior government officials including Chief Secretaries at the Centre and State level. The Prime Minister himself addressed a course of top government officials at the TMTC. Of the first training courses conducted by government at three top management centres the one that proved the most popular, according to *India Today*[1] was the one at the Tata Management Training Centre. It was conducted by the TMTC Director, Dr Francis Menezes.

Changing times have, however, brought about changing circumstances. The more leisurely pace of the 1950s has been lost perhaps forever and now Tata Directors fly in by the morning plane to Pune and fly back by the evening plane after making their speech. That richness of interaction, that mellowing of an executive through shared experience over a period of days is missing. This is the price of "progress".

*

When J.R.D. was sixty-one, he wrote to a Calcutta educationist:

> I thank you for your letter of the 6th August (1965) enquiring what have been the guiding principles which have kindled my path and my career. I do not consider myself to be an "illustrious personality" but only an ordinary businessman and citizen who has tried to make the best of his opportunities to advance the cause of India's industrial and economic development. Any such guiding principles I might unconsciously have had in my life can be summarised as follows:
>
> That nothing worthwhile is ever achieved without deep thought and hard work;
>
> That one must think for oneself and never accept at their face value slogans and catch phrases to which, unfortunately, our people are too easily susceptible;
>
> That one must forever strive for excellence, or even perfec-

CHAPTER II

The Philanthropist

A young man and a woman from a child relief organization were having tea with J.R.D. in 1988. As a colleague walked in J.R.D. said, 'You know, as they explain their plans to me, I'm trying to be as difficult as possible.' Even as the young man was trying to explain the project, question after question interrupted him. Sometimes, J.R.D. can be a very good listener and at other times his rapid-fire questions could unnerve anybody. A hundred artists had donated their paintings to the organization. The paintings were to be exhibited and sold in art galleries in Bombay, Calcutta, Delhi and Bangalore and the proceeds were to form a corpus of about thirty lakh rupees. They wanted a sponsor to meet the expenses of the exhibition. J.R.D. interjected again: 'Everybody wants a corpus. Why do you want a corpus?' And before the young man could answer: 'How much do you raise per year?' And, then again, 'The corpus' interest will be very little for your work.' A piece of advice followed, 'I'm impressed with what you raise every year. Continue to do so rather than relax with a corpus.'

J.R.D. then took off on a larger canvas. He asked 'But why are you concentrating more on children, some say it is women who need even more attention? How many children do you reach?' He listened to the answer, and then said, 'But women are important. In Kerala where there is 80 per cent literacy amongst women the birth rate is lowest. In Rajasthan and Bihar where eighty per cent of the women are illiterate they have the most children. Apart from my business, I've taken an interest in the problem of population which, I believe, is the cause of all our problems.' He turned to the young lady who had known him from childhood and said, 'You know Aunty Thelly picked up these young street boys and tried to teach them a craft, mainly weaving. She even drew a painting herself on cloth (about the boys).' Thereafter he switched to Sao Paulo in Brazil, this time to underline the importance of children. 'Sao Paulo has produced 20,000 children who cannot be controlled by anyone, even the police. They are little monsters. We in Bombay haven't come to that stage yet.'

Charming and relaxed as he was throughout the conversation it was evident that he was studying the two young people carefully, testing, even teasing them. He then spoke appreciatively of their dedication. Again in a lighter vein, he turned to me and said, 'Can you see the halos around them?'

Finally, he came down to business, studied the breakdown of their figures and said 'It shouldn't be a problem.' He said he would ask the Chairmen of his companies to contribute. 'I'll twist two or three arms!' It was clear that he'd been impressed by his visitors' work for he is reluctant to ask Chairmen of Tata companies for funding, even for projects dear to him. He once told me 'I hate to go again and ask X for money.' But philanthropy does not mean the donating of money alone. The trouble one takes over someone in need often demands more of oneself than the giving of funds. The Greeks knew that and coined the word which is derived from *Fil-Anthra-Pi* which means "Love of fellowmen".

J.R.D. was once taking endless trouble to find a job in Sweden for an Indian Army Brigadier whose Swedish wife was being treated for cancer in her own country. The Brigadier wanted to be near his ailing wife. J.R.D.'s sister, Rodabeh, wrote to him at that time, 'I like to think that fortune and influence are bestowed upon us as a trust during our lifetime. To be continuously faced with pleas for help is indeed a cross and you carry it with much humanity.'[1]

J.R.D. has never treated it as a cross, but has rather rejoiced in helping people. There was once a young man in Tatas, whose parents and brothers were settled in Canada, and who was therefore debating whether or not to emigrate. When this dilemma came to J.R.D.'s notice, he called the young man in and advised him like a father about his problem. He felt it was right that he migrate to Canada. Towards the end of their conversation he asked the young man 'Why didn't you come to me earlier? Why don't you use me while I am still there?'

The tradition of Tata philanthropy goes back to 1892. Admission to the Indian Civil Service (ICS) had just been opened to Indians by the British and Jamsetji Tata was keen that Indians took advantage of it. He was also eager that professionals, especially doctors, were trained in England. At the time Indian women did not go to male doctors and many of them died while giving birth. Given this situation Jamsetji first gave grants to two lady doctors to go abroad and specialize in gynaecology. 'I can afford to give but I prefer to lend,' he said. He gave the money on condition that it was returned to him in due course so that others could benefit from the same funds. In the next hundred years the J.N. Tata Endowment for the Higher Education of Indians was to give loans to over 2,000 students towards their studies abroad.

Later, in 1896, Jamsetji made his princely offer of fourteen of his buildings and four plots of land to establish a University of Science. The Indian Institute of Science, Bangalore, opened in 1911 and provided the manpower for the great national laboratories that came up in the 1940s and early 1950s.

The tradition of Jamsetji was continued by his sons, Dorab and Ratan, both of whom in their lifetime made various benefactions and on their death left behind their substantial fortunes in trusts—the Sir Ratan Tata Trust (1918) and the Sir Dorabji Tata Trust (1932). Through Sir Ratan's benefaction was established the Sir Ratan Tata Department of Social Sciences at the London School of Economics (LSE). Gandhiji in South Africa and G.K. Gokhale in Bombay received considerable assistance from Sir Ratan. When the Prince of Wales Museum started in 1922 the art collection of Sir Ratan Tata along with that of Sir Dorab, was the first major acquisition of the museum. The Tata Collection is displayed in a separate gallery.

J.R.D.'s own association with the Tata Foundations began with his appointment as a Trustee of the Sir Dorabji Tata Trust in 1932. He played an active role in establishing the Tata Memorial Hospital. The trust minutes of 13 November 1937 reveal that the trustees were debating whether the hospital should have thirty-two beds or fifty or more. The trustees were understandably concerned about keeping the expenses of treatment at a manageable level; they were not even thinking of research. At that stage J.R.D. came up with his own vision for the hospital 'This hospital should be able to carry out the triple objects of treatment, research and education. We should treat research almost as important as treatment.'

In 1941, when the Tata Memorial Hospital was inaugurated, Sir Roger Lumley, Governor of Bombay, said, 'This hospital will become a spearhead of the attack on cancer in this country, providing not only a centre where specialised treatment can be given, but also one from which the knowledge of new methods of treatment and diagnosis will go out to doctors and hospitals throughout the country.' His vision was prophetic. Doctors trained by Tata Memorial have gone out to hospitals throughout the country for the last fifty years.

In September 1939, when the war broke out, a brilliant scientist from Cambridge, Dr Homi Bhabha, then on holiday, was stranded in India. Tatas arranged for a chair in cosmic ray research to be established for Dr Bhabha at the Indian Institute of Science so that his talent could be fruitfully used.

Four years later, Homi Bhabha spoke to J.R.D. about his wish to establish an Institute for Fundamental Research in India. Dr Bhabha said that if such an institute was established, instead of returning to Cambridge or Princeton after the war, he would stay on in India and raise a team of highly qualified scientists. J.R.D. encouraged him and asked him to send his proposal to Sir Dorabji Tata Trust.

In March 1944, Dr Bhabha wrote to the Chairman of the trust, Sir Sorabji Saklatvala, saying that if such an Institute for Fundamental Research in mathematics and physics (including nuclear physics) was established, when the time came for nuclear energy to be applied for power production, India would not have to look abroad for its experts but would find them ready at hand. He promised to build a school of physics comparable to the best anywhere.

In April 1945, the trustees, after an interview with Homi Bhabha, accepted the proposal. This was four months before the first atomic bomb was dropped on Hiroshima and the world woke up to the power of atomic energy. The trustees recorded that the Bombay Government and the Government of India be involved from the very start with the Tata Institute for Fundamental Research (TIFR).

When Dr Bhabha was put in charge of the Atomic Energy Establishment in Trombay in 1957, he took with him forty-six of his top TIFR trained scientists. Dr Bhabha had kept his promise. That was a time when the entire staff of the fledgling Atomic Energy Establishment at Trombay was looked after by the TIFR; also its administration was looked after by the TIFR, which had the necessary infrastructure. The control system of Apsara, the first atomic reactor in Asia, was built under the auspices of the institute in a wartime hutment and many parts of the reactor were fabricated at the TIFR workshop. When the Atomic Energy Training School (now the Bhabha Atomic Research Centre [BARC] Training School) was started in 1957, to train the manpower required for the growing atomic energy programme, a major part of the teaching load in the initial years was carried by the TIFR scientists. Dr Bhabha observed at the time:

> It is not an exaggeration to say that this Institute (TIFR) was the cradle of our atomic energy programme, and if the Atomic Energy Establishment at Trombay has been able to develop so fast, it is due to the assisted take-off which was given to it by the Institute in the early stages of its development. It is equally true to say that the Institute could not have developed to its present size and importance but for the support it has received from the Government of India.[2]

Dr M.G.K. Menon who, at the age of thirty-six, succeeded Homi Bhabha as Director of the TIFR in 1966 wrote to J.R.D. in 1970: 'This Institute would never be what it is today but for the support that you gave Homi to set it up and the continuing support that you have given it ever since.'[3] J.R.D. was on the Board of the Atomic Energy Commission (AEC) from its inception up to 1977 when Morarji Desai as Prime Minister dropped him from the reconstituted Board.

It is not just the funds made available by the Tata Trust to the TIFR in the early years that made the difference. Equally important was the time and energy J.R.D. personally gave to the institute in its formative years. Some at the TIFR remember J.R.D. coming in with Dr Bhabha on Sunday mornings to discuss on site the building of the institute.

Earlier, during the Second World War, Jawaharlal Nehru had written to J.R.D. for funds to provide a cyclotron, an apparatus to accelerate sub-atomic particles, for Professor M.N. Shah, the well-known physicist of Calcutta University, and J.R.D. had arranged to do so.

In the 1960s, during his visit to France, J.R.D. was intrigued by the fact that over seventy per cent of the top jobs in the Civil Service and scientific institutions of France were occupied by men who were trained in the four Grande L'Ecole Polytechniques of France. (The first polytechnique was originally established by Napoleon to train his civil engineers.) J.R.D. invited Professor Jean Capelle, Director-General of Education, Paris, to go round India with a group of educationists from within the country in order to report whether a similar institution could work in India. He wanted to submit such a proposal to the Nehru Memorial Trust for which he had already raised a substantial sum of money. (The Trust was thinking at that time of erecting statues of the former Prime Minister and investing in some other ventures.)

J.R.D. placed this report of the French-led expert committee before the Nehru Memorial Trust. The Chairperson was Indira Gandhi, who had been elected Prime Minister some months earlier, and the former Maharaja of Kashmir, Dr Karan Singh was the Secretary. One important criterion about those who would attend the four French Polytechniques was that all the participants had to be good at mathematics because math, it was felt, gave decision-makers an added edge, clear and disciplined thinking and decision-making. The report placed before the trust naturally reflected this thinking. Dr Karan Singh, a philosopher, found this objectionable. 'Do you mean to say,' he asked J.R.D., 'that because I am not good at mathematics I cannot join the proposed Institute?' J.R.D. replied that, regrettably, he could not and explained why. Indira Gandhi turned up her nose at this and said 'Jeh, the Institute is elitist.' J.R.D. was disheartened.

Looking back he says, 'Perhaps I should have pressed ahead and got Tatas to undertake this project.'

He did not pursue the project of the educational institute perhaps because he was disheartened by the response but he did support soon after a proposal by Jamshed Bhabha, a trustee of the Sir Dorabji Tata Trust for a National Centre for the Performing Arts. When some trustees were sceptical about a charitable institute going for such activities, J.R.D. observed: 'While we want to build a prosperous society we do not want it to be merely a materialistic society.' After an animated discussion the trustees agreed to the proposal. This move has resulted not only in Indian music being preserved for posterity but it has established a most beautiful theatre—the Tata Theatre.

*

One day a book publisher expressed an interest in a portrait of J.R.D.'s hung in another Director's room. J.R.D. was not sure it was truly representative of him. 'It's not me,' he said. Then settled down on a chair and viewed it. 'It makes me look aggressive.' 'But you can be quite determined, sir.' 'Determined, yes; aggressive never,' he gently added. An instance of this determination is the revival by him of his idea of an educational institute suitable for India years after he had abandoned it.

Eighteen years later, even though the original idea had been shelved, J.R.D. came up with the idea of initiating a new institute suitable for India. On a visit to France he invited Professor P. Olmer, former Director-General of Higher Education, Ministry of Education in Paris, to form another Working Group and assess the sort of institution India needed in the 1980s. J.R.D. selected for the Working Group Professor M.G.K. Menon, Professor Satish Dhawan, Dr L.K. Jha, Dr H.N. Sethna, former Chairman of the Atomic Energy Commission, and Professor Rustom Choksi of the Sir Dorabji Tata Trust.

Unlike in 1966, by the mid-1980s, the Indian Institutes of Technology (IITs) and the Indian Institutes of Management (IIMs) had established themselves as first-rate institutions and the needs of the country obviously were different now.

The Working Group recommended an Institute of Continuing Education. This, because the fund of knowledge is reported to double every seven years and no academic institute can prepare a person for a thirty-year career. The committee recommended an institute that would have a wider range of study than the IITs and the IIMs. Those who attended it would be those who had some years of experience in business, industry, education, the civil services and the like and had shown promise of leadership.

There are two ways of establishing an institute. The first is to start an institute and then look for the Director, the other is to find the Director and later build an institute around him. With the TIFR, Tatas had found the man—Homi Bhabha—and the institute was built around him. Two decades later, his brother Jamshed Bhabha recommended the construction of the National Centre for Performing Arts and the institute was built around him.

This time, too, Tatas waited for the right man. When Dr Raja Ramanna stepped down as Chairman of the Atomic Energy Commission, and was free to head this institute, it was established in Bangalore. The Sir Dorabji Tata Trust financed the National Institute of Advanced Studies (NIAS) initially with grants totalling two crore rupees, followed by Tata Steel with one crore; some other Tata companies too are planning to join in supporting the institute.

*

For decades J.R.D. has been the President of the Governing Council of the Tata Institute of Fundamental Research and the President of the Court of the Indian Institute of Science at Bangalore. These are perhaps the two most prestigious scientific institutions of India and their standards are comparable to similar institutions anywhere in the world. J.R.D. is also Chairman of the Sir Dorabji Tata Trust.

However, in the evening of his life, what interests him most is the future of two trusts that he has personally started. In 1944, by donating his own shares of Tata Sons and other companies, he started the J.R.D. Tata Trust. It is a multi-purpose trust. From time to time he transferred more shares to it and in all this time he has been the only donor to his trust. It has given out one crore rupees to charity over the years. J.R.D. has always looked upon this trust "as a small trust". Now, with the sale of his duplex flat at Sterling Apartments, an expensive uptown Bombay residential building, his wife and he are planning a J.R.D. and Thelma J. Tata Trust.

In the earlier years his trust gave generously to causes dear to him like family planning, but a couple of years ago a small event took place that gave a fresh impetus to J.R.D.'s thinking. To the formidable list of national and international awards that he already held*, a comparatively minor one was added in 1989—the Dadabhai Naoroji Award.

J.R.D. read in a life of Dadabhai Naoroji by R.P. Masani that this great Indian political figure was one of the pioneers of female education in India. By 1850, men had started receiving higher Western education but they had no intellectual fellowship at home to cheer or inspire them. So Dadabhai, then a college student, along with some colleagues used to go from house to house to persuade parents and guardians to allow their daughters to attend the first girls' school to be started or to at least allow them to sit on their verandas where they could teach the three R's to their daughters.[4] A few irate fathers had threatened to throw them down the steps for making such a preposterous proposal but others took advantage of the programme. Within two months the energetic band of volunteers had succeeded in capturing sixty-eight pupils. Dadabhai's exploits in this

---

* Among them are the Padma Vibhushan, 1955; Knight Commander of the Order of St. Gregory the Great (a Papal Honour), 1964; Knight Commander's Cross of the Order of Merit of the Federal Republic of Germany, 1978; Commander of the Legion of Honour of the French Government, 1983; Gold Air Medal of the Federation Aeronautique Internationale, 1985; Bessemer Medal of the Institute of Metals, London, 1986; Edward Warner Award, International V Civil Aviation Organisation, 1986; Daniel Guggenheim Medal Award, 1988.

direction inspired J.R.D. and set him thinking. Soon after receiving the award he sanctioned grants from his trust—unsolicited—to eight institutions engaged in the literacy and uplift of women. Literacy results in a lower birthrate and promotes his concern for family planning. He would like to see opportunities created for the employment of women and to give them skills. He would like this to be the thrust of his and his wife's new trust. Established late in 1991 the future of this trust is of absorbing interest to him.

The book *100 Great Modern Lives*, edited by John Canning, features only two Indians, Mahatma Gandhi and Jamsetji Tata. Writing on the Tatas, the book concludes:

> Probably no other family has ever contributed as much in the way of wise guidance, economic development and advancing philanthropy, to any country as the Tatas have to India, both before and since Independence.[5]

CHAPTER III

The Citizen

=====

*Perpetual devotion to what a man calls his business is only
to be sustained by perpetual neglect of many other things.*
                                                    - R.L. Stevenson

J.R.D.was the first prominent Indian to espouse, with missionary zeal,
family planning and the first also to spell out the need for a Presidential
System of government for India. He propagated family planning soon
after independence and has been advocating the Presidential System since
1968.

After independence when J.R.D. raised with Jawaharlal Nehru the
importance of curbing the population, Nehru replied: 'But Jeh, population
is our strength!' Undeterred, in 1951, J.R.D. again spoke publicly of the
necessity of propagating the need for population control. India's population was then only 361 million. In his Chairman's statement to Tata Steel's
shareholders he said:

> There has, in the past, been an extraordinary reluctance to
> consider the population problem at all or an equally danger-
> ous tendency to oversimplify it by relating it only to food
> production. Even if the latter could be made to keep pace
> indefinitely with the rise in our population, the problem would
> be solved only in part because even the poorest of our people
> are not likely to be satisfied for long merely with a sufficiency
> of food.[1]

To meet the rising expectations of people for clothing, shelter, education, health, would be a prodigious task, he warned, even if the population

308

remained at the 1950 figure. He noted that the population would double in fifty years. He was only a bit off the mark: it doubled in forty.

He urged the government to appoint

> without delay, a high-powered commission consisting of eminent scientists, economists and sociologists to investigate the problem in all its aspects.[2]

When no action was taken by the government he returned to the suggestion again in his Chairman's statement of 1953. In keeping with his principle that if the government did not follow his suggestions he should do what he could, he set about studying the issue himself. For forty years he has kept in touch with Indian and world authorities on the population problem.

He has also taken two practical steps in the matter. In December 1954, J.R.D. encouraged Dr John Matthai, then Chairman of the Sir Dorabji Tata Trust, to propose to the Health Minister of India, Dr Rajkumari Amrit Kaur, that a School of Population Studies be started in association with the Tata Institute of Social Sciences (TISS). Such a school would train a corps of persons who would carry out population studies for the government and the universities. Dr Matthai indicated to the Health Minister that the United Nations might well be interested in participating in the venture if facilities for its participation could be made available. The following year, when Dr Rajkumari Amrit Kaur was at a UN Conference in Europe, she discussed the matter with UN officials.

Subsequently, the Union Government and Sir Dorabji Tata Trust started a Demographic Centre for Training and Research in July 1956 with UN collaboration. Later, the institute which was jointly started by the UN, the Union Government and Tatas became the independent International Institute of Population Studies (IIPS).

But J.R.D. wanted to have an institution that would come to grips with India's population problem more directly; so, in 1970, he started the Family Planning Foundation of India. His idea was to launch a private initiative to supplement the government's effort in the field of population control, particularly in relation to "path-finding scientific research". Dr Douglas Ensminger, the Ford Foundation representative in India, played a prominent role in collaborating with J.R.D. The Ford Foundation donated $200,000 to the project on condition that the Family Planning

Foundation make up the rest of the finance in the ratio of 1:3. Over the years, Tata companies have contributed fifteen million rupees and other contributions have come from DCM, Escorts and Godrej.

The Foundation has supported 250 research studies and action-oriented projects, assisted in publishing the *Population Atlas of India* and the *Atlas of the Child in India* which have been much appreciated. It publishes a quarterly journal called *Focus on Population* and constitutes a powerful lobby and think-tank in the field of population control. J.R.D. has spoken extensively about family planning both in India and abroad. Despite all this the results of his effort—in terms of population control—have been far from satisfying.

As its founder President he makes it a point to attend the Foundation's frequent meetings in Delhi and one sometimes finds him reading bulky seminar papers on the subject. 'I have so much to read,' he complained one day picking up bulky documents of a Family Planning seminar. 'Why do you have to read all that?' I inquired. He replied he had to in order to discuss the papers with one or two professors.

The first Five-Year Plan that was being drafted when he first sounded the alarm, allocated to family planning, sixty-five lakh rupees out of a total of Rs 2,069 crores. During the Fourth Plan it was 1.8 per cent of the total allocation, reduced to 0.9 per cent in 1979 and for the Sixth Plan it was one per cent. In 1981 he urged that the ridiculous low incentives for vasectomy and tubectomy of Rs 200 be raised to Rs 2,000 and Rs 5,000 per person. It costs a minimum capital investment of Rs 7,000 to provide goods and services to every additional citizen and each operation could save the birth of up to four or five children.[3] Is the expenditure not worthwhile? he inquired.

In 1983, J.R.D. was asked to address the International Consultation of N.G.O.s on population issues in Geneva, the city where Margaret Sanger held the first World Population Conference in 1927. J.R.D. observed that India's economic growth had increased 3.5 per cent per year over the previous thirty years but her population had grown annually by 2.4 per cent resulting in a miserable one per cent growth in per capita income. 'Despite $170 billion spent on planned development, the number of people below the poverty line ($100 per annum!) is higher today than it was 34 years ago.'

He sees two areas of hope for the future. The first is the propagation of family planning through TV—a process which has already begun.

The other is to step up the literacy of women. Where women's literacy is the highest—almost a hundred per cent in Kerala—the birthrate is the lowest. In Rajasthan, U.P., and Bihar—the birthrate is the highest. His latter day interest in women's literacy springs from his twin wishes to educate them as well as his belief that literacy will have an impact on population control.

*

J.R.D.'s other great concern is for a system of governance that would best suit India. His thinking on the subject actively began in the late 1960s. The 1967 elections showed a sharp decline in the popularity of the Congress. Though Indira Gandhi was elected at the Centre she needed the support of the left parties to run the country. The Congress which had provided stable governments in almost all the states was turfed out of some of the largest like U.P. A coalition of desperate former Opposition groups joined together to govern some states. Defections and instability became the order of the day. The rule of law was openly flouted in states like West Bengal. The existing scenario reminded J.R.D. of France before De Gaulle introduced a Presidential System of government that gave France a measure of stability.

In February 1968, at the Indian Merchants' Chamber, Bombay, he proposed a Presidential System for the first time. He said:

> India is one of the twentieth century's major political anach-
> ronisms. The parliamentary system which was evolved over
> a thousand years of trial and error for the Government of a
> small, occidental island, and is predicated on the existence
> and smooth working of a sophisticated two-party system
> through a single Parliament is sought to be adapted to admin-
> istering an Asian sub-continent...
>
> The British system has been worked by generations of trained
> professionals and highly skilled politicians and administra-
> tors. In contrast, most of India's politicians are untrained and
> inexpert in the complex management of a modern society,
> while the main responsibility for administering the country is
> borne by an overworked cadre of senior civil servants whose

number is grossly inadequate to cater effectively to the needs of over half a billion people.

In addition, the machine has been burdened with the most ambitious economic planning and development programmes ever attempted outside Soviet Russia and with immensely difficult problems of defence, external affairs and finance. Up to the early 1960s, the strain on the machine was hidden by the dominating personality of a great leader, while a benevolent one-party autocracy maintained a facade of political stability and democracy in action. With Nehru gone, the facade has begun to crack and the machine is showing increasing signs of breaking down.[4]

He explained his foray into a political issue by explaining:

While I have always advocated, and still do, that businessmen should not mix with politics, this does not mean that in their capacity of educated and responsible citizens they should not take interest in political matters and form rational views on them. No intelligent analysis of economic issues is possible without taking into account the dominating influences of politics.[5]

He was very clear that his concern with political stability was rooted in his desire to improve the economic climate. The Parliamentary System, he felt, is, 'unsuited to the conditions in our country, to the temperament of our people and to our historical background.'

If the President and the State government were elected for a fixed term they could summon experts to help in the governance of the nation, he stated. The legislature could still make the laws. To study the problem he pleaded for the appointment of a high-powered commission consisting of experts on politics, law, education, science and other professions. He believed that, 'the experiment of welding our people together permanently into a single united nation,' could hinge on the political will and courage to change the Constitution.

Unlike other instances, where he could pursue a private initiative in the face of government's lack of interest, here he was helpless. All he could do was to speak about it—often privately. For example, at a meeting with Indira Gandhi in April 1980, after she returned to power, he urged her to use her secure position in the Lok Sabha to introduce legislation for the Presidential System once the Upper House was reconstituted. Nothing happened, of course.

In the 1990s with governments changing every few months and with corruption, incompetence and political instability rising to unprecedented levels there are many prominent advocates of the Presidential System.

*

Jamsetji Tata built the Taj Mahal Hotel without calculating the returns on his capital because he loved Bombay city and wanted to give it a hotel that was worthy of it. J.R.D. has the same love for Bombay and has made his own contribution to it.

At a civic reception accorded to him in 1983 by the Municipal Corporation of Greater Bombay he recalled,

> the Bombay of my youth with its magnificent harbour, its shady wooded hills, its flowering trees, its then disciplined population—there were no *morchas* then to impede one's travels through the city—its virtual absence of beggars, its freedom from law and order problems, and how happy a place it was in which to live and to work, a city of which we could be proud.[6]

He now contemplates 'with sorrow, not unmixed with anger, the present scene.' In fifty years Bombay has been transformed from a beautiful, clean city into a heavily polluted one with the world's largest collections of hutment dwellers and dilapidated houses. Besides, its air is heavily polluted. As "an old citizen" J.R.D. cannot but be torn between two emotions—deep sympathy for the five million or so people who have come from the countryside to find work, and indignation at the government's and Municipal Corporation's failure to deal with the added infliction upon the city.

'You will not find a single hutment in New Delhi,' he noted. 'These millions would not have come so speedily and in such numbers to Bombay if they had found they could not occupy footpaths or public or private accept compromises and would never be a party to the political shenanigans into which Indian politics increasingly sank in spite of Jawaharlal's effort to keep the political arena clean. Jayaprakash died a sad and disillusioned man whose friendship and regard I felt privileged to have earned.[1]

J.R.D. responded to JP's occasional appeals for his social causes but their relationship went deeper than that. It even transcended their ideological differences. J.R.D. wrote a remarkable letter to JP on 4 January 1955: 'I must...confess that I do not share your understanding of the capitalist system or its place in history. With great respect, I wonder whether you are not making the mistake of viewing the capitalist system as it was many years ago and not as it is today or in the form into which it is clearly developing all over the world. It is true that such evolution is somewhat uneven and that progress has been less in economically backward countries such as ours than in the more advanced democracies of the West, but the trend is clear and unmistakeable and I am convinced that those who are today so confidently sealing the fate of the capitalist system on behalf of history are likely to change their mind in due course or to regret the change which they will have brought about. I believe that in most parts of the world the system of free enterprise, far from dying, will be given a renewed lease of life in recognition of its ability and willingness to serve the community well and also from a revulsion against the unpleasant reality—as distinct from the myth—of State socialism.'

JP did not live to see these words come true thirty-five years later with the historic reversal of East Europe to a market economy. But J.R.D did and when I showed him the above words in March 1991 he read them aloud carefully, the last lines twice over. 'Not too bad,' he said. Without claiming any credit he returned the paper to me.

Gandhiji, says J.R.D., was

> by far the greatest personality and, to this day, the most extraordinary human being I have ever met: he inspired in me, as in most people, a mixture of awe, admiration and affection combined with some scepticism about his economic philosophy despite which one would follow or support him to the end, come what may. Perhaps the most unexpected and endearing trait I found in him was his almost childlike sense

of fun to which he gave vent in a chuckle which he sometimes used deliberately to put one at ease in his presence. He was also, like Jawaharlal Nehru, the most considerate and courteous of men who would never leave a question or a letter, however unimportant, unanswered.[2]

On one occasion when J.R.D. met Gandhiji with G.D. Birla and Sir Purshotamdas Thakurdas in 1944, he found Pyarelal, Gandhiji's secretary taking notes of the conversation.[3] After the meeting he casually mentioned to Sarojini Naidu that he found it rather strange that their private conversation was being so recorded. Rather unwisely she went and repeated this to Gandhiji. Gandhiji wrote to J.R.D.:

Dear Jehangirji,

I am glad you drew Mrs. Naidu's attention to the fact that Shri Pyarelal was taking notes of our talks yesterday. It is usual with him. I had omitted to tell him that talks such as between yesterday's company were not to be taken down. I have now had them destroyed in my presence.

Please excuse the indiscretion.

Yours sincerely,
M.K. Gandhi

The same day J.R.D. replied that he was very grateful for the letter and went on to explain how the misunderstanding had happened. He said that in the course of his conversation with Sarojini Naidu 'I incidentally mentioned, partly out of curiosity and partly to make conversation, that Shri Pyarelal had throughout our interview been busily writing something. I am terribly sorry to find that Mrs. Naidu thought that my remark implied some suspicion or disapproval on my part....I beg you to believe that nothing was further from my mind. When she told me that the late Shri Mahadev Desai and now Shri Pyarelal always jotted down whatever you said in the course of your many interviews it struck me as being quite a natural, proper and useful procedure.

'I am therefore distressed to think that my remark, as conveyed to you, that I am to continue to get that tea in case you die before I do.'

In his letter of 12 March 1965 J.R.D. replied, 'I have considered your suggestion of making a provision in my will to ensure that you continue to receive that tea after my death, but as I do not expect anything to be left of my estate after paying taxes, I am making other arrangements to ensure continuity of supplies even while I am waiting for you on the other side. I only hope we will both go to the same place!'

CHAPTER V

A Friendship

'I want to give you a little trouble,' wrote Jawaharlal Nehru, 'I should like to take a parcel of mangoes with me for the Mountbattens. I suppose you will be able to get fairly good Alphonso mangoes by the time I go. I do not want to add too much to my luggage so do not give me too many.' A cyclone had damaged mangoes on the west coast and none had arrived in Bombay when the Prime Minister's letter of 4 April 1949 reached J.R.D., but he assured the Prime Minister that he would have the mangoes within a fortnight. J.R.D. wrote on 13 April 1949: 'Please do not worry about the weight on your flight as mangoes carried by the Prime Minister are of a very special kind and the laws of gravity do not apply, and which therefore weigh nothing.'

A cheerful Jawaharlal wrote back on 7 May: 'I suppose you have received a letter from Lady Mountbatten to thank you for the mangoes, which were appreciated very much. I sent some of these mangoes to the Princess Elizabeth, as it was her birthday. Apparently they reached the King (George VI) who enjoyed them thoroughly. Among other recipients of your mangoes were Attlee and Stafford Cripps. So you see they went pretty far. Later I got two baskets sent by G.D. Birla.'

The friendship of Jawaharlal Nehru and J.R.D. went back to a time when Nehru was a dashing thirty-six and J.R.D. was only twenty. On the occasion that Motilal stayed with R.D. Tata, J.R.D. remembers Motilal's two children coming to visit. With Jawaharlal came Vijayalakshmi, known affectionately as "Nan". Wrote J.R.D. about Vijayalakshmi in *Sunlight Surrounds You*, a birthday gift to Mrs Pandit from her three daughters:

> She was one of the loveliest women I have ever seen and
> I fell into an enchanted trance on the spot from which I
> never fully recovered.

When India became independent, Air-India International had not yet started so J.R.D. arranged for a domestic Air-India plane to fly Vijayalakshmi Pandit, to the Soviet Union as India's first Ambassador to that country.

Jawaharlal Nehru inspired J.R.D. Both were modern men, they both believed in the virtues of machinery and modern science. J.R.D. could find an equation with Nehru he never could with Mahatma Gandhi. Nehru represented the hopes of millions whom he had reached and addressed in the remotest villages of India. What is more, Nehru had also captured the imagination of India's elite with his books. His readers had lived with Nehru as he sat beside his dying father, Motilal, propped up in bed "like a wounded lion"; they had travelled with Jawaharlal to Lausanne and visited the bedside of ailing Kamala with him. They had stood with him in the witness box as he defended the cause of India's freedom: and, when he thundered on the public platform on the sands of Chowpatty, they, facing him, echoed in their hearts his words. They had lived through his incarcerations and seen the inside of British jails through his eyes. Nehru was conscious of his magnetic personality and once quoted T.E. Lawrence:

> I drew these tides of men into my hands and wrote my will
> across the sky in stars.[1]

*

'Bliss was it in that dawn to be alive, but to be young was very heaven!' The enthusiastic lines of Wordsworth at the start of the French Revolution found an echo in many Indian hearts as with stirring words Nehru ushered in the dawn of India's own freedom:

> At the stroke of the midnight hour, when the world sleeps,
> India will awake to life and freedom. A moment comes, which
> comes but rarely in history, when we step out from the old to
> the new, when an age ends, and when the soul of a nation,
> long suppressed, finds utterance.

Nehru said in the Constituent Assembly, 'Long years ago, we made a tryst with destiny and now the time comes when we shall redeem our pledge, not wholly or in full measure, but very substantially.'

In redeeming that pledge J.R.D. thought perhaps he too might have a small part to play, for when freedom came, at the pinnacle of politics stood Nehru and at the apex of industry stood J.R.D.

*

Full of hope and promise J.R.D telegraphed Nehru on 14 August 1947:

DEAR JAWAHARLAL: ON THIS DAY MY THOUGHTS GO TO YOU WHOSE STEADFAST AND INSPIRED LEADERSHIP HAVE BROUGHT INDIA TO HER GOAL THROUGH THESE LONG YEARS OF STRUGGLE AND SUFFERING STOP I REJOICE THAT YOU WHO HAVE ALWAYS HELD SO HIGH THE TORCH OF FREEDOM ARE THE FIRST PRIME MINISTER OF FREE INDIA AND I SEND YOU MY HEARTFELT WISHES FOR SUCCESS IN THE HEAVY TASK OF GUIDING HER TO HER GREAT DESTINY STOP

After the euphoria of independence, the reality of Partition dawned. The tremendous upheaval in Punjab and in Bengal resulted in refugees streaming into India with stories of the horror they had experienced. The Tata Institute of Social Sciences (TISS), Bombay, sent a team to assist with refugee relief. Nehru later said: 'We found the difference in their work and the work of many others who were earnest and had done their best but who did not have the training to do it well. There is a difference between the trained workers and the merely enthusiastic workers.'[2] As millions of refugees poured across the frontier, J.R.D. suggested to Nehru in October 1947, that a national fund for relief and distress should be started in the name of the Prime Minister and said that Tatas would be glad to make a substantial grant to it. 'If you have no such intention, could you advise me as to what we should do? In other words, I feel that such voluntary contributions should be used for meeting needs which cannot or would not be normally from Government help.'

On 11 December 1947 Nehru agreed and wrote: 'There has been a great delay in coming to some decision about the National Relief Fund. After consulting my colleagues here we have arrived at some decisions which are incorporated in the attached note. I hope you approve of them.' In the note the proposed "Trustees of the Fund" were as follows:

- Prime Minister
- Deputy Prime Minister
- Finance Minister
- President of the Indian National Congress
- Chief Justice of India
- A Representative of the Tata Trustees
- A Representative of Industry & Commerce

And possibly two other names of persons in their representative capacities.

In 1948 Nehru agreed to the formation of Air-India International, the first joint sector undertaking between private and public enterprise. J.R.D. pinned his hope on more such ventures with the government, but this was a dream that was to fade.

Also in 1948 J.R.D was selected by Nehru as a delegate to the United Nations' session at Paris. Vijayalakshmi Pandit was the leader of the delegation. Every morning she briefed the Indian delegation. J.R.D. notes: 'She told everyone what to say or do and told us off when she felt like it. She did it with such a lovely smile....Fortunately for me I avoided directing it to myself by the simple experiment of speaking so little as India's delegate to the Economic Committee to which she had assigned me, that she could hardly find fault with what I said.' J.R.D. says he spent a good part of his time listening to the political debates in the General Assembly which were far more exciting.

The following year (1949) Nehru invited J.R.D. to go to the United Nations again. J.R.D. felt that he had so much to do in India and felt that attending these proceedings was "such a waste of time". He, therefore, politely declined the offer. Looking back on this refusal he admits, 'Maybe I was wrong.' He felt this especially because this refusal came after another a few months earlier. Nehru had written a "Secret and Personal" letter, on 16 March 1949, inviting J.R.D. to be the Chairman of the monazite sand project in Travancore, now called Indian Rare Earths, one of the earliest public sector undertakings. India's leading scientists, Dr Homi Bhabha and Dr S.S. Bhatnagar, had requested the Prime Minister to invite J.R.D. to be the Chairman. Nehru assured J.R.D. that the actual work of the Chairman would be very little as there would be experienced executive officers in charge. J.R.D. wrote back on 7 April 1949, 'I am unfortunately one of those who cannot undertake a job without taking a deep and detailed interest in it,' and, in the absence of a Managing Director, he felt that he would have to give quite a lot of time to the activities of the company, especially in its early stages.

'I greatly appreciate the confidence evinced in me by your request. As you know, my services are always at your disposal, and if after reading this letter you still think my appointment as proposed is necessary or desirable in the interest of the project I shall bow to your wishes.' J.R.D. noted that in view of the tremendous pressure of work he faced he had already given up the Chairmanships of a number of companies and 'a couple of months ago I decided, to my regret, to give up administrative charge of Air-India and Air-India International.' J.R.D did offer that 'even if he was entirely unconnected with the company he would gladly be available for consultation and negotiations, etc.' In a P.S. to the letter, he said: 'Since dictating the above I have learnt about the appointment of (Sir Homi) Mody as Governor of U.P. As a result I shall have to take over much of his work in the firm and will be harder pressed than ever. I do hope, therefore, that you will kindly agree to release me from the proposed appointment.' J.R.D.'s deputy, J.D. Choksi, was finally selected as Chairman of Indian Rare Earths.

Despite refusing Nehru's offers, and despite his huge workload—at the peak of his career, which extended for about three decades, J.R.D. put in a seventy-five to eighty hour week, and his sister Rodabeh says that in those days when they rang him in the office he would flatly tell them he was too busy to talk—J.R.D. still found a lot of time for Nehru. His concern and affection for the man is evidenced in letters to the Prime Minister.

On the death of Vallabhbhai Patel, in December 1950, he wrote a personal letter to the Prime Minister: 'I have been appalled at the load you have been carrying and the pace at which you have been driving yourself all these years....There is, however, a limit to the strain to which any man, even one possessing your physical and nervous resources, can put himself and I hope that with the destinies of millions of people depending so much on your continuing life and health you will somehow find it possible to spare yourself the innumerable, non-essential calls on your time and energy to which you constantly allow yourself to be subjected.' He said to Jawaharlal, 'You are always in my thoughts and I pray that you will long be spared to lead our unfortunate country in these dark and evil days.'

reinforced his interest and only the rigidity and autocratic nature of the Marxist state seem to have kept him from going the whole way in his acceptance of it. In his Presidential Address to the Indian National Congress in 1936, Nehru had said, 'Socialism is a vital creed which I hold with my head and heart.' Notwithstanding his views against capitalism, prior to coming to power, Nehru did show an open friendship towards J.R.D. Once in a position of power, his ideological thinking perhaps told him not to get too close to J.R.D., lest his feeling for J.R.D. as a friend made him yield his principles.

Again, perhaps, the issue of airline nationalization made Nehru realize that a certain distance had to be kept. At the same time Nehru's sensitivity and care is apparent in the two-page letter he wrote when he found J.R.D. very upset on the issue*.

A further divide between Nehru and J.R.D. arose at the time C. Rajagopalachari started the Swatantra Party. The Communist Party of India, then united, emerged as the second largest party in parliament in the 1957 election. This development made J.R.D. very sympathetic to Rajaji's attempt to start the rightist Swatantra Party.

On 15 May 1961, C. Rajagopalachari appealed to J.R.D. to assist him. 'While you may exercise your judgement and help the party in power, I respectfully urge that in the interest of good government and parliamentary democracy, the national interest in general as well as the particular interest of those engaged in the industrial and commercial progress of the country justify adequate assistance being given to efforts calculated to build up and bring an opposition party into effective operation....Your decision will serve to give a lead to all others....'

On 15 July J.R.D. replied that his colleagues and he had been following with interest and sympathy Rajaji's efforts. 'We have also realised the imperative need to ensure that the Communist Party is not the only effective alternative to the Congress Party and that India's political life develops in a truly democratic way around two main opposing Parties, neither of which would be to the extreme Left or the extreme Right. This, in our view, is the *sine qua non* for the survival of democracy....' He assured Rajaji of support from Tatas and went on to explain that Tatas would continue to support the Congress which had given political stability although this policy might appear to some to be self-defeating.

---

* Quoted in the chapter "Nationalization of Airlines" in Part II.

In the coming weeks when he met Jawaharlal Nehru, J.R.D. thought it best that Nehru heard from him rather than anyone else about Tatas' intentions. J.R.D. told him that Tatas were planning to also support the Swatantra Party. Nehru blew up. 'You have no business to do that,' he stormed.

J.R.D. thought it best to write a letter of explanation to Nehru that the Prime Minister could read when he was in a calmer mood. Writing to him on 16 August 1961, he said: 'Even though we may not feel happy about some of the policies of Congress and Government, particularly in the economic field, we intend to continue that support,' which he reminded Jawaharlal, went back to the days of his father, Motilal Nehru. While acknowledging the stability and unity that the Congress had provided, J.R.D. said:'We have been perturbed by the total absence of any responsible and organised democratic Opposition which we feel is an equally indispensable element of any permanent democratic organisation of society. As a result, we have been increasingly worried about the future, however distant, when the strong and outstanding leadership which you have provided may no longer be there. I am one of those who believes that the single party regime under which we have lived since Independence has been up to now a good thing for the country as it has provided the stability and the means of concentrating the national energies and resources on orderly development, which would have been impossible without a strong and continuing administration. But even you will agree, I think, that if continued indefinitely this situation contains the seeds of trouble and risk in the future.'

J.R.D. pointed out the dangers of the Communist Party providing the only alternative to the Congress, and said: 'The only party which, it seems to us, offers any possibility of developing ultimately into a responsible and democratic Opposition, is the Swatantra Party which, after all, consists mainly of people who have been fostered by the Congress, have spent many years within the Congress and, while conservative in outlook, are not reactionary or communal or extreme rightists.'

J.R.D. concluded: 'I am anxious that there should be no misunderstanding in your mind as to our views, and motives. My feelings towards you and my personal devotion to you and to the cause to which you have dedicated your life remain unchanged. Please do not take the trouble to reply to this letter. I only seek your understanding.'

Nehru replied promptly on 18 August: 'Although you have asked me specially not to reply to your letter, I am sending this relatively brief acknowledgement.

'During the fairly long life that I have devoted to political and like matters, I have tried to follow, with more or less success, certain paths aiming at certain objectives. I suppose that, as was natural, I have learnt from experience and occasionally varied the policies I pursued somewhat. But basically I think I have been fairly consistent. This is so because I firmly believed in them. Naturally if I do so and think that those policies are beneficial for the people of India, I must continue to follow them.

'You are, of course, completely free to help in any way you like the Swatantra Party. But I do not think that your hope that the Swatantra Party will emerge as a strong Opposition is justified. I think that it will be disappointed at the turn of the next General Elections. It seems to me that it has no roots in the thinking of either the masses of India or the greater part of the intelligentsia. Indeed it seems to me to be cut off from modern thinking even in Europe or America. It is quite remarkably out of date and out of step with events. However, that is just my view.'

I observed to J.R.D., 'I find from your letters that you kept your personal equation with Nehru separate from your ideological differences. You not only cared for Nehru but one can say you had an affection for him, as your letters of condolence to Vijayalakshmi and Indira show.'

'There remains my affection for him,' he shot back, as if Nehru was still living. Then sadly, softly, he added: 'But there was nothing I could talk to him about. In fact I tried, I once told him, "Jawaharlal there are so many things I want to talk to you about but you never seem to have the time!" Nehru said, "All right, come and have lunch with me. Come at 12.45 p.m. so that we can have 15 minutes before lunch." So I went. I started by saying that I was not against some controls but the private sector need not be held back excess̄ . At that point Nehru did his usual trick. He looked out of the wind̄ ̄ ̄ was not interested in listening. Thereafter, he said, "Com̄ ̄ ̄h." He took me to see his giant panda. It was the sam̄ ̄ me I tried. Otherwise, as a gentleman, a considerate fellow, you coū ̄'t have had a better person.'

In a letter to me in May 1989 J.R.D. said that it had been a long-held regret that, 'sharing as I did Jawaharlal's deep sympathy for the poor and deep urge to alleviate their poverty, I was denied or denied myself, opportunities to help him in that task by my disagreement with his socialism.'

On another occasion J.R.D. said: 'I remember one of the last times I was invited for a lunch with him. There were only some of his personal guests like his sister Nan (Vijayalakshmi), but no politicians were present. He turned up quite late from some meeting and said, "Hello, hello" to us, sat down on a couch, fell back and fell asleep! So we all said softly "What do we do?" You could see the man was tired out.'

The last time J.R.D. met Nehru was in May 1964 in Bombay. J.R.D. found 'his eyelids were swollen, a very sure sign of lack of circulation.' Nehru died a week later on 27 May 1964.

Looking back on Nehru, J.R.D. says, 'I think it is fate. It is right that a man with his heart and tremendous sympathy for the poor became the Prime Minister. When a man is young he has a heart. I can understand Nehru in his younger days, excited by the Russian Revolution as most young men were, but when he did not grow out of his fascination, I think it showed a lack of realism….It is to his credit that he did not make India Communist. He accepted the private sector under sufferance. When it came to expanding Tata Steel, he did give us the permission to expand by a million tonnes. There was a mixture in him.'*

On the occasion of Nehru's death, J.R.D. wrote to Vijayalakshmi Pandit, 'Jawaharlal was that kind of man who made you feel he was a part of your own life and I grieve at his passing.'

When I observed in passing that Nehru's going was 'the end of an era for India,' J.R.D. added 'and for me too.'

And so it was that these two men, each at the top of his respective field of endeavour, once close friends, were later distanced by ideology. Had they worked together they could have given much to India. But fate decreed otherwise and India was the loser.

---

* J.R.D. said this when he was moved while talking to the author about Jawaharlal Nehru in August 1988. Earlier, in July 1986, J.R.D. mentioned in his Foreword to *Keynote* that if 'Sardar Patel was a younger man and had become the Prime Minister, India would have followed a different path and we could have been in a better economy than we have today.' The view expressed in *Keynote* is, I believe, his considered opinion.

CHAPTER VI

The Family

Nestling among trees, beside one of Bombay's tallest skyscrapers, is "The Cairn". Situated at the end of a side lane off Altamount Road in South Bombay, it is a remnant of a bygone age, a bungalow with a spacious compound, shaded trees and outhouses for servants. The bungalow goes up in tiers. The entrance is at a lower level and a winding passage interspersed with steps—fifty-three in all—takes the visitor to the rooms of J.R.D. and his wife.

Once the house was full of life, its public rooms in mint condition. For the last decade, however, since the hostess Thelma (Thelly) Tata had a stroke, the public rooms are never used. Taking a French journalist round J.R.D. said, 'I can afford to re-do these rooms, but they are not used, so why waste money.' The same evening, at supper, the journalist told me that he could not believe that J.R.D.'s counterpart in France would ever contemplate living as simply.

J.R.D. received the journalist in his study. The journalist would have been even more surprised had he learnt that the small study was also his bedroom at night. The wide grey sofa is converted into a bed. I chanced to observe that the room, about 15 x 20 feet, was too small for a person like him. He just shook his head. 'Maybe, but I prefer a small, compact room where every need is quickly and easily accessible. I don't need a big room.'

The smallness of the room is compensated for by the expanse of the bathroom with its tub and exercise equipment. The bedroom originally was much bigger but J.R.D. divided it to make a workshop.

Facing the sofa in the study are two comfortable swivel chairs and two telephones. Behind the sofa, on a desk, is a globe, two framed photos of his wife, some fresh flowers, an unused blotting pad, and two glasses filled with pencils and ballpoints. Along one wall are bookshelves with a fine

334

selection of titles that include a twenty-four volume encyclopaedia and numerous books on aviation, *The Spirit of St. Louis* by Charles Lindbergh and Sir Frank Whittle's autobiography among others. Two of the best thumbed copies in his library are the complete short stories of O'Henry and the book on Jamsetji Tata by F.R. Harris. There are also a number of thrillers and nature books, mostly American.

His wife's room, next to his own, is a rather spacious one with facilities for TV and dining. Since her first heart attack and stroke in 1980, and then a fall in 1982, she is mostly confined to bed or a wheel chair. Forty years ago, 'there were always people coming to the house and Thelly ran a beautiful home, always full of flowers,' says her friend Tina Khote. 'Jeh is finicky to the enth degree. You have to be a perfectionist if you are married to such a person....What I liked was that there was no ostentatious living, everything was done in the most simple but gracious style, interior of the house or decor, all was in good taste. In such a liveable house you could curl up in a corner.'

A Polish lady, Tina Khote came out in her twenties to marry an Indian. Realizing that the young foreigner might find life in a strange city difficult Thelly befriended her. Tina Khote remembers being taken by Thelly Tata to Anand Kendra, a centre she had created for homeless boys. 'It was,' she says, 'my first introduction to social work in this country....She was marvellous with the boys. She knew every one of them by their first name. She would tell me their stories and their background and later on she would ask us whether we could find any employment for them.'

Two days before J.R.D.'s seventy-sixth birthday, Thelly had a severe heart attack followed by a stroke. While she was ill in hospital J.R.D. spent all his spare time with her and gave up a game that was most dear to him—golf. When his wife came home from hospital he decided to spend more time with her and for this reason did not resume his weekend golfing. 'As I didn't know how much longer we would have together, I never started playing again.'

As they had no children, over the years several friends advised J.R.D. and Thelly to adopt a child but they did not. He loves children. Once an employee called Robert Lawrence, who had taken J.R.D.'s dog Knight to the vet, came to report on its health. A black Doberman, trained to keep out intruders, Knight did not take very kindly to human beings he did not know.

On this occasion Robert had brought his young daughter Chantel along.

Proudly he told J.R.D. 'You know Knight allowed her to embrace him!'
'He did!' said J.R.D, acting more surprised than he was to the amusement
of the child. J.R.D. told Chantel he could hardly recognize her after two
years.

'How old are you now?'

'Ten.'

'You are so tall! Come and play with Knight again but don't grow up
so fast.'

Knight bounded into the room.

He asked the child, 'What will Knight think of you if you grow up so
fast?' Chantel gave a charming smile; J.R.D. had got her wavelength.

At the same time as this incident I received a letter from Robert Jansen,
Air-India's former Sales Manager in Germany. In it, he related an incident
that took place in the 1950s in Dusseldorf: 'When I received him (J.R.D.)
at the airport he asked whether we could go to my house, as he wanted to
wash his hands. I knew that he wanted to see how an Air-India man lived.
He liked the house and the nice garden which was located very near the
Dus Airport. He greeted my wife and when he saw my children, aged 4
and 6, he grasped a comic book about Mickey Mouse and started reading
it to the children. And then in between laughter about Mickey Mouse he
started asking me questions, very sharp and precisely, about my target,
my revenue, sales arguments, motivation. At the same time he showed the
comic book to the kids and during his questions he never looked at me but
pointed to the comic. Somehow he was satisfied with my answers and my
wife prepared a cup of tea for him. When I dropped him at the station he
went in to a flower shop and ordered a beautiful bunch of flowers to be
sent to my wife just because she prepared a simple cup of tea.'

When I once inquired of J.R.D. whether he missed not having children,
he put on a brave face and said, 'Not really.' He went on to talk of how
disappointed an eminent businessman was with his son and daughter. On
a second occasion when the same question came up, vis-à-vis the future
of Tata Sons, I inquired, 'Do you miss not having a son?' 'I would have
liked to have a daughter, not a son,' he said, 'because I feel a son would
probably have created a problem for me. Tatas is much too big to have a
hereditary succession. Tatas is, thank God, a national institution and must
continue to be so.'

The relationship between a husband and wife is too intimate for an
outsider to understand. Auguste Comte was perceptive when he described

the female sex as affective and emotional and the male sex as effective and active. 'A woman's thoughts,' says Comte 'are less abstract than man's. A woman does not altogether understand a man's need for action. Man's thoughts travel by aeroplane. They fly above space and time. They discover wide but unsubstantial landscapes, they mistake "the straws of words for the grain of things." Woman's thoughts usually go on foot.' J.R.D. flew in more senses than one. His mind spanned national and world events. Thelly stayed on the terrestrial plane, looking after him, loving him intensely, some say jealously, being his hostess and, like him, wanting to be worthy of Tatas.

Today Thelly has a night and a day nurse. Sometimes J.R.D. reads her his favourite stories from O'Henry. He is there for dinner and after dinner they watch TV or a movie on the video. When his book *Keynote* was launched on 29 July 1986, in New Delhi, he had arrived in the capital from Europe in the morning. The function started at six in the evening and ended after 7 p.m. J.R.D. drove straight to the airport to catch the last flight to Bombay. He wanted to be home with his wife before midnight. It was his birthday.

'In some ways,' says one of J.R.D.'s colleagues 'he has everything. In other ways he has nothing.' He stays on in office working away till late in the evenings and if anyone pleads with him to go home he tends to reply 'What have I to go home to?'

J.R.D.'s closest male heir, Nusli Petit, son of his sister Sylla, applied for an Air-India job on his own. He says his uncle, Jeh 'always wanted me to get in on my own and he made it a point to show no favouritism towards me. He didn't hire me in fact. When my name came up to the Board, I believe he said "This is my nephew and it is up to you whether you want him or you don't want him. I have nothing to say about it."'

Of Nusli's mother Sylla, J.R.D. proudly says that she was the first woman to get a flying licence in India. She was also a skilled tennis player who participated in the Western India Lawn Tennis Championship. J.R.D. says that his mother and Sylla were the two great influences in his life. As children Sylla and he even looked somewhat similar.

Sylla married Fali Petit, a bright law student in England. His son relates his father's adventures as a student in the 1920s: 'My father once drove from London to Cambridge. He was stopped by the police for some offence and was given a ticket. He went on, was stopped for speeding, got another ticket. He got to Cambridge and there they (his friends) challenged him to go up a one-way street, he got another ticket. So he thought this was getting a little too hot for him. He went back to London, shipped his car back to India and left for India. Meanwhile his father died, so he became Sir Dinshaw Petit! For years the cops would go to his digs in London asking his landlady where he was. But by then he had changed his name and disappeared from the face of the earth.'

As children Nusli and his sister spent their long holidays at their home in the South of France. At other times Sylla stayed at Petit Hall, Nepean Sea Road, Bombay. Speaking about his mother, Nusli Petit says, 'People would say she was absolutely adorable in everything she did. People would say "What a marvellous person." She was gentle, she looked after the children and the servants. In their village in the South of France when she went shopping the shopkeepers and the bus drivers used to love her. She was one in a million.' When asked if she had "the peppery temper of the Tatas" he replied: 'She did have a slight temper. If she found the servants were not cleaning properly she would scream and shout but once it was done it was over. She'd still look after the guy.'

Throughout the war she played an important part in the Women's Voluntary Service—WVS—in India. 'She was the boss,' says Nusli. 'A lot of sailors would write thank you letters and she would reply to them. In a silver frame is a picture (given to her) saying in gratitude from HMS....Whenever the family went to Genoa after the war sailors would come and greet her.'

In 1962, it was diagnosed that Sylla had terminal cancer. 'She never gave up talking about life,' says Nusli. 'Even during the last month of her life she talked about changing her Mercedes Benz when she got up and I had to sit there and say "yes," knowing she was not going to get it. It was pretty tough for us. We had to keep up with her.'

Some time after this J.R.D. wrote to a friend from Geneva, in June 1963: 'I have been spending a good deal of time during the last fifteen days in Nice, where my sister Sylla is dying of cancer. The end may come any moment now. I am going back there tomorrow. I should have been back in Bombay quite a few days ago but could not tear myself away. She is by far the best of the four members of the Tata family, and certainly the bravest. She has been through three terrific operations in two years, alas to no avail.'[1]

'She could not be operated (upon) any more,' Nusli continued. 'She said "If I die, don't wait and mope around my grave." When she died I felt terrible. I broke down, crying my eyes out. I felt awful looking at the coffin at her burial place in the village and yet as they lowered the coffin in her grave, all of a sudden I felt complete relief from everything. It was as if someone was patting me on my back and saying "go on". I put in some flowers and a little earth of our property on the coffin and then I felt "you've done your duty, now carry on". Bus drivers, shopkeepers who do not attend funerals unless they know the person well, all came to the funeral.' Writing to her doctor in New York, J.R.D. said, 'To the end she was more concerned with the health and welfare of her husband, children and other members of the family than her own (self). She was truly a most exceptional person, loved by all.'

Next in line after J.R.D. was his brother Darab. Friends describe Darab as a very friendly and popular person. He was married but was childless; he adored children and every Christmas he would take a number of them from one of the children's homes in Bombay to Regal Cinema and then on to the Taj (of which he was Chairman for many years). He would give them a Christmas party, squat on the floor and play with them all. Nusli Petit says of his uncle, Darab, 'I don't think he was ever destined to be a great businessman but he was a great art lover—a lover of nature. I would say he was very gentle in that respect—more than his brother or my mother. He'd love to go into the countryside, be with the people. We both had jeeps, he loved jeeps and I would go with him. He should have been a tourist guide. That guy knew all the spots and we would go to places nobody knew. He had gone round beaches and spots which I can't find even now if I wanted to. He used to take us to a beach with black sand somewhere and the villagers knew him. He'd been there so often they would sit him down, bring him water, talk to him. Very sad what happened to him.'

'At what time of his life did his health decline?'

'I would say around 50. It suddenly came on. Sort of a nervous disorder that took over. He was never the same after that. He was always a slightly nervous person. All the Tatas are slightly hyper-strung.'

Laura, his friend, cared for Darab as did Sylla and Dabeh (Rodabeh). 'Dabeh,' says Nusli 'was the go between J.R.D. and him and she cared for him right to the end. When Darab died he was 75.' The only letter on record of J.R.D. to his brother is one dated 24 August 1983 to Darab in Geneva urging him to visit an endocrinologist in London, who had been recommended. J.R.D. thought that perhaps there was some biochemical deficiency which resulted in the disorder. 'It would be a pity,' he said 'if you came back to India with the same symptoms without having made this one attempt at tackling your problem through the chemical route...'

The relationship between Darab and J.R.D. was uneasy for a good part of their lives, though in Darab's last years it somewhat improved. The distance may have arisen from the fact that J.R.D. got all the attention and plaudits, while Darab, also a Director in Tatas, lived perpetually in his elder brother's shadow. Though the youngest of the family, Jimmy, had lived for only twenty years J.R.D. had always felt closer to Jimmy than to Darab.

Rodabeh, J.R.D.'s younger sister, is a skilled interior decorator who was responsible for some fine work at the Taj. The Sea Lounge in the hotel is her creation. She was married to Colonel Leslie Sawhney. J.R.D. admits he did consider Leslie Sawhney as the person who would possibly take his position as Chairman of Tata Sons. 'He had great leadership qualities,' says J.R.D. Unfortunately, he had a sudden heart attack and collapsed on the golf course in Bombay. Today, Rodabeh is J.R.D.'s only living relative. He makes it a point to take her out for lunch every weekend and when she is unwell he often visits her late in the evening, after a long day in the office.

Through all these trials J.R.D. has maintained a sense of humour. It is his saving grace. Once a stream of guests stayed at "The Cairn" but now the flood has dwindled to a trickle. One of the few privileged to have stayed at the house in the last few years is the remarkable Kira Banasinska. When she was past eighty, she launched a major project to have a permanent centre (the first of its kind) for training Montessori teachers in Hyderabad.

She is short, with shining eyes and a friendly smile, and one of the couple's oldest friends (from the 1930s.) Three years older than J.R.D., she comes up to him one day, a sprightly eighty-seven, and says in her Polish accent 'Jeh, the doctors say that I yem more healthy than *you* are. My heart and my pulse rate is better than yours.'

'Kira,' replies J.R.D., 'you are indestructible. Ultimately the only way to get rid of you will be to shoot you!'

CHAPTER VII

Vignettes

In July 1979, after I had done four months of research on Tatas for my book, *The Creation of Wealth*, I requested an appointment with the head of the House, J.R.D. Tata. As I entered his well-appointed office, I noticed a couple of paintings on the walls and a map of the world that spread over a substantial part of the wall behind him. The sofa set and his chair were in white leather. He was seventy-five then. He stood erect, keeping his head high, looking taller at first sight than he actually is.

Before the interview began, he questioned me on what basis I was writing the book and then answered the question himself. 'As former editor of *Himmat* you are coming to have a look at Tatas from the outside and writing about it.' He then enquired, 'Why are you writing on Tatas? Who is interested in Tatas? Who is bothered about Tatas?'

I was patient and polite at the start but when he kept up this refrain, I could see my work of the previous months being wasted if I did not secure his cooperation soon. Finally, I said rather firmly, tapping my finger on his desk, 'Sir, the hand of history has shaped the tapestry of Tatas and it is a story that has *got* to be told.' He was taken aback. He leaned back in his chair and looked at me quietly and intently for a moment, with raised eyebrows. After a pause he said, 'All right, what can I do for you?'

I replied, 'I want to know what is Tata Sons' holding in different companies.'

He rang up the Secretary of Tata Sons. The telephone rang and was not picked up. He sent a man across and the Secretary rang back. 'Why is it that when you are not in your seat, no one lifts up the telephone? You must always have someone to answer your phone,' he said into the receiver. He then asked the Secretary to give him the particulars I'd asked for and within a couple of minutes a large sheet of paper arrived on his desk. He took out his calculator and worked out the figures.

Encouraged by this triumph, my next question was more sharp: 'If the Tata Trusts hold 80 per cent of the shares, who holds the rest?'

I thought I could perhaps point out in my book that if the family owned a good chunk of the remaining twenty per cent, it would not be incorrect to say it still had a considerable private interest in the Group.

'There is an outside party,' J.R.D. replied, 'who owns a good part of it. The family owns very little.' He rang up the Secretary of the company again and asked him to send me the break-up of the holdings. We parted friends. The next day the break-up was on my desk. I felt a bit humbled. The outside party, which has no connection with Tatas, owned 17.45 per cent, the family and Tata Directors together, owned about 1.62 per cent.

Years later, when I got to know him better, I learnt that J.R.D. liked to question people sharply not because he disagreed with them but because he wanted them to clarify their own thinking and define their objectives. Once he was sure of the person's convictions he was quick to give his unwavering support.

Ratan Tata, now Chairman of Tata Sons, gives an instance of J.R.D.'s operating style. In the late 1960s he was put in charge of NELCO. The company was losing money heavily. Soon after his appointment, the subject came up at a meeting before the Tata Directors and all of them criticized the company. Ratan was naturally upset. He had had nothing to do with the past performance of the company and he was being penalized for it. 'Jeh came to my rescue,' says Ratan Tata, 'and slowly turned round the whole conversation. If you are confident, he will question you and grill you, but if you are fighting with your back to the wall, he will come and duel beside you. He had the vision even then that the future belonged to electronics and NELCO had a future.'

*

In September 1989, J.R.D. and I were to meet at his home to leave together at 10 a.m. for a function. As I was early, I waited in the silver-grey Mercedes, outside his home. His driver, Peter, who had worked with him for twenty years, was seated in the car. I enquired of Peter whether he had seen any difference in his boss over the years. He replied: 'Yes, earlier, he was more strict. But I liked that strictness because he is good-hearted. In the old days, by the time he arrived on the veranda, I had to start the engine. (He still does!) He would tell me as soon as he sat down, "*Chalo!*

Chalo!" If he was having lunch at Air-India, he would inform me or say, "If I don't come by 1.15 you can go for lunch." If he unexpectedly made a lunch date he would come down himself, 22 floors in the lift, tell me not to wait, and go to lunch. Sometimes I went to lunch and if he came down after that, he would take another car and come to Bombay House. Never once did he take it up with me.'

As J.R.D. was told I was waiting, he thought he was late for our 10 o'clock date and arrived at the car breathless. His eye fell on my watch which showed five past ten. He quickly looked at his watch. It was five *to* ten.

'Your watch is fast.'

'Yes, sir,' I said, 'I keep it fast.'

'Why do you do that? Whom are you fooling?'

He was not impressed with my explanation that I kept my watch fast in order to be punctual.

As we left the function, he saw the dilapidated houses in which people of the area lived and began speaking with passion about the conditions our people had to live in. Some time later, when driving with him at Flora Fountain, he saw a poor man crossing the road with a bundle on his head. 'This is all he possesses in life,' noted J.R.D. with feeling. Most of us manage to reconcile ourselves to the poverty, dilapidated houses, streets with potholes that we see around us. J.R.D. refuses to accept what is wrong. At eighty-seven, he rebels and wants to fight back.

*

He takes some time to give you his heart, but once he does so, you can depend on him, whether you cross his path only six times a year or sixty times. Minoo Masani, who once worked for Tatas, relates what happened when he helped Rajaji to start the Swatantra Party in 1959. Masani was then in Tatas' employment. When Colonel Leslie Sawhney informed J.R.D. of this development, J.R.D. said, 'Damn good idea, but Minoo will have to resign.'

This was conveyed to Masani. Surprised, Masani asked J.R.D.: 'But why? Under the British you never asked me to resign for my political activities.'

J.R.D. replied, 'I don't want my shareholders to suffer from the activities of my staff.' He paused and asked, 'Are you angry with me?'

'Not angry,' Masani replied, 'just disappointed.'

'I know that I am a bit fussy on this point, but I do not want it to boomerang on my shareholders.'

J.R.D. then asked, 'If you don't get elected what will you do? I don't want to throw you on the footpath! If I was Mr... I would give you Rs. 5 lakhs under the table. But I cannot give you that. So I must find an open way.'

Some time passed after this encounter, and one day J.R.D. rang up Masani: 'I am spending sleepless nights over what to do for you.' Masani replied, 'Oh, make me a consultant.' J.R.D. promptly agreed, and Minoo Masani started a company, Personnel and Productivity Services, which Tata companies consulted with. J.R.D. even spent some time designing Masani's letterhead! Technically, though, Minoo Masani was out of Tatas. Later, when J.R.D. met T.T. Krishnamachari, the Union Minister said, referring to Tatas' arrangement with Masani, 'Whom do you think you are kidding?'

When in his mid-eighties, J.R.D.'s colleague, J.D. Choksi (J.D.C.) began to lose his memory. As often happens, one by one J.D.C.'s friends found it difficult to converse with J.D.C. and stopped calling on him. But right to the end J.R.D. would call on J.D.C. every Sunday, try to recall jokes of the past and cheer him up. 'In his last months J.R.D. was the only non-family member to call on J.D.C.,' J.D.C.'s sister-in-law says.

*

D.R. Pendse, a Cambridge-educated economist working in Tatas, was invited to address an international conference in London in 1979. Pendse, who later became Economic Advisor to Tatas, was not so senior then.

When J.R.D. heard of the invitation to one of his staff, he called him and enquired, 'Where is the text of your speech?' Pendse replied that he usually spoke extempore. J.R.D. exclaimed, 'You mean you will address an international audience of five hundred people without an address in your pocket! Have you rehearsed your speech?'

'No, sir. I will do it in the London hotel, the day before my speech.'

J.R.D. saw the potential of young Pendse but figured he needed help. Like a father he advised him: 'Your audience will hear you but have you heard yourself? Keep a tape-recorder in front of you every time you rehearse at home and play it back. Then you will know how the audience is going to hear you.' J.R.D. added, 'I've no doubt you know the subject well, but I think you should talk slowly. Listen to the tape-recorder and and you will find out for yourself.' He paused. 'Do you have a tape-recorder?'

'I'll get one from some friend. No problem, sir.'

J.R.D swung around on his swivel chair, picked up his ever-ready tape-recorder, checked if there was a clean tape in it and handed it over to Pendse. 'Take this. Write out your speech. Rehearse at home. Listen. Then you can return it to me. I will find another recorder. Don't worry and good luck to you.'

Pendse followed the Chairman's advice and when, at the end of his speech, the round of applause came, Pendse says, 'It was beyond my expectations. And as I sat down acknowledging it, I could almost feel the Chairman smiling at me.'

*

On occasions he is mischievous, especially when he is in a relaxed mood. Soon after he came out of a Bombay hospital in April 1991, he was going to America for his heart treatment. On the eve of his departure to America, I called on him. As I rose to leave, he said rather plaintively: 'You know, at my age nobody is really interested in what happens to me except myself,' (pause) 'and the ladies!'

'Whose hearts you have broken?'

'No, whose favours I seek.'

*

While he can be supremely collected and patient at moments of crisis, he can be terribly impatient over trivial matters. Lily Lalvani, a former Air-India hostess, was ahead of J.R.D.'s car on a traffic-clogged Bombay street. At the traffic light, her car stalled. As cars behind her honked away, she became increasingly nervous. Then she heard someone behind her say in a loud voice. 'Oh, these women drivers!' The friend beside her turned and said, 'My, it's J.R.D. He has got out of his car.' Lily became even more nervous but finally managed to get her car out of the way.

The next day, on 9 April 1985, Lily Lalvani wrote to J.R.D. that she had been a great admirer of his and had served him on flights. However, she said her estimation of him had suffered a blow after the previous day's incident. J.R.D. was away when her letter arrived but the day he saw it, he wrote back saying he was afraid he did not recall the event but that if he had done what she'd accused him of doing, he was very sorry. He added that he'd always felt that he was chivalrous to ladies and for this reason was even more surprised at hearing about this incident.

The letter was delivered at Lily Lalvani's office along with a bouquet of flowers.

*

J.R.D. has a high threshold of endurance, and this may be partly due to his ability to cut any tension he feels through humour. One weekend he was in a reminiscent mood about his maternal grandfather, who happened to be a practical joker.

'My grandfather and a friend once went out on the streets of Paris with a rope and a red flag. They held up the traffic, pretending they were engineers with those telescope-like sights to size up the road distances...'

As he began relating this, news came in of a disturbing item about Tatas in that morning's paper. He was visibly upset. 'Did you see the earlier item on the subject?' He asked. His jaw was set. He continued, 'It created a storm! It created a storm!' It was a serious matter about differences between two of his senior colleagues.

He thought for a minute and dialled a co-Director. He was amazed when all he got was a servant. 'Think of it,' he said, 'they go away for the weekend and don't even tell the servant where they are going.' He smiled and the tension left him.

He then continued reminiscing as if nothing had happened: 'As I was saying, my grandfather and his friend would hold the rope at either end across the street and stop the traffic. Next, they would get hold of two passersby and request them to hold on to the rope, saying they had to measure distances for widening the road and urging them not to let the rope go whatever happened. Then they would melt away into the crowd. The traffic would pile up, cars would blow their horns, policemen would come on the scene and berate the hapless rope-holders. Meanwhile, they would watch the happenings from a distance and enjoy it.'

This ability to transport himself from one subject to another, from one

346

age to another with astonishing ease, enables him to break the tension at moments of crisis.

*

Biologist and Nobel Prize winner Alexis Carrell pointed out that every man has two ages, the chronological age and the biological age. Chronologically, a man may be sixty-eight, but biologically, he may only be forty-eight years old. If biologically J.R.D. appears twenty years younger than he is chronologically, it is because he has taken the trouble to order his life meticulously rather than let it be run by circumstances. He has worked hard all his life but taken time off to play. And for forty-four years, every year for three weeks, he has skiied in the Alps, be it the Swiss, Austrian or the Italian Alps.

He started his skiing career at an age when most men retire from skiing—at forty-one, after the Second World War. From then on, till the age of eighty-five, he has gone skiing. In his sixties he could ski faster than ever before, his sharp pilot's reflexes and fit physique serving him well. He won the second highest award for proficiency in skiing, the *Chamois Bronze*, one short of the *Chamois d'Or*.

In his eighties he bicycled on an exercycle with pressure adjustments. Also, at that age he could do twice as many push ups as men half his age. A couple of years before his heart operation he demonstrated to me one of the more strenuous exercises he did. He has installed a padded plank at home, rather like an extended ironing board placed at an angle of about sixty degrees to the wall. He lies on it, head down, and then he throws up his hands and levers himself up through the strength of his stomach and thigh muscles alone, until his hands almost touch his feet. Most people cannot do that lying horizontally on the floor, let alone against the law of gravity. I was aghast at his doing the exercise. Before I could say anything he had done it once, twice, thrice till he was red in the face. I cried out, 'Enough, sir! Enough!' 'Ideal for the stomach,' he said nonchalantly. Próbably, but I wondered what would happen to the heart. In fact, before his commemorative flight, Karachi–Bombay, at the age of seventy-eight, he did strain his heart with his preparations for skiing the following January in Switzerland! 'I don't believe in slow exercise,' he says. At seventy-seven, he tried his hand at hang gliding, one of the most dangerous of sports. J.R.D. believes that man is made to live till 125. 'We don't because we don't exercise.' 'Sometimes,' he admits, 'I am lazy and do not feel like it.'

*

The *Economic Times*[1] relates two incidents involving J.R.D. and L.K. Jha, a top ranking bureaucrat. L.K. Jha and Morarji Desai were going by the same Air-India flight to New York, *en route* to Washington for a World Bank meeting. Jha, the Finance Secretary, liked to sip a couple of whiskies, "away from his puritanical Finance Minister's disapproving eye". But he had reckoned without the then Air-India Chairman, J.R.D. Tata.

J.R.D. had come to the airport to see off the Finance Minister and his entourage. On checking the seating arrangement, he found that Jha and Morarji were going to sit far apart. Upbraiding his officials, Tata ensured that Jha got the seat next to Morarji. Through the entire flight, a frustrated Jha was forced to be a teetotaller and spent his time ruing the unwanted attentions of the Air-India Chairman.

On another Air-India flight, when both he and Jha were travelling together, Jha noticed that J.R.D. was suddenly missing for the best part of an hour. When J.R.D. returned, Jha asked him where he had been.

'Oh, I went to see if the toilets are clean, and that everything is as it should be,' J.R.D. replied. 'But what took you so long?' Jha asked. 'The toilet rolls were not placed properly,' J.R.D. said. It turned out that the father of Indian civil aviation had gone to each of the several toilets in the Boeing 747 and personally corrected whatever was wrong. As Jha recalled, it was this kind of attention to detail and concern for service that made Air-India, under J.R.D.'s stewardship, one of the premier commercial airlines in the world.

*

In the rush of our working life most of us say harsh things to our colleagues or subordinates, oblivious to the damage we do to their self-esteem. J.R.D. thinks twice before raising a point and often puts himself in the other person's shoes before he does so. Once I had done something he did not approve of. When I went in to see him, he said before I could take a chair: 'You will be unhappy to hear what I am going to tell you...' Without anger he made his point. It did hurt but when a man cares so, you can take it from him.

J.R.D cannot stand the suffering of people. In his eighties he dislikes movies portraying suffering and violence. 'Do we not have enough of it already?' Over the years his interests have widened rather than narrowed down—especially his interest in people. He has never stagnated. 'Man ascends,' says Professor Bronowski in *The Ascent of Man* 'as he discovers the fullness of his gifts.' And these gifts are not only of the head but of the heart.

Once he heard that one of his officers undergoing chemotherapy was reacting too violently. 'I would like to send you to America,' he volunteered to the officer one day. The next morning he told his secretaries that the priority that day was to make the officer's travel arrangements. He issued instructions for the necessary funds and asked for a Reserve Bank form to personally obtain foreign exchange for the officer. When J.R.D.'s secretary informed the officer of J.R.D.'s plans, the officer rang J.R.D. and said he was thankful but he could manage to deal with the Reserve Bank himself.

J.R.D. also spoke personally to the Chairman of Indian Hotels to arrange for the officer's stay at the Lexington Hotel in New York. When the officer arrived in New York for medical consultations, J.R.D. happened to be there and invited him for tea in his suite.

'I do hope they won't give you any more shots,' he said.

After seeing the doctor, the officer rang J.R.D. to say that the consultant at Memorial Sloan-Kettering hospital had recommended that the chemotherapy be stopped. 'You don't know how happy you have made me today,' replied J.R.D.

A couple of days later, as J.R.D was leaving for Geneva, the officer and his wife happened to spot him in the Lexington Hotel's foyer. The officer's wife said, 'Mr. Tata, I want to thank you for all that you have done for my husband.'

'Don't thank me, my dear, thank his faith and his God.'

CHAPTER VIII

Eventide

---

*I have scaled the peak and found no shelter in fame's bleak
and barren height. Lead me my guide, before the light fades,
into the Valley of Quiet, where life's harvest mellows into
Golden Wisdom.*

> - Rabindranath Tagore

When "fast falls the eventide" and life's harvest mellows, even practical and earthy persons take time off to think of the hereafter. This is true of J.R.D. as well.

For decades he has been vigorously against organized religion, including Zoroastrianism, to which as a Parsee he is born to. He has at the same time appreciated the ethos of the religion of his ancestors that has enabled his tiny community to make a remarkable contribution to the enlightenment and advancement of the Indian nation. At a reception given by the Parsee Panchayat to eighteen distinguished Parsees in various fields of life, J.R.D. said: 'As an Indian, I am proud to be a Parsee.' However, he has found some of the customs of the Parsees, like their funeral rites and their "senseless exclusiveness" irksome and outmoded. In 1943, when Piroja Nanavutty sent him a translation of Zoroastrian prayers he said in reply, 'While I am certainly not an atheist or even an agnostic I do not believe in the outward, priest-created manifestations of religion and in fact hold the view that they have been for centuries and are still one of the principal causes of disunity and backwardness amongst people, particularly in our unhappy country.'[1] When the question of Parsees' rites and customs came up in a recent conversation he said: 'More than rites I am against their insistence in not accepting into the Zoroastrian faith children of Parsee girls marrying outside the community on grounds of racial purity. Partly as a consequence of this, the community is dying out.'

In a letter to Noshir Pavri on 21 April 1984, JRD wrote: 'The Zoroastrian philosophy of hard work, honesty and charity is one that the world should know today.'

He is even-handed in his treatment of all religions. For instance, in 1949, he wrote a letter to the General Manager of the Taj Mahal Hotel, Bombay, L.P. Vachek, stating it was not appropriate to play hymns on New Year's eve by the hotel band. He, however, did attend in 1952 several meetings of the Moral Re-Armament (MRA) when Dr Frank Buchman, its initiator, came with 200 people from around the world with MRA plays. Perhaps the idea that people of different faiths could associate together in a moral movement with the common denominator of listening to inner voice appealed to him. In fact he was the group's main host in Bombay. Later his interest in the movement flagged, though he maintained his personal friendship with people in the movement. Since the 1970s, J.R.D. has showed and retained a keen interest in Auroville. I asked him whether the teachings of Shri Aurobindo had an influence on him. He replied, 'I can't claim I have any real knowledge of the philosophy and the life of Aurobindo, though I have read what Nani (Palkhivala) has written about him. Shri Aurobindo was a man who could see into the future. Auroville is one of the outcomes of this and the moving spirit behind it was the Mother. My interest of the Auroville lies in the fact that it is a commune. I've always been interested in communes. They have all failed. I find Auroville is one which has survived.'

He is impressed by the sacrifices that skilled foreigners have made to the building of the place. 'I am an internationalist and a world citizen, and here is a project that interests me.' He spoke of the greenery of what was once a barren area and the enthusiasm of the local people. But his interest in Auroville is not rooted in any religious teachings or faith.

J.R.D. has served on the Auroville International Advisory Council, since 1981. There were four members. Two of these were P.V. Narasimha Rao and the Director-General of UNESCO.

Earlier in 1954, he had written to a person who invited him to a spiritual meeting, saying, 'In my present life on earth at least, I seem to be destined more for the rough and ready type of life in which action predominates over meditation. Perhaps, when I retire from my existing activities, I may find some inclination in the direction in which you evidently wish to lead me today.'

Nearing retirement, J.R.D. met Rev. Fr. M.M. Balaguer, S.J., an active prelate of almost ninety. Fr. Balaguer had been principal of St. Xavier's College, Bombay, the Head of Jesuit Education in India and the President of the Eucharistic Congress when Pope Paul VI came to this country in 1963. The two octogenarians exchanged notes on how age affects the mental processes. J.R.D. said that one's memory tended to fade with age. Fr. Balaguer replied: 'Retention of all facts is not expected of you. Judgement is expected.' J.R.D. replied: 'I suppose so!' When the conversation came around to creativity, J.R.D said: 'I've been given credit for a lot that I do not deserve.' Fr. Balaguer observed that he had done a good deal in his lifetime to which J.R.D. replied: 'Yes, but one is never satisfied.'

When the subject turned to God and religion, J.R.D. agreed that God is love, but expressed his dislike of organized religion that had prompted wars, including the Crusades. Fr. Balaguer suggested he should read more of religion and find out for himself. 'You haven't given enough time,' said Fr. Balaguer. J.R.D. agreed that he had not taken the trouble to read enough, adding, 'At 20 one should find out which religion will answer all one's needs. Otherwise, like myself, you end up believing in nothing. To me religion is service—one must play one's role.'

J.R.D. was interested in finding out how Fr. Balaguer joined the church. Fr. Balaguer spoke of how, at the age of fourteen, his father explained to him through a book what the vow of chastity would involve. At this point I, who was also present, asked Fr. Balaguer whether, in his seventy-five years of priesthood, the vows of chastity, poverty and obedience had ever bothered him.

'Not once,' said Fr. Balaguer. There was a moment's stunned silence.

'You are lucky,' I said.

'Maybe because it was God's calling for you,' observed J.R.D.

J.R.D. then talked about his belief in reincarnation, 'I hope to be re-born in this life,' he told the Jesuit. Fr. Balaguer replied: 'You may not necessarily be here.'

'You mean I will be in some other part of the universe?'

'Not necessarily,' said Fr. Balaguer. 'God is without time or space.'

'Then where will I be?'

'We shall either be with God or not with God,' replied the Jesuit calmly. One sensed J.R.D was searching for something deeper in his life.

J.R.D. conceded towards the end of the meeting that if there was no

God, 'too much is inexplicable.' As is usual with J.R.D., when a visitor leaves, he enquired whether the Father had transport and whether his car could drop him home.

After he had bid J.R.D. goodbye, the priest and I walked down the corridor silently. I thought of J.R.D.'s comment, "Where will I be?" and his interest in reincarnation, and realized that in this man whose life had been one of action, was a pilgrim soul he seldom revealed.

Turning to Fr. Balaguer I remarked: 'I wish before he moves on he could come closer to God.'

'Don't you think he is already?' replied Fr. Balaguer.

*

Though J.R.D. can recite by heart English and French poems about death, the pursuit of the higher realm has not been a preoccupation in his life. His worldly interests are far more pressing. Apart from his business and his enduring interest in family planning, and education he is concerned with what goes on in the nation. He is not preoccupied with petty and ephemeral issues, the political antics which characterize the Indian political scene. But he is alert to the trends in the country. He lives in the stream of national life and when the torrent is overpowering, like the rest of us, he is carried over the precipice. But he is unlike most of us who drift with the mainstream. He is like a trout that defies the current, tries to jump vertically up the waterfall, yearning to get back to its homewaters.

He feels that quite a few of our problems will be resolved including relations with Pakistan and the situation in Kashmir and Punjab, if the government and the leaders dare to go against the torrent and decide to remove the roots of conflict.

At a lunch meeting with Field Marshal Sam Maneckshaw in September 1986 he related how as a teenager in Paris he was worked up about Alsace-Lorraine being taken away from France by Germany in 1870. He rejoiced when France, by the Treaty of Versailles in 1919 got them back from Germany. He recalled the bitterness between France and Germany before and during the First World War which was compounded by the tragic events of the Second World War. Said J.R.D.: 'Today you could hardly imagine two nations more unlikely to go to war than France and Germany.'

He paused and then shot his question: 'Sam, how long will it take for

India and Pakistan to arrive at such a relationship?' The hero of the war against Pakistan in 1971 was somewhat taken aback at the question.

At eighty-seven, when he could rest on his laurels, he still keeps thinking about the nation and the Kashmir issue is very much on his mind, which if unsolved, could well result in a war. 'People don't realise what war means in today's world.' He is sad that the press has not for four decades created the necessary public opinion which would have enabled us to at least discuss the issue with an open mind. Be it the situation in Kashmir, Punjab or Assam he believes that India's continuing political turmoil is substantially due to the insistence of politicians that the Union Government has a right to dominate and control all political and economic actions in the country and therefore the life of the whole nation. 'There is no justification for that.'

J.R.D. goes on to give the example of how successive Union Governments have circumvented directives inscribed in the Constitution that industry was a State subject except for a few strategic industries which parliament chose to assign to the control of the Union Government. By getting parliament over the years to pass laws, over ninety per cent of industries including the manufacture of such "strategic" items like razor blades and tennis shoes and others were transferred from State to Centre although that wasn't the real intention of the Constitution makers. This J.R.D. feels was, 'bad enough in that it resulted in a serious infliction of unwanted socialism on the country, but it also encouraged the Union Government's claim to exercise Central control over virtually all activity in the country.'

J.R.D. is convinced that most if not all the unrest and violence in the country could be eliminated if the government and the political parties recognize that in a federal type of democracy such as ours, people should be allowed to exercise their inalienable right to choose the form of administration they want and to manage and control the economy. The minimum number of subjects should be left for the Centre to handle. He has no doubt that if that were to happen all talk of secession would disappear.

*

As stated earlier, at the age of sixty-eight, J.R.D. voluntarily shed the Chairmanship of TELCO in favour of Sumant Moolgaokar. He was a sparkling seventy-eight when he gave the Chairmanship of Tata Chemicals to Darbari Seth and two years later that of Tata Steel to Russi Mody. He thus demonstrated that men who contribute substantially to the progress of a firm deserve the highest office. The old have to graciously make way for the new. I mentioned, 'Some of your colleagues say you stepped down from the Chairmanship too soon.'

'No,' he replied, 'I don't think so. If anything I think I should have stepped down earlier.' And he quoted Russi Mody as an example of a man in whose favour he should have done so sooner.

A key question throughout the last decade has been the succession to the Chairmanship of Tata Sons. Though weakened by the abolition of the Managing Agency System, it was still the pivotal company started by Jamsetji Tata, with holdings in every Tata company.

In 1981, J.R.D appointed Ratan Tata*, son of his distant cousin Naval Tata as Chairman of Tata Industries, a hundred per cent owned subsidiary of Tata Sons. Trained as an architect at Cornell University, USA, Ratan Tata was happily settled in the States. He had a job he liked, he had a nice flat in an apartment complex with a swimming-pool, and looked forward to furthering his career there. At the age of twenty-four, due to the pleading of his grandmother, Lady Navajbai, Ratan Tata returned to India. On his return J.R.D. invited him to join Tatas. In 1962 the enthusiastic young man wrote a letter to J.R.D. where he said 'Words could never adequately express my sincere gratitude and appreciation for your decision—I shall attempt to express my thanks by serving the firm as best as I can, and to do all I can to make sure that you will not regret your decision.'[2]

Ratan Tata was sent for training to TELCO and TISCO in Jamshedpur. After he completed his training he was given two sick companies to run, first NELCO and later Empress Mills. The next step was the Chairmanship of Tata Industries. 'The Chairmanship of Tata Industries,' says Ratan Tata 'was a titular one. Tata Industries had a great aura about it but it was only a Rs. 60 lakh company. I had no plan at the start. It was a soul-searching time to begin with. The first decision was to reduce the ownership of Tata Sons and sell 70 per cent of it to the rich operating companies and then to have a rights issue to raise the holdings.' Then, as S.A. Sabavala says, 'He took Tatas into the hi-tech areas and introduced strategic thinking.' His first strategic plan was unveiled in 1983. Its accent was on electronics and

---

* Also see brief family tree at the beginning of the book.

bio-technology and gave the shell company, Tata Industries, a fresh purpose and a role. In the 1980s Tata Industries launched five hi-tech industries and one finance company—Tata Honeywell, Tata Telecommunications, Hitech Drilling Systems, Keltron Telephones and Tata Finance.

What was a shell company became the cutting edge of Tata enterprises. J.R.D. watched all this carefully, though he gave little indication of his thinking on the succession issue vis-a-vis Tata Sons. Whenever asked, he would say it was for the Directors to decide when he gave up office. In November 1987, he spoke to me of how after the abolition of the Managing Agency System, the other big industrial houses had either split or drifted apart and proudly added, 'But we have not.'

'So long as you are there,' I observed. 'After you?'

'Ratan is aware of that.'

His words were a clear pointer to his thinking, but J.R.D. had still not settled the succession issue. When he was in his mid-eighties there was understandable concern amongst his colleagues on when they would know the identity of the new chief. Though clear in his own mind, J.R.D. would still not put up a name. A skiing friend of J.R.D. says that he is

a little like a fellow who is holding in his hands a grenade with the pin removed.[3]

However, it will be wrong to consider that he is rash in the slightest way. Often when I go enthusiastically to him with an idea, he will caution 'Take care.' There is a balancing factor to his daring which prevents foolhardiness and stops him from being impetuous. In fact he spoke to me of two senior Managing Directors in Tatas, both highly competent. Of one he said, 'He is impetuous but he consults me *before* he does something.' Of the other he said, 'He too is impetuous but he informs me *after* he does something.' Needless to say he took adequate time before he gave the final authority to both these outstanding executives.

J.R.D. was clearing the ground by making Ratan Tata, still young in his forties, the second most important person in key Tata companies like Tata Chemicals, Tata Electric and Tata Steel. When Sumant Moolgaokar stepped down, he wanted Ratan Tata to be the next Chairman of TELCO. In 1989, Ratan became Deputy Chairman of Tata Sons, at the age of fifty-one. J.R.D.'s thinking had crystallized.

TELCO was the first major operating company Ratan Tata had to handle.

Soon after taking over as its Executive Chairman, Ratan Tata was faced with an explosive labour situation in the Pune plant which he was not responsible for. He showed guts and grit in handling it.

A year-and-a-half later, on 28 February 1991, J.R.D. had a very exhausting day in the office ending with a public function in the evening at which he spoke at length. The next morning just before flying to Jamshedpur, he had anginal pains. He still flew. He attended the Founder's Day function on 3 March and the inauguration of a Foundation for Business Ethics at a college there. The pains recurred. Even so, with the help of pain-killers, he went through a gruelling schedule, rural visits and all. On his return from Jamshedpur, he was persuaded to stay in Breach Candy Hospital for a check-up for five days. He left hospital on a Saturday, and to the surprise of everyone, on Monday morning he was back at his desk. A Tata Director out of courtesy phoned to say: 'I am glad, sir, you have come to office today.'

'What,' he shot back, 'you mean to say that you are not glad when I come to work on other days!'

The following week on Wednesday, 27 March, the usual meeting of the Board of Tata Sons was to be held. Its date was advanced to Monday, 25 March. The agenda was a well-kept secret between J.R.D. and his colleagues on the Board. J.R.D. started the meeting by speaking about his sixty years in Tatas and his experiences. It was a moving occasion. 'I wish we had a tape-recorder,' said one of the Directors. J.R.D. told his colleagues that the time had come for him to handover charge and he proposed the name of the gentleman sitting on his right, Ratan Tata, for the Chairmanship. Pallonji Mistry, a non-Tata Director, seconded the proposal. The Directors were conscious that they were witnessing the end of an era. The helmsman who had steered the ship for fifty-two years rose and with courtly dignity offered his chair to Ratan Tata, who was born just a year before J.R.D. assumed the Chairmanship in 1938.

There was exceptional coverage in the Indian press when J.R.D. stepped down.

One item noted that when he had assumed charge of Tatas, its turnover was Rs 280 crore. When he stepped down, it was Rs 10000 crore. Many papers noted that his was certainly the longest corporate reign of any industrialist in India.

More important than the number of years one puts in, is what one puts into the years. Several papers lauded his contribution.

When asked how he would like to be remembered he replies, 'As an honest man who tried to do what he thought was right.'

In that summing up is an echo of the Emperor Darius, who, influenced by the teachings of Zarathustra, said: 'What is right, that is my desire*.'

Some worship at temples. Others steer clear of temples, and in spite of frailties and failures, attempt to live out the principles of religion in their lives. J.R.D. belongs to the latter category.

*

Three weeks after stepping down, J.R.D. was in San Francisco for an operation to open a blocked heart artery. On the first day they opened up one artery and on the second day another.

Feeling very fit after the angioplasty, he inflicted a punishing schedule on himself. He flew to New York, then on to Cleveland, for the hundredth birthday celebrations of his good friend Fred Crawford and the following day flew back to New York. He came back to Bombay in May and continued working full throttle as if nothing had happened.

In June 1991, two months after his angioplasty, he flew to London for a meeting of Tata Limited. After the meeting he felt breathless and rang up the Royal Brompton Hospital asking for an appointment the following day. The doctor replied quickly: 'You are not coming tomorrow. You are coming *tonight*.' At an emergency operation the right coronary artery was subjected to angioplasty. His personal physician, Dr Gool Contractor, was rushed to London. Soon after at another angioplasty on the same artery a stent was inserted.

On his return to Bombay in July 1991, although his mind was as alert as ever, he was forced to cut down the long hours at the office. Group Captain G. Leonard Cheshire, who knows him well, says:

> I think that zest is the key to understanding J.R.D's remark-
> able achievements, zest for whatever he is engaged in, and
> zest for life in general... Most of us have it in one form or
> another when we are young, but if we are not careful it can
> begin to fade as we get older, especially if we let ourselves
> get too immersed in routine. Zest for life carries one over

---

* Quoted in the chapter "The Heritage" in Part I.

difficulties and setbacks, even over disaster, it carries us forward into new fields and new interests and is a truly life-giving force. But if once it disappears we are in danger of disappearing too. Look at Jeh, skiing up to the age of 85, and still vitally and constructively interested in everything around him.[4]

One might add, 'Not only in everything but also in everyone around him.' For example, when I met him after his second heart procedure, he said: 'I suppose you will write another book after this biography. You know I am fond of history and I think that the book that needs to be written is on the Death of a Religion.' He paused, 'The Death of a Religion in 70 years—Communism. I remember my school days in France when most of our teachers were Communists...To them Communism was a religion. Where is it today?'

In August 1991 he had been unwell for a couple of days—a problem unrelated to his heart ailment. I attempted to cheer him up by pointing out at the various things he could look forward to. I need not have taken the trouble. He said: 'I will tell you one thing. In November I am going to Disneyworld and I am going to take somebody with me to see it—and I am not going to tell you who!' He then went on to speak of Disneyworld's ingenuity and the perfectionism that appealed to him on his last visit. 'Disneyworld is worth visiting for grown-ups too. You should go there.'

For him there is always tomorrow. In the five years I have spent working on this book I have got to know him more intimately. In this short span of time I have been able to see how his sympathies have widened. His sensitivity to other people has sharpened as one year yields to the next. The body weakens as it must, but the spirit shines brighter. And as the evening mellows and the shadows lengthen, somewhere above in the sky, in an invisible Puss Moth, is a voyager still pressing ahead to cross beyond the last blue mountain where a glorious sunset awaits him.

APPENDIX A

J.R.D. Tata's Journey Logbook*

Logbook 1

Journey from Karachi to Jask

Date	Places of Departure, of intermediate landings and of arrival	Times** Arrival	Departure	Incidents and observations
3/5/30	Karachi		6.15	Head winds all the way. Bad visibility first hour due to dust haze. Flew all through at 6 to 7,000 ft. where winds were reported less strong. Stopped 1 h at Gwadar for petrol and food. Calm. Machine and Engine O.K. Customs at Jask. Leftwing incidence corrected Karachi O.K.
	Gwadar	10.15	11.15	
	Jask	15.05		
				Spent night at Mr. W. Jones' house (Indo European Telegraph)
	Total: 7 h–50			

* J.R.D. Tata flew in a Gipsy Moth G-AAGI for the Aga Khan Prize Flight, 3 May–12 May 1930.
** All Karachi time.

Logbook 2

Journey from Jask to Basra (Shaiba)

Date	Places of Departure, of intermediate landings and of arrival	Times		Incidents and observations
		Arrival	Departure	
4/5/30	Jask		5.20	Head winds all the way, especially strong between Bushire and Basra. Visibility fair. Calm upto Bandar Dilam. Bumpy upto Basra. Drifted inland slightly from Quishen island covered with thick layer of clouds @ about 2,000 (flew all through at about 6,000). Had to double back from Salt marshes north of Lingoh. Took petrol at Lingoh and Bushire. Delayed 2 hours at Bushire for petrol. Aircraft and engine O.K. Spent night at Imperial Airways rest house.
	Lingoh	8.45	9.20	
	Bushire	12.55	14.55	
	Shaiba (Basra)	18.15		
	Total: 10 h–20			

Journey from Basra (Shaiba) to Rutbah Fort

Logbook 3

Date	Places of Departure, of intermediate landings and of arrival	Times Arrival	Departure	Incidents and observations
5/5/30	Basra (Shaiba)		6 a.m.	Started from Basra on compass course but after about 120 miles ran into sand storm or clouds. Doubled back to Euphrates river. Sand at 9,000 ft. visibility very bad all the way. Calm upto Baghdad. Baghdad to Rutbah; good visibility. Very bumpy. Followed track across desert. Strong head winds right through from Basra. Machine and engine O.K. Spent night in Rutbah Fort.
	Baghdad	10.15	11 a.m.	
	Total: 7 h–35	2.20 p.m.		

Logbook 4

Journey from Rutbah to Cairo (Heliopolis)

Date	Places of Departure, of intermediate landings and of arrival	Times Arrival	Departure	Incidents and observations
6/5/30	Rutbah		5.50	From Rutbah flew on compass course. Either compass inaccurate or strong southerly winds. Flew north of correct course and reached coast at Haifa about 90 miles north of Gaza. Landed on Haifa landing ground (disused) for name of place. No one available for 3/4 h. Took petrol at Gaza and examined and set plugs, filter, etc. Slight misfiring before reaching Gaza. Singh* in Moth landed after me India bound. Visibility good all through. Rough after Haifa. Very rough from canal to Cairo.
	Haifa	9.30	10.05	
	Gaza	11.15	12 p.m	
	Cairo (Heliopolis)	14.40		
	Total: 7 h–50			

* Manmohan Singh who also competed with J.R.D. Tata for the Aga Khan Prize Flight.

363

Logbook 5

Journey from Cairo to Benghazi

Date	Places of Departure, of intermediate landings and of arrival	Times Arrival	Departure	Incidents and observations
7/5/30	Cairo		5.40	Easterly winds. Landed at Aboukir for compass swinging. Delayed there for 2 1/2 hours. Good weather. Slightly bumpy. Took petrol at El Sellum. Machine and Engine O.K. Compass was badly inaccurate (25° off). Met Aspy Engineer on Aga Khan flight opposite direction. Exchanged life jacket for set of plugs.
	Aboukir	7 a.m.	10.30	
	El Sellum	2.10	2.30	
	Benghazi	6.50		
	Total: 9 h			

Logbook 6

Journey from Benghazi to Tripoli

Date	Places of Departure, of intermediate landings and of arrival	Times Arrival	Departure	Incidents and observations
8/5/30	Benghazi		6.45	Southerly winds resulting in sand upto 10,000 ft. Landed at Sirte for weather report. Ran suddenly into low clouds near Hones. Flew to Tripoli at 100 ft. to 500 ft. Bad visibility owing to sand in the air. Calm. Intended to proceed to Gabes or Tunis but petrol not arrived at the aerodrome. Machine and Engine O.K.
	Sirte	10.20	11.10	
	Tripoli	14.10		
	Total: 6 h–35			

365

Journey from Tripoli to Catania and Naples

Date	Places of Departure, of intermediate landings and of arrival	Times		Incidents and observations
		Arrival	Departure	
9/5/30	Tripoli		6.50	Trouble in starting at Tripoli, found one push rod slipped out of rocker car and latter bent. Repaired. Good trip across Mediterranean; passed over Malta after 2 1/2 h. run. Further hour upto Sicilian coast. Visibility good upto coast. Then thick haze and very bumpy. Landed Catania for petrol and customs. Clouds and intermittent rain between Catania and Naples. Head winds.
	Catania	11.10	1.30 p.m.	
	Naples	5.45 p.m.		
	Total: 8 h–55			

Journey from Napoli to Roma Littorio   Logbook 8

Date	Places of Departure, of intermediate landings and of arrival	Times		Incidents and observations
		Arrival	Departure	
10/5/30	Naples		9.30	3/4 Head winds. Calm. Landed Rome for petrol and customs.
	Rome	11.00		
	1.30			
	C.F.			

Logbook 11

Journey from Paris to Lympne, Croydon

Date	Places of Departure, of intermediate landings and of arrival	Times		Incidents and observations
		Arrival	Departure	
12/5/30	Le Bourget		8.30	Westerly winds bad weather upto coast. Low ceiling. Improving over Channel. Good upto Croydon.
	Lympne	10.40	11.30	
	Croydon	12.15		
	Total: 2 h–55			

On The Night Air Mail[*]

I was most grateful for the opportunity you gave me at the end of last month to place before you our point of view in connection with the unfortunate controversy which had arisen over Civil Aviation.

In the light of your remarks and advice, I had hoped that the controversy would cease forthwith, and I certainly intended to take no further part in it myself. I did not, however, expect that Mr. Kidwai would make so unfair and unreasonable a speech in the Assembly. This speech, which I regret to say contains many incorrect statements and allegations, has naturally received a tremendous amount of publicity throughout the country.

If we give no reply at all, we shall stand doubly condemned in the eyes of the public, and much permanent damage will have done to the good name and reputation of Air-India, and, for the matter of that, of Tatas. I have received innumerable enquiries from Share-holders, employees, friends, etc. regarding Mr. Kidwai's charges, and in each case the hope is expressed that we shall at least state the facts where they have been incorrectly given by the Hon'ble Minister.

There is also the Legislature to consider. It is natural that schemes for speeding up mails and lowering fares should be politically popular, irrespective of their technical merits or demerits, and that for this reason the members of the Legislature should have some bias in favour of the Hon'ble Minister's policies. I could have no complaint on that ground, but the effect of Mr. Kidwai's speech, although partially corrected by your dignified and constructive intervention, for which I am deeply thankful, has gone much further. By resorting

---

[*] As expressed in a letter J.R.D. wrote to Pandit Jawaharlal Nehru on 8 December 1949.

to incorrect and unfair statements and allegations, he has convinced the majority of the members of the Assembly, who were naturally not in a position to differentiate between what was true and untrue in his remarks, that the airlines in general, and Air-India in particular, are dishonest and greedy, cannot be trusted, and fully deserve their present plight.

In the circumstances, we have had no option but to prepare a memorandum answering the more important statements and charges contained in Mr. Kidwai's speech. For ease of reading, we have had it printed and I enclose three copies herewith for yourself and your Secretariat. After you have glanced through it, I am sure you will agree that we had no option but to publish it and I hope you will forgive me for having taken this step notwithstanding the advice you kindly gave me in Delhi last month.

I was heartened to learn from your statement in the Assembly that Government would, outside the context of the recent debate, consider submitting the case of the air transport industry to an impartial board of enquiry. I do hope that this will be done soon and that in view of the heat generated by the present controversy, the composition of the board will be entirely independent of the Ministry of Communication as well as of the air transport industry.

I am sorry to have been partly responsible for adding further to your worries. I would not have done so had I not felt that wrong policies were driving an important national industry to the wall. I do not ask for any special treatment or protection but only plead that the case of the air transport industry be made the subject of an impartial enquiry by the Tariff Board or a similar body.

In conclusion, I beg you to believe that I shall never seek your help or intervention to serve personal ends nor ever betray such faith as you may have in me.

APPENDIX C

On Nationalization Of Airlines[*]

My dear Jehangir:

I was very sorry to notice your distress of mind when you came to lunch with me the other day. You told me that you felt strongly that you or the Tatas, or at any rate your air companies had been treated shabbily by the Government of India. Indeed you appeared to think that all this was part of a set policy, pursued through years, just to do injury to your services in order to bring them to such a pass that Government could acquire them cheaply.

You were in such evident distress at the time that I did not think it proper to discuss this matter with you. Nor indeed am I writing to you today with any intention to carry on an argument. But I feel I must write to you and try in so far as I can, to remove an impression from your mind which I think is totally wrong and is unjust to Government, to me as well as to you.

I cannot of course deal with any individual acts of discourtesy that might have occurred in the Secretariat here or of any attitude adopted which was not becoming. That might well have happened.[**]

All I can say is that I regret that exceedingly and, whenever any such act has been brought to my notice, I have immediately taken steps. The machinery of Government functions in a peculiar way which is not to my liking. I have been six years here and still feel rather like

* As expressed in a letter Pandit Jawaharlal Nehru wrote to J.R.D. Tata on 10 November 1952.
** When J.R.D. was shown this letter and asked if anyone had behaved in an unbecoming manner he replied: 'No—I never complained. I could never have discussed that the officers were rude because everybody was always kind.'

a fish out of water. I have been impressing upon our officials and others that the old type of superior governmental behaviour is objectionable.

So far as the Tatas are concerned, you know my own high appreciation of the record of this outstanding firm in India which has pioneered so many projects. I think in this matter I reflect the general views of most of my colleagues. I have not heard at any time any adverse comments in regard to the Tatas, although there is plenty of criticism of others here. Of course, there may be disagreement in regard to policies and we may look in a somewhat different direction sometimes.

But the charge you made the other day which amounted to a planned conspiracy to suppress private civil aviation and, more particularly, Tata's air services, astounded me. I could not conceive of it and I am sure that nobody here could do it. This matter of air services in India has been discussed by us in Cabinet and outside on a very large number of occasions. As a matter of general policy, we have always thought that transport services of almost all kinds should be State-owned. Indeed, so far as the Congress is concerned, we laid down this general policy about twenty years ago. It is true that the policy could not be implemented for various reasons and we gave it no high priority. But the matter was discussed on many occasions. It was chiefly the lack of finances that prevented us from going ahead.

For a considerable period, Rafi Ahmed Kidwai was in charge of the Communications Ministry. Your grievance is that a large number of services were permitted to operate and these tended to eat each other up. I am not competent enough to judge this policy. But obviously all of us are anxious to develop air services in India. It may be that we went too far. It may be also that the Tatas were too cautious in some matters, such as the night air mail. I am not dealing here with the merits of the particular policy, but with your charge that a policy was pursued by the Government of India with the deliberate intention of injuring Tata's air services. That, I am quite sure, is completely unjustified. Rafi Ahmed Kidwai used to discuss air services with me frequently. He always had a good word for your air services, except for the fact that he thought you were too cautious and he did not like the idea of high fares. He may have been right or wrong. But I have no doubt in my mind that any desire to injure your air services was never present in his mind. So far as his subordinate officials were concerned, they may have behaved or misbehaved occasionally, but Kidwai did not give them too much rope and even in their case it seemed to me incredible that, secretly in their minds, they were

374

pursuing this long-distance policy of causing harm so as to make acquisition easy later. I doubt if they even thought of acquisition.

During the last few months, that is since Jagjivan Ram has been Communications Minister, this matter came up before the Cabinet on several occasions. We examined it thoroughly. We were driven to the conclusion that there was no other way out except to organise them together under the State. I remember that even then stress was laid on the excellence of your services, and more particularly, Air-India International. We did not wish to touch the Air-India International. We appointed a Committee of the Cabinet to go into this matter. Their report was that it would be difficult in the circumstances to isolate Air-India International.

The purpose of my writing to you is to remove the impression from your mind that any policy has been pursued by us with the deliberate intention of acquiring them later after their value came down. Both from the civil and the defence point of view, we have naturally been anxious to develop aviation in this country. Our eagerness to do so may have gone too far. A situation arose ultimately when we were driven to a certain conclusion.

I do not want you to carry in your mind the impression you gave me when you came here. We want your help in this and other matters and it is a bad thing to suspect motives and nurse resentment. Coming from an old friend like you, this distresses me greatly.

Yours sincerely

sd/-
Jawaharlal Nehru

*

SHRI JAWAHARLAL NEHRU[*]
PRIME MINISTER OF INDIA
NEW DELHI

NOVEMBER 12, 1952

I DEEPLY APPRECIATE YOUR LETTER OF THE 10TH JUST RECEIVED AND AM MOST GRATEFUL FOR THE ASSURANCES OF YOUR REGARD FOR THE

---

[*] J.R.D. Tata's telegram to Pandit Jawaharlal Nehru on 12 November 1952.

ORGANISATION I REPRESENT
STOP

AM SORRY IF YOU FEEL THAT IN MY DISTRESS I HAVE DONE INJUSTICE TO
ANYONE BUT MY ONLY IDEA WAS TO CONVEY TO YOU FRANKLY MY VIEW
OF THE POLICIES AND ACTIONS OF GOVERNMENT WHICH HAVE BROUGHT
ABOUT THE PRESENT SITUATION IN THE AIR TRANSPORT INDUSTRY
STOP

LEAVING ASIDE THE PAST HOWEVER I AM SINCERELY CONVINCED THAT
THE MINISTRY'S PRESENT NATIONALISATION SCHEME IS NOT SOUND
AND WILL NOT RESULT IN THE CREATION OF AN EFFICIENT AND SELF
SUPPORTING AIR TRANSPORT SYSTEM
STOP

IF AS APPEARS FROM YOUR LETTER GOVERNMENT HAVE ALREADY
DECIDED UPON THE ADOPTION OF THIS SCHEME I CAN ONLY DEPLORE
THAT SO VITAL A STEP SHOULD HAVE BEEN TAKEN WITHOUT GIVING US
A PROPER HEARING
STOP

MODY AND I WERE CALLED BY MR. JAGJIVAN RAM ONLY TO BE IN-
FORMED OF GOVERNMENT'S DECISION TO NATIONALISE THE INDUSTRY
STOP

ALTHOUGH I TOLD HIM THAT I HAD PREPARED AND BROUGHT WITH ME
AN ALTERNATIVE SCHEME WHICH IN MY HUMBLE JUDGEMENT WAS
BETTER CALCULATED TO ACHIEVE GOVERNMENT'S OBJECTIVE THE
MINISTER SOUGHT OUR ADVICE ONLY ON QUESTIONS OF COMPENSA-
TION AND THE LIKE
STOP

IF HOWEVER IN VIEW OF WHAT I HAVE SUBMITTED ABOVE YOU FEEL
DISPOSED TO KEEP THE MATTER OPEN FOR DISCUSSION I WOULD BE
GRATEFUL FOR THE OPPORTUNITY AND WOULD BE ONLY TOO HAPPY TO
PLACE MY VIEWS AND PROPOSALS BEFORE YOU MR. JAGJIVAN RAM AND
SUCH OF YOUR OTHER COLLEAGUES AS YOU MAY LIKE TO HAVE AT AN
INFORMAL CONFERENCE
STOP

SHOULD YOU DECIDE TO GIVE US A HEARING MODY AND I ARE AT YOUR
DISPOSAL ANY TIME YOU WANT US TO COME TO DELHI
STOP

TO SAVE TIME MAY I SUGGEST THAT YOUR REPLY BE COMMUNICATED
TO ME THROUGH PAN
STOP

I BEG YOU TO BELIEVE THAT I AM MOTIVATED BY NO SELF INTEREST IN
THIS MATTER
STOP

MY ONLY ANXIETY IS TO SEE A STRONG AND EFFICIENT INDIAN AIR
TRANSPORT SYSTEM BUILT UP AND AT THE SAME TIME TO SEE JUSTICE
DONE TO INVESTORS AND STAFF WHO HAVE SUFFERED HEAVILY.

(FROM) J.R.D. TATA

Dismissed From Air-India[*]

February 26, 1978

My dear J.R.D.,

Please refer to your letter of the 13th February 1978, which has caused me some surprise and some distress.

I shall first clarify the point to which you refer in the penultimate paragraph of your letter. My reference to the efficient working of the two airlines obviously referred to the need for having a common Chairman for the two airlines. It had made no reference to the past performance of the two airlines. I can only say that out of understandable sensitiveness over the manner in which the two Boards were reconstituted, you were not able to appreciate the reference in its true context. In fact, if I were not fully cognisant of the efficient management of Air-India under your Chairmanship, I would not have said what I did in the second paragraph of my letter of the 4th February.

From your letter I have not been able to judge whether you feel aggrieved about the termination of your chairmanship of Air-India. Frankly speaking I thought that after so many years of useful services to the Corporation and the manner in which you had brought it up to its present stage of remarkable distinction, you would yourself offer to place the responsibilities on younger shoulders. You did not offer to do so which is why I did not broach the subject. We had felt, however, that it would be better to have the two airlines combined under one wholetime Chairman. There was, therefore, no alternative for us but to make the change. In doing so I naturally presumed that with your preoccupations and commitments it would not be possible for you to undertake the combined wholetime Chairmanship.

* As expressed in a letter Morarji Desai wrote to J.R.D. Tata on 26 February 1978.

378

You have also made a grievance of the fact that you have been excluded from the reconstituted Boards of the two Corporations of which you have been a member since their inception. I think I am correct in saying that never in the history of the two Boards have the outgoing Chairmen been nominated as Members of the succeeding Boards. The reason is clear. The successor would be put in an embarrassing position in case he has to review any of the previous decisions or practices approved by his predecessor. Secondly, the staff of the organisation would also be overawed somewhat by his presence on the Board and may not be able to express themselves freely on matters in which the previous Chairman had taken a decision or any particular view. Moreover, I am sure you will agree that there is some advantage in bringing to bear on the affairs of an institution a fresh outlook uninfluenced by past associations. In the circumstances I assumed that you yourself would find it embarrassing to continue on the Boards.

I do hope that in the light of what I have said you would appreciate better the reasons for which we made a change and in any case I can assure you that no extraneous consideration whatsoever was involved in bringing about this change.

With kind regards,

Yours sincerely

sd/-
(Morarji Desai)

Mission To The West[*]

<hr/>

I have read with pain and concern Mahatma Gandhi's statement published in the papers yesterday regarding the forthcoming visit of Indian industrialists to the U.K. and the U.S.A. As one of those concerned, I feel it my duty to say that Gandhiji has done us a grave injustice in issuing a public statement impugning the motives of those undertaking the visit. I emphatically repudiate the suggestion that, in visiting Britain and America for the purpose of investigating the difficult problems with which Indian industry is faced at this critical juncture, we are out to enter into 'shameful deals' behind the country's back. In controversies of this kind Gandhiji usually refrains carefully from ascribing motives to those with whose action he disagrees and takes pains to state their case with scrupulous fairness.

It seems that some people have made it their business to misrepresent to Gandhiji the objects of the visit, and it is a thousand pities that he should not have sent for some of us and obtained a clarification of our aims before giving public expression to his views. We could have convinced him that we are not going on any mission or deputation under the auspices of the Government of India.

The facilities extended to us by Government, and for which we are grateful, are the granting of air passage priorities, both to England and America, and the release of dollar exchange without which a visit to the United States would be impossible. We are visiting these two countries as nothing more than a group of individuals, each travelling and living at his own expense, with the sole object of surveying the economic and industrial developments which have taken place in those countries since the war and of acquiring from them such knowledge and experience as will enable us better to play

<hr/>

[*] J.R.D. Tata's rejoinder to Mahatma Gandhi's statement to the press, 8 May 1945.

our part in the future economic development of the country.

It is somewhat difficult to see how a business trip of this nature, undertaken with such objectives, can be considered detrimental to the interests of the country. India cannot afford to stand still while other nations, great and small, are forging ahead. The fact that India has not been able to secure political freedom is one which industrialists and businessmen deplore as much as any other class, and I cannot see how our visit can be construed as detracting in any way from the national demand either for the release of the Congress Working Committee members or for complete political freedom.

Jamsetji Tata[*]

Jamsetji Tata died over half-a-century ago. Few people remain today who had personal contact with him or remember meeting or seeing him. His sons and lieutenants, his political and business associates, his staff and his workers have passed from the scene. It is probably true to say that to most of the present generation, outside those who work within the House of Tata, he is primarily known as a wealthy industrial pioneer and philanthropist of the last century who gave his name to a group of industrial concerns and endowment trusts. Yet few men have played so vital a part in the renaissance of India and there is much to be learned even today from the life and work of this remarkable man.

It was Tata who laid the foundations of modern Indian industry, without which the present programme of industrial development would be difficult of achievement or have been long delayed. Amidst the present welter of conflicting ideas and philosophies in the economic and industrial field, many of them half-baked and half-digested, we could do with some of the clear thinking, the courage and reliance on basic truths and values which the reader will find in abundance in the life of J.N. Tata.

Much has happened since Jamsetji Tata died. Here in India, as elsewhere, the extent and character of the changes have no parallel in previous history. Even more remarkable has been the rate of change. Between Jamsetji's death in 1904 and the publication of Harris' book in 1925, the changes, marked as they were, appear inconsiderable in the light of subsequent developments in the political, industrial and economic life of the country. If we are to view Jamsetji's life-work and ideas in their proper perspective, we should do so in the light of these subsequent changes and against the background of the conditions which prevailed at the time in which he lived.

Jamsetji was born and lived the earlier part of his life during one of the darkest periods in the long and chequered history of India. He was eighteen at the time of the abortive Revolt against foreign rule which took place just a hundred years'ago. As he began his career, the grip of colonial rule and exploitation reached its peak. Independent thought or action in any field other than the literary, spiritual or

---

[*] J.R.D. Tata evaluates Jamsetji Tata in the Foreword to F.R. Harris' *J.N. Tata: A Chronicle of his Life,* Bombay: Blackie and Sons (India) Limited, 1958.

religious was frowned upon. The people of India, their spirit broken, stagnated politically, socially and economically, and the elite sought solace in religious meditation or sterile contemplation of the country's past glories.

Then came on the scene a succession of remarkable men. Beginning with Dadabhai Naoroji and ending with Jawaharlal Nehru, men like Pherozeshah Mehta, Tilak, Tagore, Motilal Nehru, Gokhale, C.R. Das, Vallabhbhai Patel and above all Mahatma Gandhi, lighted the darkness, reawakened the political and social consciousness of the people and in a unique struggle, peaceful and constructive, the like of which has never been seen before, won back the people's self respect and finally their independence. J.N. Tata has an honoured place amongst these giants for, as we look back upon his life, it is clear that the part he played in his own field was as important to India's rebirth as that of the others.

Indian industry did not begin with Jamsetji Tata. In fact no date or period can be laid down for such a beginning. India was, for thousands of years, a highly industrialized country in the sense that it produced with great skill and in large quantities goods and products which not only met its own requirements but were exported to half the world, from Europe to China. Electricity, railways, and the telegraph came to India in the nineteenth century, pretty well at the same time as they were introduced in the countries of their origin. The power-driven cotton textile industry, in which Jamsetji Tata played such a prominent role, was itself established well before J.N. Tata entered the field. The genius of Jamsetji, however, and the essence of his contribution to the later growth of Indian industry is that he alone in his time understood the full significance of the industrial revolution in the West and of its potentialities for his own country. Where others thought primarily or exclusively in terms of political action, he saw clearly that India's freedom could not be achieved or maintained by political means alone, that freedom without the strength to support and, if need be, defend it, would be a cruel delusion and that the strength to defend freedom could itself only come from widespread industrialization and the infusion of modern science and technology into the country's economic life.

It is a measure of the man that despite the changes which have since taken place, his ideas and industrial philosophy remain today as fresh and vital as they were in his time. Jamsetji Tata dreamt of an industrialized and prosperous India. Under conditions which would have appalled and discouraged lesser men, he set about to breathe life into his dream. He realized that if he was to achieve anything worthwhile in his lifetime, he must not fritter away his resources by attempting to do too much over too wide a range of activities. He wisely concentrated his energies, knowledge and wealth on three schemes which would not only be within his own resources to bring to life but which, he saw, were essential to any future progress and, once established, would set the pattern and the pace for future development. The three basic ingredients of economic progress, as he saw them, were steel, electric power and technical education combined with research. The soundness of his reasoning may seem pretty obvious to us today, but under the conditions prevailing in India some hundred years ago when all progress, other than that geared to satisfying colonial needs, was firmly discouraged or ridiculed,

383

it required an exceptionally lucid and well-furnished mind, backed by courage and determination, to see the basic issues so clearly and to plan in detail the programme of action which he adopted.

He died before the Steel Works at Jamshedpur, the Hydro-Electric Power Plants near Bombay and the Institute of Science at Bangalore were established, but, fortunately, the enthusiasm and devotion which he inspired amongst his sons and lieutenants and their faith in his ideals were so great that his plans for the three great projects were pursued with unabated vigour and determination after his death and, in due course, brought into being. They were, in fact, rapidly followed by many new ventures in the fields of engineering, chemicals, civil aviation and insurance among others, all, with rare exceptions, inspired by the same driving purpose of promoting and accelerating the industrial development of the country. It is impossible, within the compass of a brief foreword, to attempt to assess with any precision the effect which the creation of these key industries and the Indian Institute of Science at Bangalore had on subsequent events in the country. There can be no doubt, however, that they did in fact largely achieve the purpose for which J.N. Tata and his successors conceived them. They not only produced and supplied essential materials and products to other industries but also created amongst Indians of all classes a resurgent belief and confidence in themselves and in their ability to stand on their own feet and, by making them impatient of existing conditions, they strengthened the freedom movement and quickened the country's march towards independence.

However we may judge the part that Jamsetji Tata played in the national events of the last hundred years, there can be no question about his role and influence in the life and growth of the Firm which bears his name. To this day, the memory of Jamsetji and what he stood for burns brightly for all within the Organization. Although time has blurred the edges, his presence, wise and benevolent, is still felt in many ways. The traditions which he set, the values which he established are still cherished and defended. On innumerable occasions, I have been struck by the devotion to his ideas and ideals expressed by individuals, high and low, in the Firm. Even simple workers in Tata Companies seem to have been touched by the magic of his name and to have realized intuitively how much they owed to him. Of all the tributes paid to Jamsetji in his lifetime and since his death, none, I think, has been more striking or would have touched him more than the voluntary demonstration of affection and gratitude on the part of thousands of workers in the Tata Textile Mills at Ahmedabad and Bombay who, on their own initiative, recently raised from amongst themselves the funds to erect bronze or marble statues of the Founder. Every year at Jamshedpur and elsewhere on the anniversary of his birth, thousands of workers garland his statue in grateful homage to his memory.

It is, however, within the counsels of the top management of the Firm that his continuing influence and leadership are most felt, and half-a-century after his death, important decisions and policies are still measured against his ideals and the traditions set by him. If to this day his policies continue largely to be followed, it is not merely out of personal loyalty to the memory of a great leader but because, despite the passing of time, his basic philosophy and his ideas on industrial management have been found to be sound and applicable even to the changed circumstances of the present day. His anticipation of modern thinking is a continuing source of surprise and admiration to the present generation of Tata management; his emphasis on the application of science and modern techniques and methods in industrial management; his generous and yet realistic understanding and acceptance of the needs and rights of workers at a time when they were frequently exploited in the West as well as in India; his insistence on absolute standards of integrity; his sense of trusteeship and his realization that to survive and prosper, free enterprise must serve the needs of society, are all remarkably in tune with modern thinking and with the ethical and social standards of the most advanced societies of today.

If the House of Tata survived some exceptionally difficult times and prospered and continued to grow both in size and stature, I sincerely believe it is because throughout, it held fast to the principles and ideas of J.N. Tata. I further believe that if a greater proportion of the private sector of business and industry in India were to adhere more closely to like principles, the prestige of free enterprise and public confidence in it would be higher than it is and it might suffer less from restrictive legislation and controls than it does today.

In 1895, when a new extension of the Empress Mills was opened, Jamsetji Tata said: 'We do not claim to be more unselfish, more generous or more philanthropic than other people. But we think we started on sound and straightforward business principles, considering the interests of the shareholders our own, and the health and welfare of the employees the sure foundation of our prosperity.' That is not an original or a remarkable statement but it will be of interest to contrast it with an equally frank statement made by someone else ten years later. During a conference in London with Tatas' representatives in 1904–5 regarding concessions in freight rates for bulk conveyance of raw materials, iron and steel, the Managing Director of the Bengal–Nagpur Railway said: 'It does not appeal to us at all if you can only show that in an indirect and remote way this would be for the benefit of India. The only appeal that can be made to us is that we can make money out of it. This Railway Company, you must always bear in mind, is a commercial undertaking, and must not be actuated by anything like patriotic or philanthropic motives; it must only be actuated by commercial motives. We must not consider the advantages to India and, figuratively, we do not care a snap of the fingers about the advantages to India. What we have to think of is the interests of the Company as a commercial concern.'

It is not difficult to judge which view has prevailed in both Western and Eastern practice. "Sound and straightforward business principles," they go to the root of the matter. Jamsetji Tata spoke then, as he spoke always, with modesty, directness and robust good sense. Not the least of his remarkable gifts was a way of going straight to the heart of a matter. Yet this practical man and realist was always to take the far view, to seek not quick returns but the ultimate benefit of India in his schemes and plans; to have in his make-up a good deal of the visionary and dreamer. Men of business are not often at home in the world of ideas; it was

Jamsetji's distinction that he lived in both worlds, the world of ideas and the world of action.

It was because Jamsetji Tata lived in the world of ideas and had imagination that he played the role of pioneer in India in another direction—that of constructive philanthropy. Alms-giving and doing good to others are as old as human history, but new ideas, current in the latter half of the nineteenth century directed much of such activity towards higher education and the spread of education. Men began to recognize the importance of science and engineering for answering the needs of a rapidly changing world. The new philanthropy turned in every conceivable direction to pioneer schemes of human welfare and to increase human knowledge. It led, in the United States and in Europe, to the establishment of great foundations whose names are household words.

Amongst the great pioneers was Andrew Carnegie, whose life and outlook bear a resemblance to those of Jamsetji Tata's and who was roughly his contemporary, though the Scotman lived to a much greater age. I believe Tata would have heartily subscribed to the social philosophy of Carnegie which is embodied in the following words: "This, then, is held to be the duty of the man of wealth; to set an example of modest, unostentatious living, shunning display or extravagance; to provide moderately for the wants of those dependent upon him; and, after doing so, to consider all surplus revenues which come to him simply as trust funds, which he is called upon to administer...the man of wealth thus becoming the mere trustee and agent for his poorer brethren."

Jamsetji started sending brilliant young men abroad for higher education in 1892. In 1898 he made an offer of three million rupees for an institution of higher learning in India. At the value of the rupee in those days, this was a considerable sum by any standard. Nothing of this magnitude had been known in the country but what was truly significant was the breadth of outlook that characterized the intended gift. Notwithstanding the substantial philanthropic grants he made in his lifetime, Jamsetji left a considerable fortune to his sons Dorab and Ratan. It grew further in their hands as, with vigour and ability, they brought to life the great projects planned by their father. They too made large grants in their lifetime and, on their death, left the whole of their fortune in philanthropic and charitable endowments which, today, hold about three-fourths of the capital of the parent firm, disburse funds over a wide range of objects and in their policies bear ever in mind the spirit and outlook of the Founder. The wealth gathered by Jamsetji Tata and his sons in half-a-century of industrial pioneering formed but a minute fraction of the amount by which they enriched the nation. The whole of that wealth is held in trust for the people and used exclusively for their benefit. The cycle is thus complete; what came from the people has gone back to the people many times over.

To sum up in a few words the life of J.N. Tata and the lessons to be drawn from it is a task I shall not attempt. That he was a man of destiny is clear. It would seem, indeed, as if the hour of his birth, his life, his talents, his actions, the chain of events which he set in motion or influenced, and the services he rendered to his country and to his people, were all predestined as part of the greater destiny of India.

NOTES

Part I

Chapter I

1. Sir Dorab Tata's words—Sir Dorab was educated at Cambridge.
2. Lala, R.M., *The Creation of Wealth: A Tata Story*, Bombay: IBH Publishing Company, 1981, p. 25.
3. ibid.
4. Fraser, Lovat, *Iron and Steel in India*, Bombay: The Times Press, 1919, pp. 103–04.
5. Harris, F.R., *J.N. Tata—A Chronicle of his Life,* Bombay: Blackie and Sons (India) Limited, 1958, p. 126.
6. A Souvenir on the occasion of the India Steel Industry's Golden Jubilee, Jamshedpur 1958, p. 126.
7. Harris, F.R., *J.N. Tata—A Chronicle of his Life*, Bombay: Blackie and Sons (India) Ltd., 1958, p. 260.
8. ibid., p. 274.
9. ibid., p. 275.

Chapter II

1. Clarke Moma, *Light and Shade in France,* London: John Murray, 1939, pp. 2,7,8,14 & 15.
2. ibid., p. 14.
3. ibid., p. 14
4. Maurois, André, *A History of France,* London: Jonathan Cape, 1949, pp. 469–70.
5. Churchill, Winston, *Great War Speeches,* London: Transworld Publishers Limited, 1957, p. 107.

Chapter IV

1. Churchill, Winston, *Great Contemporaries*, London: Thornton Butterworth Limited, 1937.

2. ibid., pp. 302–03.
3. *Partha,* Bombay, September 1988.

Chapter VI
1. Legislative Assembly Debates, Vol. 4, Part IV, 1924.
2. Verrier, Elwin, *The Story of Tata Steel,* Bombay: TISCO, 1958, p. 61.

Chapter VII
1. Harris, F.R., *J.N. Tata—A Chronicle of his Life,* Bombay: Blackie and Sons (India) Ltd., 1958.
2. Verrier, Elwin, *The Story of Tata Steel,* Bombay: TISCO, 1958, p. 32.
3. Fraser, Lovat, *Iron and Steel in India,* Bombay: The Times Press, 1919; quoted in Harris, F.R., *J.N. Tata—A Chronicle of his Life,* p. 14.
5. Ashby, Lilian Luker & Whately, Roger, *My India: Recollection of Fifty Years,* Boston: Little Brown and Company, 1937, p. 373.

Chapter VIII
1. Morgan, Ted, *Winston Churchill 1878–1915,* London: Jonathan Cape, 1982, p. 84.
2. Russel, Janney, *The Miracle of Bells,* New York: Prentice-Hall Incorporated, 1946.

Chapter IX
1. Nanavutty, Piloo, *The Parsis,* Delhi: The National Book Trust, 1977, p. 107.

Chapter X
1. Verrier, Elwin, *The Story of Tata Steel,* Bombay: TISCO, 1958, p. 36.
2. Harris, F.R., *J.N. Tata—A Chronicle of his Life,* Bombay: Blackie and Sons (India) Ltd., 1958.
3. Lala, R.M., *The Creation of Wealth,* Bombay: IBH Publishing Co., 1981.
4. ibid.
5. Harris, F.R., *J.N. Tata—A Chronicle of his Life,* Bombay: Blackie and Sons (India) Ltd., 1958.
6. The *Tata Monthly Bulletin,* March 1946, p. 11.

Part II

Chapter I
1. Hoare, Sir Samuel, *India by Air,* London: Longmans, Green and Company Limited, 1927.
2. ibid.
3. ibid.
4. ibid.

Chapter IV
1. Tata, J.R.D., *Keynote,* Bombay: Tata Press, 1986, p. 20.

Chapter V
1. Quoted in Sen, A., *Five Golden Decades of Aviation,* Bombay: Aeronautical Publications of India, 1978, p. 44.

Chapter VI
1. *The Magic Carpet,* October 1977.
2. *The Magic Carpet,* March 1968.
3. *The Magic Carpet,* August 1968.
4. *The Magic Carpet,* October 1972.

Chapter VII
1. Brecher, Michael, *Nehru: A Political Biography,* London: Oxford University Press, 1959, pp. 609–10.

Chapter VIII
1. J.R.D.'s speech at the Annual General Meeting of the shareholders of Air-India Ltd., 22 June 1953.
2. G.D. Girla to J.R.D., 23 September 1946.
3. Sardar Patel's speech in the Central Legislative Assembly, November 1946.

Chapter IX
1. Peter Menezes to J.R.D., 28 July 1953.
2. J.R.D. to Floyd Blair, 21 May 1953.

Chapter X
1. *The Journal of the Royal Aeronautical Society,* January 1961.
2. J.R.D. to M.M. Philips, 4 July 1963.
3. J.R.D. to Sir Frederick Tymms, 5 August 1985.
4. J.R.D. to Pandit Nehru, 26 June 1954.
5. Karnik, A.S., *Kashmir Princess,* Bombay: Jaico Publishing House, 1953.
6. J.R.D. to Aspy Engineer, 19 October 1957.
7. J.R.D.'s speech on civil aviation in India, 28 August 1957.
8. J.R.D.'s message to Air-India staff, 23 October 1962.
9. J.R.D. to N.S. Kajrolkar, 12 November 1962.

Chapter XI
1. Sir William Hildred to J.R.D., 4 January 1954.
2. Sir William to J.R.D., 13 January 1954.
3. J.R.D. to Sir William, 14 March 1955.
4. Sir William to J.R.D., 18 March 1955.
5. Pandit Nehru to J.R.D., 2 August 1957.

6. Sir William to J.R.D., 27 February 1978.

7. J.R.D. to Sir William, 3 June 1977.

Chapter XIII

1. J.R.D. to Sir Frederick Tymms, 18 January 1967.

2. J.R.D. to Rose Benas, 15 June 1971.

3. J.R.D. to Sir William Hildred, 13 January 1971.

Chapter XIV

1. Sampson, Anthony, *Empires of the Sky,* London: Hodder & Stoughton, 1984.

Part III

Chapter I

1. Quoted in J.R.D.'s Foreword to Harris, F.R., *J.N. Tata—A Chronicle of his Life,* Bombay: Blackie and Sons (India) Ltd., 1958.

2. ibid.

3. Tead, Ordway, *The Art of Leadership,* New York: McGraw Hill Book Company, 1935.

Chapter II

1. Sir Homi Mody to J.R.D., 26 December 1955.

2. Tata, J.R.D., *Keynote*, Bombay: Tata Press, 1986, p. 153.

3. ibid.

4. ibid.

5. ibid.

6. Lala, R.M., *The Creation of Wealth*, Bombay: IBH, 1981, p. 101.

Chapter III

1. Halifax, Lord, *Fullness of Days*, London: Collins, 1957, p. 199.

2. ibid., p. 196.

3. ibid., p. 205.

4. ibid., p. 204.

5. Munshi, K.M., *Pilgrimage to Freedom,* Bombay: Bharatiya Vidya Bhawan, 1967, p. 54.

6. ibid., p. 55.

7. Moraes, Frank, *Jawaharlal Nehru: A Biography,* New York: Macmillan, 1956.

8. J.R.D.'s Convocation Address to the TISS, 1962.

9. Wavell, Lord & Moon, Penderel, eds., *A Viceroy's Journal,* London: Oxford University Press, 1973, p. 446.

10. Ward Mowbrays, Benedicta, trans., *Saying of the Desert Fathers,* London: OUP, 1975, p. 69.

11. Wavell, Lord & Moon, Penderel, eds., *A Viceroy's Journal,* London: OUP, 1973, p. 452.

12. ibid., p. 52.

Chapter IV

1. Hanson, A.H., *Process of Planning: A Study of India's Five Year Plan 1950–1964,* London: Oxford University Press, 1966.

2. *Transfer of Power,* London: H.M.'s Stationary Office, 1973, Vol. 4, item 344, p. 672.

3. Iengar, H.V.R., *Planning in India,* Delhi: Macmillan India Ltd., 1974, p. 8.

4. Hanson, A.H., *Process of Planning: A Study of India's Five Year Plan 1950–1964,* London: OUP, 1966.

5. Tyson, Geoffrey, *Nehru: The Years of Power,* London: Pall Mall Press, 1966, p. 26.

Chapter V

1. G.D. Birla to J.R.D., 1 May 1945.

2. Lala, R.M., *The Creation of Wealth,* Bombay: IBH, 1981, p. 141.

3. The *New York Times,* 25 February 1945.

Chapter VI

1. *India Today,* 8 April 1987.

2. Mehta, Ved, *Mahatma Gandhi and His Apostles,* Harmondsworth: Penguin Books, 1977, pp. 61–67.

3. Jaju, Ram Niwas, *G.D. Birla: A Biography,* Delhi: Vikas Publishing Co., 1985.

Chapter VII

1. For a more detailed list of TISCO's firsts in labour welfare, see Lala, R.M., *The Creation of Wealth,* Bombay: IBH, 1981; Appendix A.

2. J.R.D.'s speech to the Lion's Club, Jamshedpur, 22 August 1963.

3. J.R.D.'s speech at the Michael John Memorial Lecture, Jamshedpur, 2 March 1985; quoted in Tata, J.R.D. *Keynote,* Bombay: Tata Press, 1986, p. 101.

Chapter VIII

1. Chairman's (J.R.D.'s) address to the Board, 18 October 1945.

2. J.R.D. to George Woods, 27 April 1959.

3. The *Financial Express,* 4 July 1989.

Chapter IX

1. For more details, see Lala, R.M., *The Creation of Wealth,* Bombay: IBH, 1981; chapter "Treasures from the Sea".

Chapter X

1. The *Sunday Times,* London, 15 September 1951.

2. *Harijan,* April 1953; *Modern Review,* October 1953.
3. J.R.D. to Dinshaw Daji, 12 May 1953.

Chapter XI
1. Air Chief Marshal Arjan Singh to J.R.D., 27 September 1966.
2. J.R.D. to Arjan Singh, October 1966.
3. Chairman's (J.R.D.'s) statement to Tata Steel, 15 July 1968.

Chapter XII
1. Indira Gandhi to J.R.D., 5 July 1973.
2. Tata, J.R.D., *Keynote,* Bombay: Tata Press, 1986; Foreword p. xv.

Chapter XIII
1. Tata, J.R.D., *Keynote,* Bombay: Tata Press, 1986, p. 50.
2. ibid.
3. Chairman's (J.R.D.'s) statement to Tata Steel, August 1956.
4. Tata, J.R.D., *Keynote,* Bombay: Tata Press, 1986.
5. ibid., p. 77.
6. ibid.
7. ibid., p. 43.
8. ibid., p. 44.

Part IV

Chapter I
1. *India Today,* 28 February 1986.
2. The *Tata Review,* October 1965.

Chapter II
1. Rodabeh to J.R.D., 8 September 1969.
2. Lala, R.M., *Heartbeat of a Trust,* Bombay: Tata McGraw-Hill Publishing Company Limited, 1984, p. 95.
3. Prof. M.G.K. Menon to J.R.D., 5 April 1970.
4. Masani, R.P., *Dadabhai Naoroji,* Bombay: Publications Division, [reprinted] 1975, p. 13.
5. Canning, John, ed., *100 Great Modern Lives,* London: Odhams Books Limited, 1965, p. 166.

Chapter III
1. Tata, J.R.D., *Keynote,* Bombay: Tata Press, 1986, p. 109.
2. ibid.
3. ibid., p. 133.

4. ibid., p. 125.
5. ibid., p. 124.
6. ibid., p. 115.
7. ibid., p. 117.

Chapter IV

1. Tata, J.R.D., *Keynote,* Bombay: Tata Press, 1986; Foreword, pp. xiv & xv.
2. ibid., pp. xii & xiii.
3. Notes taken by Pyarelal usually giving Mahatma Gandhi's views on various subjects in conversation with visitors were published in the latter's paper, *Harijan.*
4. Tata, J.R.D., *Keynote,* Bombay: Tata Press, 1986; Foreword, pp. xiii & xiv.

Chapter V

1. Lawrence, T.E., *Seven Pillars of Wisdom.*
2. J.R.D.'s Convocation Address to the TISS, 1954.
3. James, Arthur, *Every Day with William Barclay.*
4. *The Illustrated Weekly of India,* 14 April 1988.
5. M.O. Mathai confirms this in *My Days With Nehru,* Delhi: Vikas Publishing Company, 1979, p. 90. The author says, 'One of the words Nehru disliked most was "lucrative". The nearest word he could suffix was "profitable".

Chapter VI

1. J.R.D. to R. Dennis, 14 June 1963.

Chapter VII

1. The *Economic Times,* 9 September 1991.

Chapter VIII

1. J.R.D. to Piroja Nanavutty, 22 December 1943.
2. Ratan Tata to J.R.D., 8 August 1962.
3. *Forbes,* 5 October 1987.
4. Group Captain Leonard G. Cheshire in *Aircraft and Engine Perfect.*

Index